AF760096

MATERIALS SCIENCE AND TECHNOLOGIES

POLYMER PHASE BEHAVIOR

MATERIALS SCIENCE AND TECHNOLOGIES

Additional books in this series can be found on Nova's website
under the Series tab.

Additional E-books in this series can be found on Nova's website
under the E-books tab.

MATERIALS SCIENCE AND TECHNOLOGIES

POLYMER PHASE BEHAVIOR

TIMOTHY P. EHLERS
AND
JAMES K. WILHELM
EDITORS

Nova Science Publishers, Inc.
New York

Copyright © 2011 by Nova Science Publishers, Inc.

All rights reserved. No part of this book may be reproduced, stored in a retrieval system or transmitted in any form or by any means: electronic, electrostatic, magnetic, tape, mechanical photocopying, recording or otherwise without the written permission of the Publisher.

For permission to use material from this book please contact us:
Telephone 631-231-7269; Fax 631-231-8175
Web Site: http://www.novapublishers.com

NOTICE TO THE READER

The Publisher has taken reasonable care in the preparation of this book, but makes no expressed or implied warranty of any kind and assumes no responsibility for any errors or omissions. No liability is assumed for incidental or consequential damages in connection with or arising out of information contained in this book. The Publisher shall not be liable for any special, consequential, or exemplary damages resulting, in whole or in part, from the readers' use of, or reliance upon, this material. Any parts of this book based on government reports are so indicated and copyright is claimed for those parts to the extent applicable to compilations of such works.

Independent verification should be sought for any data, advice or recommendations contained in this book. In addition, no responsibility is assumed by the publisher for any injury and/or damage to persons or property arising from any methods, products, instructions, ideas or otherwise contained in this publication.

This publication is designed to provide accurate and authoritative information with regard to the subject matter covered herein. It is sold with the clear understanding that the Publisher is not engaged in rendering legal or any other professional services. If legal or any other expert assistance is required, the services of a competent person should be sought. FROM A DECLARATION OF PARTICIPANTS JOINTLY ADOPTED BY A COMMITTEE OF THE AMERICAN BAR ASSOCIATION AND A COMMITTEE OF PUBLISHERS.

Additional color graphics may be available in the e-book version of this book.

Library of Congress Cataloging-in-Publication Data

Polymer phase behavior / editors, Timothy P. Ehlers and James K. Wilhelm.
p. cm.
Includes bibliographical references and index.
ISBN 978-1-61324-336-7 (hardcover)
1. Polymers--Solubility. 2. Polymers--Mixing. 3. Polymers--Separation. 4. Phase rule and equilibrium. I. Ehlers, Timothy P. II. Wilhelm, James K.
QD381.9.S65P638 2011
547'.70454--dc23

2011012566

Published by Nova Science Publishers, Inc. + New York

CONTENTS

PREFACE

This book examines the phase behavior of polymers. The authors present topical research in this field. Topics discussed include the phase behavior of PVP as compared with that of poly(N-vinylcaprolactam); the applicability of lattice cluster theory to the calculation of miscibility; Raman study of the pressure and temperature induced transformations in crystalline polymers of C60; polymer phase behavior in nanocomposites; phase inverting polymer systems in drug delivery medicine and the correlation between stereochemistry and phase behavior.

Chapter 1 - Newly developed hyperbranched polymers possess a compact, highly branched, three-dimensional structure, which has a high density of functional end groups and inherently low viscosity. The combination of these two properties, low viscosity and high reactivity, makes them attractive candidates for an overwhelming variety of applications. The experimental and theoretical investigation of the phase behavior of hyperbranched polymer systems is a crucial requirement for a successful introduction of new applications to highly competitive markets. In this context, thermodynamic models, which accurately account for the impact of polymer branching on the phase behavior of polymer systems, play a very important role.

The lattice cluster theory (LCT) is an extension of the well-known Flory-Huggins theory, especially in the calculation of the entropy of the lattice. Whereas the Flory-Huggins theory is limited to linear chains the LCT can be applied to arbitrary chain architecture. This situation permits the incorporation of the architecture in the thermodynamic functions useful for phase equilibrium calculations. The polymer architecture plays an important role in the physical properties of hyperbranched polymers. Additionally, the combination of the LCT with the density gradient theory allows the theoretical investigation of the interfacial properties between the demixed phases.

Chapter 2 - High hydrostatic pressure causes a number of effects in fullerene polymers; the most interesting of them being further pressure-induced polymerization and subsequent structural phase transitions in partially polymerized fullerenes. The behavior of the phonon modes of the polymeric one-dimensional orthorhombic 1D-O phase, the two-dimensional tetragonal 2D-T and rhombohedral 2D-R phases of C_{60} have been studied as a function of pressure, up to ~30 GPa, at room temperature. The 1D-O polymeric phase, even at small pressures, undergoes pressure-enhanced photo-induced transformation to a new polymeric phase characterized by twinned polymeric chains. The photo-transformed 1D-O polymer and two-dimensional 2D-R and 2D-T polymeric phases undergo irreversible structural transformations at different pressures to new cross-linked three-dimensional polymeric structures. The phonon spectra of the high-pressure phases provide strong indication that the fullerene molecular cage is preserved in the recovered phases. The decomposition of the 2D-R polymer of C_{60} during high temperature treatment leads to the initial face centered cubic structure of the fullerene C_{60} monomer.

Chapter 3 - Polymer phase (polymer matrix) behavior in nanocomposites in many respects defines the behavior of nanocomposite as a whole independently from the used nanofiller type (disperse particles, organoclay, nanotubes and so on). In connection with this it is necessary to account for polymer matrix structure changes at nanofiller introduction in initial matrix polymer. These changes can be realized with the aid of different processes, namely, crystallization, amorphous polymer phase structure change, interfacial regions formation. In its turn, such factors as polymer matrix chain flexibility, interfacial adhesion level, nanofiller particle shape and so on influence on characteristics and realization possibility of the mentioned processes. Hence, at polymer nanocomposites structure formation complex dynamics of polymer phase behavior in them is observed, that defines in the long run a nanocomposite properties. In this aspect particularly important is the role of interfacial regions, which are the same reinforcing element of structure in polymer nanocomposites as actually nanofiller. For semicrystalline polymer matrix nanofiller can play a nucleator role changing in reality the indicated matrices crystallinity degree. In the present review the quantitative relationships of the initial polymer characteristics and their modification at nanofiller introduction and their influence on nanocomposite final structure are considered.

Chapter 4 - Phase inverting polymer systems are primarily utilized in industrial applications such as the microfiltration of bacteria and reverse osmosis, but their use has been rapidly expanding in other areas. In the medical field the predominant role of these systems has been in development of new biomaterial matrixes for drug delivery and tissue engineering. The use of phase inverting systems for the controlled release of therapeutic agents is of interest due to the injectable nature of the implants, which provides a less invasive means of physically placing the implant at or near the site of action. The goals of this chapter are to: provide a basic description of the phase inversion process involved in medical implants, describe factors that affect the phase inversion and drug release processes, overview the techniques used to characterize these systems, and provide insight into the *in vivo* behavior to include biocompatibility and deviations from *in vitro* behavior. *In situ* forming implant systems are an exciting field of study, and have been successfully used to treat diseases that range in severity from prostate cancer to periodontitis. These systems provide a compelling alternative to preformed polymer implants, and may prove to be paramount in overcoming the intrinsic obstacles of the physical targeting of polymer implants for the local delivery of therapeutic agents.

Chapter 5 - Nowadays the urgency for solving plastic waste problems is inducing academic and industrial research to develop novel environmentally friendly polymers, i.e. materials produced from alternative resources, with low energy consumption, non-toxic to the environment, and biodegradable. These biopolymers should have also good physical performances. In the field of aliphatic polyesters, novel (co)polymers, containing 1,4-cyclohexylene units, appear very promising materials, which are obtainable from biomass, potentially biodegradable and characterized by good mechanical properties. Moreover, these polyesters have the interesting peculiarity that their phase behavior is strictly connected to the ratio of the two possible configurations, *cis* and *trans*, of the cyclic units. Indeed, the *trans* isomer is more rigid and symmetric than the *cis*. Highly symmetrical units tend to improve the chain packing with a consequent increment in crystallinity and crystalline perfection. On the other hand, the *cis* isomer introduces kinks into the main chain, which hinder the formation of stable crystals. Thus, at high *trans* content the polyesters are characterized by relative high degree of crystallinity, whereas at low *trans* content the polymers are amorphous. Therefore, accordingly to the final *cis/trans* ratio, the phase behavior of the homopolymers and copolymers significantly changes and the stereochemistry of the cycloaliphatic units result to be a key factor to tailor the final thermal properties of the material. In this paper the properties of some homopolymers and copolymers, containing the 1,4-cyclohexylene units with different *cis/trans* ratio, are discussed just in terms of the correlations between stereochemistry and phase behavior.

Chapter 6 - Partitioning in aqueous two-phase systems (ATPS) is a proved technology for separating and purifying of enzymes. The goal of this study was to evaluate the applicability of polymer-salt ATPS based on polyethylene glycol (PEG)/K_2HPO_4-KH_2PO_4 as a putative method to isolate and recovery of recombinant amino acid dehydrogenases (AADHs). The partition behaviors of three models of AADHs namely phenylalanine dehydrogenase (PheDH), proline dehydrogenase (ProDH) and Leucine dehydrogenase (LeuDH) in two-phase partitioning systems prepared by PEG-4000/K_2HPO_4-KH_2PO_4 were investigated. The influence of different process parameters such as polymer molecular weight, type and concentration of salt, pH, phase volume ratio (V_R), tie-line length (TLL), type and concentration of inorganic salts, temperature, and cell extract loading on system phase behavior and extraction behavior were evaluated. Furthermore, the efficiency of partition behaviors was analyzed by SDS-PAGE method. The best optimal system for model AADHs with regard the partition coefficient (K_E), recovery (R%) and yield (Y%) was: 9.0% (w/w) PEG-4000, 18.0% (w/w) K_2HPO_4-KH_2PO_4, 8% (w/w) NaCl and a TIL of 52.3% (w/w). The partition parameters were as follows; PheDH (K_E=51.4, R=84.7%, Y=92.5) LeuDH (K_E=81.8, R=94.5%, Y=95.34) and ProDH (K_E=73.4, R=91.6%, Y=94.83). Three target enzymes showed to be partitioned in favor of the PEG-4000 rich top-phase. PEG-4000 proved to have a stabilizing effect on the enzymes of interest. K_2HPO_4-KH_2PO_4 was selected as the phase forming salt because of its ability to enhance the hydrophobic difference between the phases. It was found that the partitioning was not affected by V_R, while PEG-4000 concentration and K_2HPO_4-KH_2PO_4 concentration had significant effects on separation behavior. Longer TLL and higher pH resulted in better partitioning into the top phase. Addition of sodium chloride to the ATPS proved to be suitable to increase the recovery of target enzymes. Collectively, the observed partition behaviors of the model AADHs showed that developed ATPS can be a promising system for partitioning and potential recovery of recombinant AADHs.

Chapter 7 - A blend of poly(ε-caprolactone) (PCL) and poly(styrene-co-acrylonitrile) (SAN) containing 27.5 wt% of acrylonitrile having the critical composition (80/20 PCL/SAN) was studied. This PCL/SAN blend having a lower critical solution temperature (LCST) phase boundary at 122°C offered an excellent opportunity to investigate, firstly the kinetics of phase separation above LCST (125-180°C), and secondly the kinetics of phase dissolution below LCST (50-115°C). The blend underwent a temperature-jump above LCST where spinodal decomposition (SD) proceeded, yielding a regularly phase-separated structure (SD structure). Then, it was quenched to the temperatures below LCST when the phase dissolution proceeded. Optical microscopy was used to observe the spinodal decomposition qualitatively while light scattering was used to characterize the phase separation and phase dissolution quantitatively. It was found that during phase dissolution the peak maximum moved towards a smaller angle (wavelength of concentration fluctuations increases) while the peak intensity decreased. This behavior was explained by a model. Also it was found that the fastest phase dissolution kinetics at 80°C, which was characterized by an apparent diffusion coefficient, was about 10 times slower than the kinetics of phase separation at 180°C. Crystallization after various levels of spinodal decomposition was observed by optical microscopy. Order parameter of the lamellae inside the spherulites was evaluated with the help of H_v light scattering. Transmission electron microscopy revealed interesting lamellar structure after spinodal decomposition.

Chapter 8 - In recent years, a number of polymers that undergo phase separation in water solutions on temperature rising are studied. These polymers are characterized by lower critical solution temperatures (LCST). Poly(*N*-vinylpyrrolidone) (PVP) is not thermo- or pH - sensitive under usual conditions. However, since this polymer is widely used, especially in medicine, several studies are dedicated to the problem of making this polymer stimuli-responsive, too. In the review, the phase behavior of PVP in water solutions under various conditions is covered. The phase behavior of PVP-containing copolymers and hydrogels are described. The effect of the addition of salts, including transition metal ones, on the PVP phase separation temperature is considered, the attention being paid to the different influence of anions and cations on this value. It is known that PVP readily forms complexes with many organic and inorganic compounds. Examples of such complex formation effects on cloud points of the polymer solutions are given. The phase behavior of PVP is compared with that of poly(*N*-vinylcaprolactam), a PVP close analog, which is a well-known thermosensitive polymer.

In: Polymer Phase Behavior
Editors: T. P. Ehlers and J. K. Wilhelm, pp. 1-73
ISBN: 978-1-61324-336-7
© 2011 Nova Science Publishers, Inc.

Chapter 1

APPLICATION OF LATTICE CLUSTER THEORY TO THE CALCULATION OF MISCIBILITY AND INTERFACIAL BEHAVIOR OF HYPERBRANCHED POLYMER CONTAINING SYSTEMS

S. Enders [1,*] ***and T. Zeiner*** [2,•]
[1]TU Berlin, Fachgebiet Thermodynamik und
Thermische Verfahrenstechnik,
Berlin, Germany
[2]TU Dortmund, Lehrstuhl Fluidverfahrenstechnik,
Dortmund, Germany

ABSTRACT

Newly developed hyperbranched polymers possess a compact, highly branched, three-dimensional structure, which has a high density of functional end groups and inherently low viscosity. The combination of these two properties, low viscosity and high reactivity, makes them attractive candidates for an overwhelming variety of applications. The experimental and theoretical investigation of the phase behavior of hyperbranched polymer systems is a crucial requirement for a successful introduction of new applications to highly competitive markets. In this context, thermodynamic models, which accurately account for the impact of polymer branching on the phase behavior of polymer systems, play a very important role.

The lattice cluster theory (LCT) is an extension of the well-known Flory-Huggins theory, especially in the calculation of the entropy of the lattice. Whereas the Flory-Huggins theory is limited to linear chains the LCT can be applied to arbitrary chain architecture. This situation permits the incorporation of the architecture in the thermodynamic functions useful for phase equilibrium calculations. The polymer architecture plays an important role in the physical properties of hyperbranched polymers. Additionally, the combination of the LCT with the density gradient theory

* Straße des 17. Juni 135, D-10623 Berlin, Germany.
• Emil-Figge-Str. 70, D-44227 Dortmund, Germany.

allows the theoretical investigation of the interfacial properties between the demixed phases.

INTRODUCTION

The development of linear polymers and their impact on all aspects of modern life is one of the major achievements of the last century. As this field has matured it is increasing apparent that further developments will likely arise, not from the synthesis of totally new linear polymers, but from more accurately controlling the architecture of polymers from currently available monomers. The production and processing of polymers are influenced by the presence of phase separation and segregation, which may be either necessary or highly undesirable. For example, proper orientation and crystallization conditions are needed to secure useful fibers and films; on the other hand, segregation of highly viscous phases during a polymerization process may lead to catastrophic consequences like plugged lines or overheated reactors [1, 2]. Partial miscibility also plays important roles in biology and medicine, as for example in the formation of cataracts in eye lenses [3], protein separations [4], and fibril formation related to Alzheimer's disease [3]. Polymer solutions, which are mixtures of high-molecular weight compounds (solutes) and low-molecular weight solvents, have been examined experimentally, since the early 1930s. These studies revealed some abnormalities when the results were compared with those for low-molecular weight solutions e.g. an unexpectedly small vapor depression, small boiling point elevation, small osmotic pressure and extremely high solution viscosities. These abnormalities became a strong motivating force in the search for a thermodynamic theory for polymer solutions. For polymer solutions, where the difference in molecular size of the components is very large, miscibility gaps become highly asymmetric. A useful approximate model describing the thermodynamic properties, including the phase behavior, in polymer systems is the well-known Flory-Huggins lattice theory [5]. In the lattice model the mixture is represented by a number of regularly arranged lattice sites, each of the same size. The lattice is thought to have a definite coordination number z, which however; does not remain as a relevant variable in the simplest version of the theory. Each polymer molecule is considered to be composed of a number of segments, and the entropy of mixing is evaluated by counting the number of distinguishable ways of placing the molecules on the lattice. Flory and Huggins [5] in their initial works were mainly concerned with the effects of varying molecular size on the thermodynamic properties. In the framework of this theoretical approach no information about the architecture of the monomers and the resulting polymers are involved. The architecture of the polymer is often determined by the functionality of the monomers from which it is formed. This property of a monomer is defined as the number of reaction sites at which may form chemical covalent bonds. The basic functionality required for forming even a linear chain is two bonding sites. Higher functionality yields branched or even crosslinked or networked polymer chains. Modern polymerization strategies such as dendritic macromolecular chemistry involve the formation of large multiples of covalent bonds between homogeneous monomers to produce large molecules or infinite networks with a broad range of structure control [6].

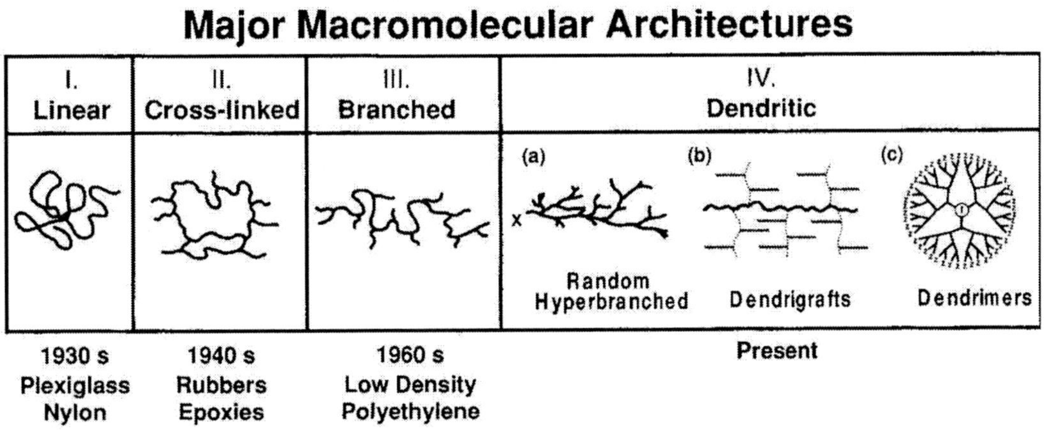

Figure 1. Representation of the four major classes of macromolecular architectures [6].

Historically, each of the three macromolecular architectural classes, i.e. (I) linear, (II) crosslinked, and (III) branched, has spawned rich polymer science (Figure 1). These architectural discoveries have been characterized by the emergence of new syntheses, structures, phenomena, properties, and products that have dramatically improved the human condition. Nanotechnology initiatives have focused on new synthesis strategies, structures, phenomena, and properties associated with length scales of 1-100 nm. These dimensions encompass biological building blocks (protein, DNA, RNA, etc.) and abiotic application areas (nanophotonics and nanoelectronics) [7]. Dendritic polymers are recognized as the fourth major class of polymeric architecture consisting of three subsets that are based on degree of structural control, namely: a) random hyperbranched polymer, b) dendricraft polymers and c) dendrimers [6, 7].

The outstanding polymer chemist Flory [8] was the first to hypothesize concepts, which are now recognized to apply to statistical, or ‘random hyperbranched’ polymers. However; the first synthesis of hyperbranched polymer could be realized 35 years later [9]. Webster and Kim [10] coined the popular term ‘hyperbranched polymer’ that has been widely used to describe this type of dendritic macromolecules. In theory, all polymer-forming reactions can be utilized for the synthesis of hyperbranched polymers, however; in practice some reactions are more suitable than others [6]. Hyperbranched polymers are typically prepared by polymerization of AB_X monomers, where x is two or more. They are produced by the one-pot polymerization of AB_X monomers or macromonomers involving polycondensation [11, 12, 13, 14, 15], ring opening [9, 16, 17, 18], or polyaddition [19, 20] reactions hence the product usually consist of broad statistical molecular weight distribution [21, 22, 23, 24]. Frechet et al. [25] presented the first example of a hyperbranched vinyl polymerization initiating the birth of a ‘second generation’ of hyperbranched polymers. Frechet and Chang [26] reported proton-transfer polymerization as a versatile route to hyperbranched polymers. Remarkable progresses have been made in recent years in the exploration of metal-mediated and metal-free click polymerization [27, 28] systems and in the syntheses of linear and hyperbranched polytriazoles with regioregular molecular structures and advanced functional properties [29]. Hyperbranched polymers are often modified to tailor their properties for a specialized purpose. Five modification methods have been developed: 1) end-capping with short chains or organic molecules; 2) terminal grafting via living polymerization; 3) growing hyperbranched polymers on the surface, or grafting from/onto the surface; 4) hypergrafting to obtain hyperbranched polymers with a linear macromolecular core; 5) blending or crosslinking.

Star polymers are three-dimensional hyperbranched structures in which linear arms of the same or different molecular weights emanate from a central core. The existence of numerous functional groups in a small volume makes these polymers important for use in biological and pharmaceutical applications. Biologically active molecules can be immobilized on the surface of the polymer gel or incorporated into the network.

Unique architecturally driven properties that may be expected from hyperbranched polymers will be largely derived from their a) amplified number of terminal functional groups, b) new rheological properties based on less chain entanglement, c) new architectural arrangements that may modulate crystallinity, flow characteristics and glass transition properties in designed systems and d) the formation of micelles.

The degree of branching is one of the most important molecular parameters of hyperbranched polymers since it determines many physical properties including the phase behavior of these polymers. This contribution focuses on the incorporation of the architecture of the polymers in the thermodynamic equations necessary for phase equilibrium calculations.

PROPERTIES AND APPLICATIONS OF HYPERBRANCHED POLYMERS

The unique properties of hyperbranched polymers are mainly manifested by their intrinsic globular structure and large number of terminal functional groups. Unlike dendrimers, however; hyperbranched polymers have elements of conventional polymers, namely molecular weight dispersity, isomerism, and geometrical shapes. Because the addition of each monomer takes place randomly, a large number of geometrical isomers can be formed even for a given molecular weight and branching degree. The polydispersity of the hyperbranched polymers is expected to increase to infinity at infinite polymer molecular weight. Besides the complex structure, also molar mass determination for hyperbranched polymers is far from trivial. It is obvious that molar mass determination by GPC lacks the fact that linear standards are not suitable for calibration. The application of light scattering and viscosity detector in the GPC as mostly done for branched polymers improves the results but still the broad molar mass distribution and the large number of polar end groups might cause problems [30, 31].

Some of these polymers exist as unusual colloid-like aggregates and form unimolecular micelles. Unimolecular micelles are defined as a class of macromolecules, wherein an interior hydrophobic core is surrounded by a hydrophilic surface layer. These structures closely resemble the shape of classical micelles except that they are static, in contrast to the dynamic nature of micelles, with all end-groups attached to the central core. The ability of guest molecules to penetrate the lipophilic interior can be used as drug or gene nano-carriers [10, 32, 33, 34, 35, 36]. In addition dendrimers can be surface engineered to release the drug at desired site, that is, as targeted drug delivery. This property along with the solubilisation behavior could improve the bioavailability of drugs. Hyperbranched polymers should be an attractive candidate for a gene- or drug-delivery system in aqueous media and could provide the phase-transfer carriers between water and organic media. Recently [34, 37, 38], biodegradable unimolecular reversed micelle consisting of a hyperbranched hydrophilic core and hydrophobic shell, were introduced. Liang et al. [39] demonstrate the possibility that a unimolecular micelle can simultaneously deliver both polar and apolar guests. In a biphasic water/chloroform mixture, the nanocapsule can transfer anionic, water-soluble guest from an aqueous phase to the chloroform phase; while when dissolved in water, the nanocapsule can efficiently capture both ionic and apolar solutes. Release of the guest can occur under the stimulus of pH or the switch of medium.

Above their glass transition temperature, these hyperbranched polymers are considered to minimize their free energy by lowering their free volume via a conventional nematic mesophase which is generated by a conformational change of their structural units from gauche to anti [40]. Percec et al. [41] were the first to report on flexible, non-spherical polymers with AB_2 mesogens in the branches that exhibit nematic and smectic behavior. On the other hand, aromatic polyamides exhibit polymer aggregation in the absence of a complexing salt resulting in a mesomorphic phase [42].

In general, hyperbranched polymers and dendrimers are more soluble in the same solvents than their linear analogous [43]. For higher generation dendritic polymers, solubility characteristics depend predominantly on the properties of their functional endgroups. As an example, dendrimers with hydrophobic interiors such as polyethers and polycarbosilanes can be made water soluble by introducing hydrophilic groups. Oppositely, water soluble dendrimers can be generated to be hydrophobic by converting their functional groups into hydrophobic units [44]. Furthermore, end group modification allows ideally to optimize their properties for special applications and to fine tune e.g. miscibility, melt rheology, surface and optical properties as well as biocompatibility.

Numerous applications in the field of coatings and blends, nanoscience, microelectronics, information technology, optics, lithography, organic light-emitting diodes, solid electrolytes, photoresponsive materials, chemical- and biosensors, and medicine (drug delivery systems, tissue engineering, magnet resonance contrast agents) have been suggested for hyperbranched polymers but only a few have yet reached full commercial exploitation. Several reviews [6, 7, 15, 45, 46, 47, 48, 49, 50, 51, 52, 53, 54] about this issue were publicized in the past. In this contribution only a few of them will be highlighted.

Application in Medicine

After drug administration, the drug may pass through different physiologic barriers and/or pathways, decreasing the actual amount of drug that reaches the site. Tissue specificity, product stability and solubility all desirable characteristics of drug, but are not always attained. Therefore, the need of develop a drug carrier system with such characteristics is of great importance. In the last 25 years [55], there have been numerous efforts focused on the development of the drug carrier systems. Investigators have made attempts to develop a specific drug carrier system, which can maintain continuous drug levels in a desired range, reduce side effects by improved tissue or organ specificity.

Many polymeric carriers have been investigated for therapeutic applications [55]. However; only a few polymers such as linear poly (lactide-co-glycolide), polyethylene glycol, and acrylic-based polymeric carriers have been introduced at commercial scale for controlled-release applications. Müller et al. [56] indicated that most polymeric drug delivery systems suffer from two major drawbacks, i.e., the cytotoxicity of polymers and the lack of processability. Hence, the successful formulation of a polymeric drug delivery system requires a system solution, i.e. a) a carrier with a narrow molecular mass distribution that fulfils nontoxicity and controlled-release criteria, b) sufficient loading capacity for the desired applications and c) processability by means of a commercial encapsulation technique. Even after considering the aforementioned system solution, factor such as the bioavailability, biodegradability, biodistribution, and drug efficacy can still limit the effectiveness of a polymeric drug delivery system. The processing of many therapeutic formulations has been suggested with commercial encapsulation techniques employ volatile organic compounds for the processing of polymeric carriers, where the residuals leading to the major health concern. However; most polymeric carriers such as polyesters and polyamides are very difficult to process in the absence of volatile organic compounds, because of high melting points, limited solubilities in supercritical gases, or high solution and melt viscosities are nonprocessable in most encapsulation techniques. Furthermore, harsh operating conditions and the cost effectiveness of the encapsulation technique may also limit the commercialization of the polymeric drug delivery system.

In the search for an ideal carrier system, the hyperbranched polymers may have significant potential. Hyperbranched polymers and their substitutes can be used as nanomaterials for host-guest encapsulation for several molecules such as dyes, pharmaceuticals, cosmetics, fragrances, catalysts and pollutants as well as the fabrication of organic – inorganic hybrids, and even directly as nanoreactor [37] for some reactions. As carriers, hyperbranched macromolecules can offer their interior or peripheral functional groups to covalently fix bioobjects, or depending on their core-shell architecture, to sequester guest molecules. The information stored at a molecular level plays a key role in this process. For a controlled release application, a change in pH, temperature, pressure, or a bacterial, enzymatic, or catalytic activity disintegrates the carrier system, leading to the release of the encapsulated active substance. The loading capacity and the release kinetics of carriers based on hyperbranched polymers are dependent on the polymer backbone, the number and type of functional groups, the molar mass, the polydispersity, and the amphiphilicity of the macromolecule. In controlled-release applications, hyperbranched polymers are functionalized in such a way that the shell of the hyperbranched carrier not only protects the encapsulated guest molecules, but also responds to the target environment via specific physiological change such as temperature, pH, chemical, and/or enzymatic reactions [57, 58].

Well-characterized, hyperbranched polymers were subjected to functionalization for preparing drug delivery systems of low toxicity, high loading capacity, ability to target specific cells and transport through their membranes [57]. They open new routes for the development of controlled-drug delivery systems with the potential to inhibit microbial adhesion to host tissues. In past years, the scope of encapsulation has expanded from pharmaceutical applications to cosmetics and agrochemical products [51]. The common goal for these applications is a high degree of control over the release mechanism for the encapsulated active substances.

As hydrophobic drug substances are difficult to introduce in the human body, many studies focus on increasing the solubility of drugs in the aqueous phase. As an example, tamoxifen, which is a hydrophobic breast anticancer drug, was encapsulated by Tziveleka et al. [59] in hyperbranched polyglycerols modified with polyethylene glycol. The results demonstrate that molecular encapsulation, based on hyperbranched nanocarriers, allows transporting 12 times more tamoxifen into an aqueous phase, compared to the solubility of the pure drug. The release of tamoxifen observed upon addition of sodium chloride is, in most of the cases, significant only at concentrations exceeding the physiological extracellular concentration [59].

Hyperbranched polymers such as polyesters, polyethyleneimine, polyglycerol, poly(ethylene glycol) and different copolymers have been investigated as potential polymeric vehicles for drug delivery applications and have proven to be highly biocompatible, thermally and chemically stable [57]. Moreover, although most dendritic polymers show cytotoxicity, low-molecular mass hyperbranched polyglycerols have proven to be nontoxic. The encapsulation of therapeutic agents such as ibuprofen [60, 61, 62, 63, 64], acetaminophen [65], tamoxifen [59, 66], ketoprofen [63], diflunisal [63], naproxen [63], paclitaxel [67, 68, 69, 70, 71, 72, 73], docetaxel [73], glimepiride [74], doxorubicin hydrochloride [75, 76], methotrexate [75], sodium ibandronate [75], cisplatin [77], chlorambucil [78], cytochrome c [79], amphotericin B [80], nimodipine [81], indomethacin [82], doxorubicin [76, 83], and docetaxel [70] were studied. The in vitro release of ibuprofen from drug-dendrimer complex is appreciably slower compared to pure ibuprofen. The complex drug enters A549 cells much more rapidly than pure drug suggesting that dendrimers may be able to carry the complex drug inside cells efficiently [64]. Hyperbranched Polyol with 128 OH end groups appears to encapsulate approximately 24 drug molecules [60].

The use of oral antidiabetic drugs for management of type 2 diabetes increases rapidly caused by the discovery and approval of several new types of oral antidiabetic drugs with different mechanism of pharmacological action [74]. Many of these drugs show poor solubility, slow dissolution rate in water, pH-dependent solubility and high permeability. Several approaches to improve water solubility include prodrugs, complexation, cosolvency, solid state modifications, surfactants were investigated. Among these the addition of cosolvents, the formation of cyclodextrins or micellar inclusions and the preparation of solid dispersions are the most commonly used. Many of these solubilisation techniques have their own limitations, toxicity, nephrotoxicity. The transformation of the drug to its amorphous form is often desirable since the solubility increases from a few to many-fold. The presence of hydrophilic compounds in close contact with the drug molecules increases the solubility by maintaining the drug in a molecular state and maximizing the surface area of the compound. The polymeric molecules also act as crystallization inhibitors and preserve the drug in its amorphous state [74].

Polymeric biomaterials are also of particular interest for the parenteral administration of peptides and proteins [84]. In an initial phase, release occurs predominantly by pore diffusion through an interconnecting network formed by the dissolving drug substance itself. The second release phase is governed by polymer degradation. Polyphasic drug release profiles can be overcome either by formulation approaches or by modification of the biodegradable polymers. While the release properties of biodegradable microspheres can be modified only in a limited sense, polymer modifications provide a broader spectrum of possibilities. Hydrophilic multi-arm polyethylene oxides seem to be promising candidates for parenteral protein delivery systems, allowing a synchronization of both pore diffusion and polymer erosion for the controlled release [84].

Boltorn hyperbranched dendritic polymers functionalized with mannose have been used to inhibit DC-SIGN-mediated infection in an Ebola-pseudotyped viral model [85]. Their physiological solubility, lack of toxicity and especially their low price suggest the application of these glycodendritic polymers for possible formulation as microbicides [85].

Gene delivery is a relatively new discipline of drug therapy and is defined as the incorporation of foreign nucleic acid that can be mediated by viral and nonviral methods into the host cells. An ideal gene delivery method must meet three major criteria: a) it should protect the transgene against degradation by nucleases in intercellular matrices, b) it should bring the transgene across the plasma membrane and into the nucleus of target cells, and c) it should have no detrimental effects. Nucleic acids are conventionally bound to viral vectors, but polymeric carriers such as polyethylenimine (PEI) have also been used. Linear PEI shows high transport efficiency, because of the relatively better loading capacity. However; an increase in the loading capacity of PEI occurs with an increase in the molar mass, which, in turn, increases cytotoxicity. Moreover, factors such as nondegradability, cytotoxicity, and short circulation time in bloodstream hinder the desired effects of the therapy. A common feature of the hyperbranched polymers related to gene therapy is the exhibition of polyvalent interactions, while for multifunctional derivatives, a number of targeting ligands determine specificity, and another type of group secures stability in biological milieu and prolonged circulation, while others facilitate their transport through cell membranes [58, 86]. Furthermore, polymers employed for gene delivery should be or become cationic in the biological environment for the formation of complexes with the negatively charged genetic material [87]. Hyperbranched polyethylenimine and their modifications [72, 88, 89, 90, 91, 92, 93], polyethylene glycol based cationic hyperbranched polymers [94, 95], hyperbranched polysiloxysilane [96], hyperbranched Boltorn H® with introduced tertiary amines [97], hyperbranched poly(amido amine)s [98, 99, 100, 101], and poly(amino ester)s [90, 102] have been investigated as carriers for gene delivery applications. It was found that hyperbranched polyglycerols were nontoxic to mice, even after the dosage injection of 1g per kg body weight [103]. Hyperbranched poly(amido amine) a highly efficient non-viral vector for gene delivery into numerous cell lines in vitro and in vivo [90, 104]. However; results obtained from in vitro studies do not always correlate to similar observations obtained in the in vivo experiments [6, 87].

The products for skin, hair or body care are generally referred to as "personal care products". Many of them contain biologically active substances, such as vitamins, supplements, or fragrances. In a recent study, it was shown that hyperbranched polymer molecules having primary amine groups react with aldehyde-based fragrances to form a fragrance delivery system [105]. Such fragrance moieties can e hydrolyzed to release fragrance molecules such as aldehydes and/or ketones via a reverse Schiff base reaction mechanism.

Dendritic substances have made crucial advances and have already been tested in preclinical studies, particularly in the field of contrast media for magnetic resonance. The relaxation time of the water protons are significantly shortened by application of the contrast media, such as Magnevist (gadolinium salt of diethylenetriamine pentaacetic acid) into the organ that is to be visualized and leads to a picture with a very good signal-to-noise ratio. However; clinically used contrast media reveal a disadvantage. Immediately after intravenous application they diffuse into the extravenous area as low-molecular compounds. Do prevent the diffusion higher-molecular weight compounds have been developed in which several gadolinium complexes are covalently bound to albumin, dextran, and polylysine. Unfortunately, no preparation is on its way to clinical studies, because of bad renal elimination of the compounds. The most promising solution of this problem consists in the use of a new compound [106] consists of a trimesinic acid central building block, to which second generation lysine dendrons with a total of 24 complexed gadolinium ions are anchored. Animal tests showed the quantitative renal elimination and the higher intravascular retention time, which ensures an excellent signal-to-noise ratio [106].

Tissue engineering has recently attracted a great deal of attention for its potential in repairing, reconstructing, regenerating, or replacing tissues in damaged or diseased organs. A number of tissues and organs have already been investigated using this approach, including cartilage, liver, skin, bone, tendon, ureter, intestine, pancreas, and blood vessels. One approach in tissue engineering employs a hybrid system combining biomaterials as structural scaffolds to organize living cells into a desired structure in vitro or in vivo. Polymers for these scaffolds require good biocompatibility, suitable biodegradability, and the ability to interact specifically with appropriate cells. The control of protein adsorption is a key issue in many tissue engineering. All this desired properties can be found for hyperbranched polymers [107, 108, 109, 110, 111, 112, 113].

The escalating global incidence of bacterial infection, particularly in chronic wounds, is a problem that requires significant improvements to existing therapies. For example, Shepherd et al. [114] have developed hyperbranched poly(N-isopropylacrylamide) functionalized with the antibiotics vancomycin and polymyxin-B that are sensitive to the presence of bacteria in solution. When the polymers were removed from the infected skin, either in a polymer gel solution or in the form of hydrogel membranes, they removed bound bacteria, thus reducing the bacterial load in the infected skin model [114]. Because poly(ethylene glycol) is a nontoxic, water-soluble polymer that resists recognition by the immune system, hydrogels prepared from poly(ethylene glycol) star polymers are excellent candidates as biomaterial.

Ever increasing demands on the healthcare industry offer challenging opportunities for producing products that are not only more cost effective and outperform existing products, but also provide ready access to a variety of new biological and analytical reagents. Recently, Xu et al. [115] synthesized a new dendritic core-shell architectures with pH-labile linkers based on hyperbranched polyglycerol cores and biocompatible poly(ethylene glycol) shells which encapsulate the anticancer agent doxorubicin and a dye for near-infrared imaging.

Application in Chemistry

Metal nanoparticles continue to evoke great current interest due to their tremendous potential in designing smart materials for a wide variety of applications (for instance silver [116, 117, 118, 119] or gold [120, 121] with antibacterial activity [122, 123] for drug delivery purpose or antibacterial coating, platinum [124, 125], gold [126, 127] or titania [128, 129] for catalysis, iron for introduction of magnet properties [130]). Much emphasis has been placed lately in developing methodologies that could modulate the size and shape of these metal particles. Dendrimers that are monodisperse in nature with a regular and highly branched three-dimensional architecture, provide a useful platform to accomplish this goal. These hyperbranched macromolecules have been widely explored as templates in the construction of metal nanoparticles [131, 132]. In general, these metallodendrimers are thought to be macromolecular materials that can combine the advantage of both homo- and heterogeneous catalysts [133, 134, 135, 136]. Because of their ‘pseudo’-spherical nature and their resultant conformations the metal sites in these well-defined polymeric catalysts should be easily accessible for substrate molecules and reagents, and therefore exhibit characteristics usually encountered in homogeneous catalysis such as fast kinetics, specificity and solubility. Owing to their precise persistent nanoscale size they may be easily removed from product streams, e.g. by means of ultrafiltration techniques.

Gladitz et al. [137] produced a long-term-stable silver nanoparticle dispersions with narrow size distribution using amphiphilic-modified hyperbranched polyethyleneimines and investigated the antibacterial performance and morphology of thin silver-loaded hyperbranched polymer coatings on poly(ethylene terephthalate) prepared by different wet coating techniques. Thermosensitive gold nanoparticles with tunable lower critical solution temperature have been prepared by coating the nanoparticles with a thermo- and pH-responsive hyperbranched polyelectrolyte [120]. The globular, polyfunctional structure of the hyperbranched polymers will offer new options to control phase-transition temperatures of the nanoparticles.

Membrane-based gas separations have attracted much attention in the past because they offer many significant advantages over traditional separation processes such as low energy consumption, low capital investment cost, simple and easy operation. A large number of polymeric materials have been studied for this application. Hyperbranched polymers [138] show an average cavity size of 0.5 nm, which is larger than that of common gas molecules such as H_2, O_2, CO_2, N_2 and CH_4. Rölker et al. [139] examined the application for gas separation such as CO_2 absorption from flue gas. For example, one amine-terminated hyperbranched polyimide membrane crosslinked with terepththaldehyde exhibited fairly high CO_2/N_2 separation performance and good selectivity [49]. The gas permeability of different gases such as CO_2, O_2 and N_2 for hyperbranched polyimide containing trifluoromethyl groups [140], for star-like poly(ethylene oxide)s [141], for polyimide - silica hybrid membranes [142, 143] for polyimide [144, 145], and Boltorn-modified polyimide [146] was investigated. Sterescu et al. [147] describes the preparation, characterization and the permeation properties of poly(2,6 dimethyl-1,4-phenylene oxide) dense polymer films containing aliphatic hyperbranched polyesters, Boltorn (H20, H30, and H40). The gas permeability of poly(2,6 dimethyl-1,4-phenylene oxide) with 1.0 wt % of Boltorn is 2-3 times higher than the pure polymer, while at higher concentration (9.1 wt %) of Boltorn the permeability becomes almost 50% of the pure polymer [147]. The gas pair selectivity, however; stays constant. The increase in permeability at low concentration of Boltorn is due to the increase of the free volume, probably due to hydrogen bonds between Boltorn and the oxygen of poly(2,6 dimethyl-1,4-phenylene oxide) backbone, whereas the decreased permeability containing higher concentration of Boltorn (9.1 wt %) is due to two reasons: decrease in free volume as determined by positron annihilation lifetime spectroscopy as well as phase separation [147]. The hyperbranched polyesters form aggregates that migrate to the top surface of the membranes.

The applications of highly crosslinked thermoset polymers are limited by their intrinsic brittleness. To increase their toughness, different solutions, most often based on the addition of toughening or plasticizing components, have been developed. When the resulting blended system is homogeneous, the increase in toughening is generally provided by plastification of the thermoset matrix or reduction of the crosslink density. Moreover, other properties such as the Young's modulus or glass-transition temperature are usually strong affected. To limit the effect of these drawbacks, the pure resin system can be blended with phase separating rubbery modifiers, which will remain only partially dissolved in the thermoset. Phase separation may be promoted during processing either by temperature change or by chemically induced phase separation. In both cases, one resin-rich phase and a second phase rich in modifier are formed. The second phase appears as small particles whose diameter may range from a few nanometers to some tens of microns. Boogh et al. [148] suggested the application of hyperbranched polymers as reactive modifiers. They [148] have been shown to have a marked toughening effect without loss of thermomechanical properties and processability.

The large number of terminal functional groups allows the use of hyperbranched polymers as crosslinking agents.

The peculiarities of dendritic structures offer new possibilities to manipulate the properties of surface by coating them. In the formation of these macromolecules at interfacial areas, either gas-water or gas-solid transitions, a change in conformation compared to that in solution has been noticed [47].

Hyperbranched polymers can also be utilized in nanoimprint lithography. A fascinating application of the nanoimprint lithography technique is to fabricate so-called quantum magnetic disks, comprising single magnetic domain bits surrounded by a non-magnet material. Pattering of polymer films at micron or even submicron resolutions is of critical technological importance in the microelectronics industry. Because hyperbranched polymers have many functional groups, moieties with interesting optical, electrochemical, biological, and mechanical properties can be incorporated into the hyperbranched polymer films. Related techniques including template-based approach and photolithography have been developed by Crooks and coworkers [149].

Based on their unique properties, hyperbranched polymers can be applied as toughener for thermosets, curing, cross-linking or adhesive agents, dye-receptive additives for polyolefins, compatilizers, dispersers, processing aids, and rheology modifiers or blend component. The Boltorn^{TM} three-generation hyperbranched polyesters can act as outstanding toughener in epoxy matrix composites, and they can include more than a two-fold increase in the critical strain energy release rate for both low and highly cross-linked matrices, without affecting either the viscosity of the uncured resins or the thermomechanical properties of the cured material [49]. In the tubular film blowing process of linear low density polyethylene (LLDPE), Boltorn H30 hyperbranched polyester successfully acted as a processing aid, and sharkskin was eliminated with no significant influence on the other physical properties of LLDPE films [150, 151]. In processing, the hyperbranched polymer has a tendency to migrate to the surface, leading to a hyperbranched polymer-rich surface, which creates the opportunity to tailor the surface properties for various potential applications.

Polypropylene (PP) is a versatile and widely used polymeric material, however; PP does not accept dyes, owing to its non-polar structure as well as its high crystallinity, resulting from the high stereo-regularity of PP. The dyeability of PP with Dispere Blue 56 can be markedly enhanced through the incorporation of hyperbranched macromolecules (Hybrance PS 2550) into PP prior to fibre spinning [152, 153].

Perstorp demonstrated that hyperbranched aliphatic polyols can be used, for example, for hardcoat applications, because they allow for a unique combination of coating properties such as high hardness, scratch resistance, and flexibility while ensuring at the same time low viscosity and low shrinkage.

Several pharmaceutical, electronic and optical manufacturing application require “ultrapure” water that meet stringent limits concerning the dissolved and suspended solids, organic carbon, dissolved gases, and biological organisms. In this context, hyperbranched polymers have been suggested for the extraction of model impurities, such as polycyclic aromatic hydrocarbons via molecular encapsulation. Recently [154, 155], it was suggested unimolecular micelles based on hyperbranched polymers as "nanosponges" to extract pollutants from an aqueous phase followed by their subsequent release in an organic phase.

Application in Energy Technique

A solid polymeric electrolyte should meet the requirements of 1) being amorphous, 2) having a high solvating power for appropriate ions, 3) good ion transport, and 4) electrochemical stability. It is well known that polyglycerols segments satisfy the last three requirements, and hyperbranched polymers are usually amorphous. Thus, hyperbranched macromolecules possessing ethylene glycol chains have been designed, prepared and used as novel polymeric electrolytes or ion-conducting elastomers [156, 157, 158]. Novel polymer electrolytes have been prepared with hyperbranched polymers as the host and lithium trifluoromethanesulfonimide, $LiN(CF_3SO_2)_2$, as the ion source. Analysis of the salt concentration dependence of the glass transition temperature indicates that intermolecular interactions occur in the polymer electrolytes and that these hyperbranched polymers can function as a 'solvent' for the lithium salt [159].

Hyperbranched polymers can also be used to developed fuel cell membrane material [160, 161, 162, 163]. It was found that the molecular weights of the hyperbranched polymers do not affect significantly the ionic conductivity, but the molecular weight distribution might affect it, and also further branching at the terminals of the hyperbranched polymers led to a decrease in the ionic conductivity [160]. Additionally, it was demonstrated that applying the concept of dry polymer system to proton conduction is one possible approach toward high-temperature fuel cells [161, 164, 165, 166].

In recent years, the search to develop large-area solar cells at low cost has led to research on photovoltaic systems based on nanocomposites containing conjugated polymers. These composite films can be synthesized and processed at lower costs and with greater versatility than the solid state inorganic semiconductors that comprise today's solar cells [167, 168, 169]. However; the best nanocomposite solar cells are based on a complex architecture, consisting of a fine blend of interpenetrating and percolating donor and acceptor materials. Cell performance is strongly dependent on blend morphology, and solution-based fabrication techniques often result in uncontrolled and irreproducible blends, whose composite morphologies are difficult to characterize accurately. The incorporation of three-dimensional hyperbranched colloidal semiconductor nanocrystals in solution-processed hybrid organic-inorganic solar cells yields reproducible and controlled nanoscale morphology.

Phase change materials have attracted lots of interest because of their high storage density and constant temperature during phase change process. Their applied field is extensive, such as solar energy storing, smart air-conditioning, buildings, agricultural greenhouse, temperature-regulating textiles, heat management of electronics, telecommunications and microprocessor equipment. Solid-solid phase change materials have some advantages, for instance no liquid or gas generation and small volume change. Cao and Liu [170] prepared hyperbranched polyurethane for thermal energy storage as a novel material. The phase transition enthalpy was more than 100J/g with a transition point at 67°C. Doped hyperbranched polyurethane leads to a material with a transition enthalpy of 125.0J/g and the main decomposition temperature of 437°C [171, 172].

However; one should recognize that extensive research of hyperbranched polymers is still in its infancy, and the main object to apply hyperbranched polymers in industry fields is still a dream with few exceptions. DSM introduced hyperbranched polyesteramides to oil field applications, because of their ability to suppress the crystallization of gas hydrated from water-hydrocarbon mixtures [51]. BASF demonstrated that hyperbranched polymers can be used as cross-linkers for coatings [51]. Evonik Degussa uses hyperbranched polymers as performance additives and enzymatically degradable carriers [51].

THERMODYNAMICS OF HYPERBRANCHED POLYMERS

As polymer dissolution plays a crucial rule in many applications. The paper describes theoretical approaches to quantify this complex phenomenon. The phase behavior of concentrated solutions of hyperbranched polymers is of great importance for their application in the field of chemical engineering (membranes, extractive distillation, extraction, absorption, adsorption, etc.) or in the field of medicine (drug delivery systems, tissue engineering, development of sensors, etc.). Kim [173] to point out "Better understanding on physical properties of these polymers, such as solubility and miscibility of these polymers in solvents or with polymers, and functional group dependency to the thermal relaxation process are needed for further development of the subject.".

ROLE OF BRANCHING

Branching of polymer chain plays an important role in liquid-liquid phase equilibria (LLE) in polymer solutions and influences quantities like separation temperature and/or separation pressure of a polymer solution. Kleintjens at al. [174] showed for the system polyethylene + diphenyl ether that the two-phase region of a branched polyethylene solution may be shifted by more than 10°C compared with that of a linear polyethylene sample of about equal number and mass average molar mass.

De Loos et al. [175] measured cloud-point curves and critical curves in the system ethylene + branched polyethylene and concluded that the cloud-point curves of a system with a branched polyethylene are at significantly lower pressure (10 – 40 MPa) that the cloud-point curves of a system with a linear polyethylene with comparable molecular-weight distribution.

Striolo and Prausnitz [176] measured the osmotic second virial coefficients for linear and star poly(ethylene oxide) in water and propan-1-ol. They [176] figured out that at fixed temperature, the osmotic second virial coefficient for star poly(ethylene oxide) is lower than that for the linear polymer of comparable molecular weight.

Kennedy et al. [177] generalized the Flory-Huggins free enthalpy of mixing expression to solution of branched polymers, assigning different interaction energies to solvent-end segment and solvent-middle segment contacts. In 1980, Kleintjens and Koningsveld [178] realized "To data, polymer solution theories have failed to describe quantitatively the effect of polymer chain branching on liquid-liquid phase relationships in solutions". They tried to improve this situation using an adopted lattice treatment to incorporate this phenomenon [178]. A branched polymer molecule has been presented as a chain, composed of end, linear middle, and branched middle segments, each interacting through a specific surface area with the immediately surrounding molecules or segments. The expression obtained for the energy of mixing contains five parameters. The model described spinodal, critical, and cloud-point data within experimental error over practically the full measured concentration range. Phase compositions, however; could only be reproduced qualitatively.

On the basis of experimental batch-cell data, increasing short-chain branch density was found to reduce the cloud-point pressure of poly(ethylene-1-butene) solution in propane, by as much as a factor of 3 [179]. This trend was captured by a SAFT (Statistical Association Fluid Theory) approximation via an effective segment energy that depends on the branch density. However; it was not possible to accomplish quantitative agreement with the performed experiments.

Blas and Vega [180] presented an extension of the SAFT for branched chain molecules using Wertheim's first- and second-order thermodynamic perturbation theory with a hard-sphere reference fluid. Molecules are formed by hard spherical sites which are tangentially bonded. Linear chains are described as freely jointed monomeric units, whereas branched molecules are modeled as chains with a different number of articulation points, each of them formed by three arms. The theory was applied to the critical temperature of light branched alkanes, whereas the results show that the theory qualitatively predicts the experimental data, although the proposed molecular description was too simple to account for the finer details of the molecular architecture of branched chains [180]. Additionally, the branching term accounts for the branching units that are given as rigid tetramer units within a hyperbranched macromolecule and hence, only branching points with four bonds can be described.

A similar SAFT-approach was used by Kozlowsky et al. [181] for the calculation of phase behavior of polymer solutions containing hyperbranched polymers. The vapor-liquid equilibrium could be modeled very close to the experimental data over a large pressure range [181]. Unfortunately, the calculation of the liquid-liquid phase split was not successful [181].

The experimental investigation of the phase behavior of hyperbranched polymer solutions is a crucial requirement for a successful introduction of new applications to highly competitive markets. Several studies dealing with the liquid-liquid equilibria in binary [181, 182, 183, 184, 185, 186, 187, 188] and ternary [181, 182, 184, 189, 190, 191] systems can be found in the literature. For instance, Boltorn U-3000 is completely soluble in hydrocarbons (hexane, heptane, benzene, and toluene) and ethers (methyl-tert-butyl ether or ethyl-tert-butyl ether) but only partially miscible with alcohols, where the solubility in alcohols increases with an increase of the chain length of the solvent [183]. Boltorn H3200 shows a liquid-liquid phase split with lower critical solution temperature that interferes with the solidification surface of Boltorn H3200 [185].

Lieu et al. [192] as well as Mio et al. [193] presented vapor-liquid equilibrium data for binary solutions of dendrimers in a variety of organic solvents. These authors figured out that the generation number and chemical composition of the dendrimers have a significant effect on solvent-dendrimer compatibility.

Domanska and Zolek-Tryznowska [194] measures the solid-liquid and the liquid-liquid phase diagrams at ambient pressure for the system Boltorn W3000 with alcohols, or with ethers, where the liquid-liquid equilibrium shows a LCST behavior. Ko et al. [195]

investigated the influence of salt on the solid-liquid phase behavior of hyperbranched polymers.

Using inverse gas chromatography the infinite-dilution activity coefficients were measured [196, 197, 198, 199, 200, 201]. Solvent activity coefficients of amine-terminated poly(propylenimine) of generation 2 to 5 in different solvents (methanol, ethanol, ethyl acetate, toluene, and THF) change with respect to the dendrimer generation number, reaching a minimum at generation 4. Because the dendrimers are basic, the solvent quality is higher for slightly acidic solvents, such as methanol and ethanol [196].

Other properties, like associations [202], densities and viscosities [203, 204] or osmotic compressibility [205], osmotic virial coefficients [176] of hyperbranched polymer solution were also measured.

Recently, Cheng et al. [206] reports for the first time the influence of degree of branching (DB) on the thermoresponsive phase transition behaviors of hyperbranched multiarm copolymers. Two series of PEHO-star-PEOs (series A) and PEHO-star-PDMAEMAs (series B) with the hydrophobic DB-variable PEHO core and different kinds of linear arms (PEO arms or PDMAEMA arms) were synthesized. It was found these two series demonstrate thermoresponsive phase transitions with the LCST [206]. The studies on the LCST transition mechanism indicate that series A belongs to the thermoresponsive polymer system with LCST transition based on hydrophilic-hydrophobic balance, while series B belongs to the thermoresponsive polymer system with LCST transition based on coil-to-globule transition [206]. For series A, the LCST phase transition is highly dependent on the DB of the PEHO core in copolymers, whereas for series B, the LCST phase transition is independent of the DB but dependent on solution pH [206]. Compared with linear thermoresponsive polymers hyperbranched ones having spheroid-like structure exhibited an unusual salt effect: a non-linear LCST decrease upon increasing the concentration of various salts such as NaCl, KCl or Na_2SO_4 has been observed [207].

Using small-angle neutron scattering Martter et al. [208] measured bulk thermodynamic interaction parameters for blends of anionically polymerized star with different number of arms and linear polybutadienes of well-defined architecture and molecular weight. Comparison of these measured values with results from comparable polystyrene blends suggests the existence of nonuniversal aspects in the thermodynamic interaction due to entropic contributions [208]. While the interaction parameter for polystyrene star/linear blend increases monotonically with number of arms in the star, the value in the polybutadiene star/linear blends does not [208].

Massa et al. [209] provide a survey of the phase behavior of blends of hydroxyphenyl- and acetoxyphenyl-terminated hyperbranched polyesters with polymers such as polycarbonate, polyester, and polyamides. The hydroxyterminated hyperbranched polyester blend miscibility was identical to that of poly(vinylphenol); this suggests that strong interactions due to hydrogen bonding, more so than chain architecture, dominate the blend miscibility [209]. Miscibility was found in blends of the two hyperbranched polyesters and in those of poly(acetoxystyrene) with poly(vinylphenol) but none of the blends of linear analog with hyperbranched polymer was single phased [209].

All these experimental observations demonstrate clearly the role of the architecture of the polymer and the interaction between different segments present in the system.

THERMODYNAMIC MODEL – LATTICE CLUSTER THEORY

Classic theories of polymer systems, as typified by the widely used Flory-Huggins (FH) theory [1, 5], treat individual monomers as single entities, devoid of any chemical structure. In the framework of the FH theory the polymers are taken to have random configurations, which arise from placing individual monomers at single lattice sites, FH-theory subject to strict constraint applies to solvent molecules when they are also present. Flory and Huggins [5] employed a very simple mean-field approximation that effectively ignores the details of the polymer chain connectivity and, therefore, that cannot distinguish between linear, star, branched, comb polymer architecture. The FH theory [5] is based on the polymer chain cohesive interaction energy parameter. Because most of the interaction between the hyperbranched polymers resides at their chain ends, rather than the chain interaction, the conventional Flory-Huggins theory may not be able to predict the miscibility of hyperbranched polymers.

While the lattice cluster theory (LCT), originally developed by Freed and coworkers [210, 211, 212, 213, 214, 215], extends well beyond FH theory first by endowing individual monomers with explicit molecular structures that occupy several lattice sites. Moreover, the LCT develops a vastly superior solution of the lattice model to include important contributions to thermodynamic properties from packing and interaction induced local correlations. The LCT describes the thermodynamic consequences of the existence of short-range correlation that arise from packing constraints and the different monomer-monomer interactions. Because only short-range correlations are included, the LCT technically is also a mean-field theory. The accuracy of the LCT has been tested against Monte Carlo simulation of various lattice model polymer systems [216, 217, 218, 219, 220, 221, 222] demonstrating that the LCT provides an excellent estimation of all thermodynamic properties in the mean-field regions, with the exception of the specific heat at lower temperatures.

Our application of the LCT to hyperbranched polymers are designed to probe the limitations of the LCT as a guide to future efforts for rectifying these limitations, to determine the qualitative features governing the thermodynamic properties of these fascinating systems [188, 223, 224, 225, 226, 227].

The partition function for the extended lattice model of polymer systems is evaluated within the LCT by introducing approximations for the bonding constraints and for the excluded volume and van der Waals interactions [215]. We assume that all chains are fully flexible. The theoretical description of polymer thermodynamics at a fixed pressure requires treating the system as compressible. The compressibility of the system can be introduced into the extended lattice models by allowing for the presence of empty lattice sites, called voids. The voids do not represent a species of particles in the thermodynamic sense, nor do they have interactions; they are merely a simple bookkeeping device for keeping track of the excess free volume. When computations are performed for constant pressure systems, the excess free volume fraction is determined numerically from the equation of state, derived us usual by defining the pressure as the derivative of the free energy with respect to total volume.

The structure of the polymer enters the LCT free energy through a series of geometrical indices $N_{j,i}$ that enumerate the number of distinct sets of j sequential bonds in a single chain of species, i.

$$N_{1,i} = M_i - 1 \tag{1}$$

where M_i is the number of monomers in a polymer chain of type i (segment number).

$$N_{2,i} = M_i - 2 + b_{3,i} + 3b_{4,i} \tag{2}$$

where $b_{3,i}$ is the number of branching points of degree 3 of the polymer chain of type i in which 3 bonds meet, and hence $b_{4,i}$ is the number of branching points of degree 4 of the polymer chain of type i in which 4 bonds meet. The quantity $N_{2,i}$ describes the number of two consecutive bonds in a polymer chain, i. The number of three consecutive bonds in a polymer chain, i, $N_{3,i}$ is given by:

$$N_{3,i} = M_i - 3 + 2b_{3,i} + 6b_{4,i} \tag{3}$$

The number of ways in which three bonds meet at a lattice site for a polymer chain, i, $N_{\perp,i}$ can be calculated using:

$$N_{\perp,i} = b_{3,i} + 4b_{4,i} \tag{4}$$

The number of distinct ways of selecting two non-sequential bonds on the same chain, i, can be expressed by $N_{1,1;i}$:

$$N_{1,1;i} = \frac{1}{2} N_{1,i} \left(N_{1,i} - 1 \right) - N_{2,i} \tag{5}$$

Finally, the number of distinct ways of selecting two sequential bonds and one non sequential bonds on the same chain, i, reads:

$$N_{1,2;i} = N_{1,i} N_{2,i} - 2N_{2,i} - 2N_{3,i} - 3N_{\perp,i} \tag{6}$$

The geometrical coefficients, used in Eqs. 1 - 6, can be calculate from the chemical structure of arbitrarily chosen hyperbranched polymer knowing the quantities M_i, $b_{3,i}$ and $b_{4,i}$. Let us assume the system consists of two different components. For simplification we introduce the quantities K_i for component 1 and L_i for component 2 in the following manner:

$$K_1 = N_{1;1} / M_1; \quad K_2 = N_{2;1} / M_1; \quad K_3 = N_{3;1} / M_1;$$
$$K_4 = N_{\perp;1} / M_1; \quad K_5 = N_{1,1;1} / M_1; \quad K_6 = N_{1,2;1} / M_1; \tag{7}$$

and

$$L_1 = N_{1;2} / M_2; \quad L_2 = N_{2;2} / M_2; \quad K_3 = N_{3;2} / M_2;$$
$$L_4 = N_{\perp;2} / M_2; \quad L_5 = N_{1,1;2} / M_2; \quad L_6 = N_{1,2;2} / M_2; \tag{8}$$

The lattice theory of a binary blend represents each component related to LLE of hyperbranched polymer solutions. ($i = 1,2$) as n_i monodisperse polymer chain placed on a regular array of N_L lattice sites and coordination number z. In LCT the Helmholtz free energy is expanded in a double power series of $1/z$ and $\varepsilon_{LCT} / (k_B T)$, where k_B means the Boltzmann constant. The interaction energy ε_{LCT} is defined by

$$\varepsilon_{LCT} = \varepsilon_{ii} + \varepsilon_{jj} - 2\varepsilon_{ij} \tag{9}$$

where ε_{ii} is the energy of a non-bonded polymer-polymer of polymer type i, ε_{jj} is the same quantity for polymer type j, and ε_{ij} is the energy of a contact between polymer of type i and polymer of type j. ε_{LCT} is related to the usual χ-parameter of the Flory-Huggins theory by $\chi = z\varepsilon_{LCT} / (2k_B T)$. To obtain an expression for the Gibbs energy of mixing from the partition function, we assume that, at the pressure of interest here, the change in molar volume upon mixing is negligible. In this limit of an incompressible lattice the Helmholtz free energy is equal to the Gibbs free energy. The Gibbs energy of mixing for a binary blend emerges from the LCT as [223, 227]:

$$\frac{G_{Mix}}{N_L k_B T} = \frac{\phi_1}{M_1} \ln \phi_1 + \frac{(1-\phi_1)}{M_2} \ln (1-\phi_1) + \sum_{i=1}^{6} \alpha_i (\phi_1)^i \tag{10}$$

where ϕ_1 is the concentration of component 1, expressed in segment fraction, and the α_i-values are:

$$\alpha_1 = \frac{\varepsilon_{LCT}}{k_B T}\left(\frac{z}{2} - K_1\right) + \frac{(K_1 - L_1)^2}{z} + \frac{\varepsilon_{LCT}}{z k_B T}\left(\begin{array}{l} 2K_2 + K_3 + 3K_4 - 4L_1K_5 + K_6 - 2K_1L_1 \\ -4L_1K_2 + 2L_1K_1^2M_1 - K_1K_2M_1 \end{array}\right)$$

$$+\frac{1}{z^2}\left(\begin{array}{l} -4L_1^4M_2 - 4L_1L_2 + 2L_1^2L_2M_2 + 2K_1^4 + K_2^2 + 8K_1^3/3 + L_2^2 - 4K_1K_2 - 2K_1K_3 \\ -2K_1K_6 - 2K_1^4M_1 + 4K_5K_1^2 - 6K_1K_4 + 2K_1^2K_2M_1 + 4L_1K_2 + 4K_1L_2 - 8L_1^2K_1 \\ +6L_1^4 + 16L_1^3/3 + 2K_1^2L_1^2M_1 + 4L_1^3K_1M_2 + 2L_1K_3 + 2L_1K_6 - 2K_2L_2 - 8L_1^3K_1 \\ -2K_1L_1L_2M_2 - 2K_1L_1K_2M_1 - 4K_5L_1^2 + 6L_1K_4 + 2K_1L_3 - 6L_1L_4 - 8K_1L_5L_1 \\ +8L_5L_1^2 + 2K_1L_6 - 2L_1L_3 - 2L_1L_6 + 6K_1L_4 \end{array}\right) \quad (11)$$

$$\alpha_2 = \frac{1}{z^2}\left(\begin{array}{l} -L_2^2 - 8L_1^3 - 2L_1^2L_2M_2 + 6L_1^4M_2 - 8K_1^2L_1 + 2K_1^2L_1^2M_2 + 2L_1K_1L_2M_2 \\ +2L_1K_1K_2M_1 - 12K_1^2L_1^2 + 8K_5L_1^2 + 4L_1L_2 + 2K_2L_2 - 8K_1K_5L_1 \\ +16K_1L_1^2 + 4K_1^3L_1M_1 - 4K_1L_2 - 4L_1K_2 - 2L_1K_3 - K_2^2 - 6L_1K_4 \\ -2L_1K_6 - 12L_1^4 - 2K_1^2K_2M_1 + 24K_1L_1^3 + 4K_1K_2 + 2K_1K_3 + 2K_1K_6 \\ +6K_1K_4 - 4K_1^2L_1^2M_1 - 8K_1L_1^3M_2 - 2K_1L_3 + 6L_1L_4 + 16K_1L_5L_1 \\ -12L_5L_1^2 - 2K_1L_6 + 2L_1L_3 + 2L_1L_6 - 4L_5K_1^2 - 6K_1L_4 \end{array}\right)$$

$$+\frac{\varepsilon_{LCT}}{z k_B T}\left(\begin{array}{l} -2K_6 - 6K_4 + 12L_1K_5 + 10K_1L_1 - 4L_1K_1^2 - 4L_1^3 + 2K_1^3M_1 - 4K_1^2 \\ +2L_2 - 2K_3 + 8L_1^2K_1 + 2L_1^3M_2 + 2K_1K_2M_1 + 8L_1K_2 - 4K_1K_5 - 4K_2 \\ -6L_1K_1^2M_1 - 4L_1^2 - L_1L_2M_2 - 4K_1L_2 + L_3 + 3L_4 + L_6 - 4L_1L_5 \end{array}\right)$$

$$+\frac{\varepsilon_{LCT}}{k_B T}\left(2K_1 - L_1 - \frac{z}{2}\right) + \left(\frac{\varepsilon_{LCT}}{k_B T}\right)^2\left(-\frac{z}{4} - K_1^2(1 + 2M_1) + 3K_2 + 4K_5 + 3K_1 - \frac{L_1}{2}\right)$$

$$-\frac{(K_1 - L_1)^2}{z} \quad (12)$$

$$\alpha_3 = \frac{1}{z^2}\left(\begin{array}{l} K_1L_1\left(8K_1 - 8K_1^2 + 24K_1L_1 - 24L_1^2 - 8L_1\right) - 8\left(K_1^3 - L_1^3\right)/3 \\ +(K_1 - L_1)^2\left(-2L_1^2M_2 + 2M_1K_1^2 - 4K_5\right) - 8\left(K_1^3 - L_1^3\right)/3 + 4L_5K_1^2 \\ -8K_1L_1L_5 + 4L_1^2L_5 + 8L_1^4 + 4L_5K_1^2 - 8K_1L_1L_5 + 4L_1^2L_5 \end{array}\right)$$

$$+\frac{\varepsilon_{LCT}}{z k_B T}\left(\begin{array}{l} 20L_1K_1^2 + 8K_5K_1 - 12L_1K_5 + 8K_1^2 - 4K_1^3 + 2K_2 + K_3 + 3K_4 + K_6 - 2L_2 \\ +L_1M_2(L_2 + 2K_1L_1 - 4L_1^2) - 16K_1L_1 + 12L_1^3 - 28L_1^2K_1 + 4\left(-L_1K_2 + K_1L_2\right) \\ +K_1M_1\left(6L_1K_1 - 4K_1^2 - K_2\right) - L_3 - 3L_4 - L_6 + 8L_1L_5 - 4K_1L_5 + 8L_1^2 \end{array}\right)$$

$$+\left(\frac{\varepsilon_{LCT}}{k_B T}\right)^2\left(\frac{z + 5L_1 - 13K_1}{2} - 3K_2 - 6K_5 + 3K_1^2M_1 - L_2 + 6K_1^2 + 2L_1^2 - 6K_1L_1\right)$$

$$+\frac{\varepsilon_{LCT}(L_1 - K_1)}{k_B T} \quad (13)$$

$$\alpha_4 = -2\frac{(K_1 - L_1)^4}{z^2} + \frac{\varepsilon_{LCT}}{zk_BT}\begin{pmatrix} -2K_1^2L_1M_1 - 12L_1^3 - 4K_1K_5 + 4L_1K_5 + 8K_1^3 - 2K_1L_1^2M_2 \\ -28L_1K_1^2 - 4(K_1 - L_1)^2 + 2K_1^3M_1 + 2L_1^3M_2 + 32K_1L_1^2 \\ -4L_1L_5 + 4K_1L_5 \end{pmatrix}$$

$$+\left(\frac{\varepsilon_{LCT}}{k_BT}\right)^2\begin{pmatrix} -\frac{z}{4} + L_2 + 4K_5 - 2K_1^2M_1 + K_2 + 6K_1 - 4L_1 + L_1^2M_2/2 - 7L_1^2 - 12K_1^2 \\ +18K_1L_1 - L_5 \end{pmatrix} \quad (14)$$

$$\alpha_5 = -4\frac{\varepsilon_{LCT}(K_1 - L_1)^3}{zk_BT} + \left(\frac{\varepsilon_{LCT}}{k_BT}\right)^2\begin{pmatrix} \frac{K_1^2M_1 - L_1^2M_2}{2} - 18K_1L_1 - 2K_1 + 2L_1 + 8L_1^2 \\ +10K_1^2 + L_5 - K_5 \end{pmatrix} \quad (15)$$

$$\alpha_6 = -3\left(\frac{\varepsilon_{LCT}}{k_BT}\right)^2(K_1 - L_1)^2 \quad (16)$$

Using standard thermodynamics Eq. 10 can be applied to liquid-liquid phase equilibrium calculations of a binary system. In principle, the binary system can be made from a) two solvents, b) two linear polymers, c) a linear polymer and a solvent, d) two linear polymers, e) hyperbranched polymer and a solvent, f) two hyperbranched polymers, g) a linear and a hyperbranched polymer. The parameters ($M_i, b_{3,i}, b_{4,i}$) occurring in Eqs. 11 - 16 estimate the type of the component. Using Eq. 10 the spinodal condition can be derived, resulting in:

$$0 = \frac{1}{\phi_1M_1} + \frac{1}{(1-\phi_1)M_2} + 2\alpha_2 + 6\alpha_3\phi_1 + 12\alpha_4\phi_1^2 + 20\alpha_5\phi_1^3 + 30\alpha_6\phi_1^4 \quad (17)$$

Additionally, the critical condition based on Eq. 10, reads:

$$0 = -\frac{1}{\phi_1^2M_1} + \frac{1}{(1-\phi_1)^2M_2} + 6\alpha_3 + 24\alpha_4\phi_1 + 60\alpha_5\phi_1^2 + 120\alpha_6\phi_1^3 \quad (18)$$

Liquid-liquid phase separation can occur either upon heating or cooling, corresponding to lower (LCST) and upper critical solution (UCST) phase boundaries, respectively. The principle shape of the LLE behavior depends on the temperature dependency of the Gibbs energy of mixing. Caused by the above mentioned series expansion until the second order with respect to the interaction energy, Eq. 10 can also be written in the following manner:

$$\frac{G_{Mix}}{N_Lk_BT} = \delta_0 + \frac{\delta_1}{T} + \frac{\delta_2}{T^2} \quad (19)$$

Using the following abbreviations:

$$v_1 = b_{31} + 3b_{41} \qquad v_2 = b_{32} + 3b_{42} \tag{20}$$

δ_0, δ_1 and δ_2 - values in Eq. 19 are:

$$\delta_0 = \frac{-1}{3M_1^4 M_2^4 z^2}\left[\begin{array}{l}
6\phi_1^4 (M_2 - M_1) \\
+2\phi_1^3 M_1 \left(\begin{array}{l} 3M_2(M_2 - M_1)\left[\begin{array}{l} -12M_1 + M_1M_2 \\ +2(M_2 - M_1)(v_2M_1 - v_1M_2) \end{array}\right] \\ +\left(M_1^3 - M_2^3\right)(M_2 - M_1) \end{array}\right) \\
+3\phi_1^2 M_1^2 \left[\begin{array}{l} M_2^2(M_2v_1 - M_1v_2)^2 + 4v_2M_2\left(-M_2^2 - 3M_1^2 + 4M_1M_2\right) \\ +2(M_1 - M_2)\left[\begin{array}{l} \dfrac{zM_2^2}{2}(M_1 - M_2) + 4v_1M_2^2 \\ +\left(2M_2^2 - M_2 + 6\right)(M_1 - M_2) \end{array}\right] \end{array}\right] \\
+\phi_1 \left[\begin{array}{l} +2M_1^4\left(-6M_2^2 + 2M_2 - 9\right) + 2M_2^2\left(-6M_1^2 + M_1 - 3\right) \\ +M_1M_2\left[\begin{array}{l} -3M_1^3\left(-8v_2 + M_2v_2^2\right) + 6M_1^2M_2\left((M_2v_1 - 4)v_2 - 2v_1\right) \\ -3M_2^3v_1(M_1v_1 - 4) \end{array}\right] \\ +6M_1^3M_2\left(4 + 4M_2^2 - M_2\right) - 3zM_1^2M_2^2(M_1 - M_2)^2 \end{array}\right]
\end{array}\right] + \frac{\phi_1 \ln(\phi_1)}{M_1} + \frac{(1-\phi_1)\ln(1-\phi_1)}{M_2} \tag{21}$$

$$\delta_1 = \frac{-\varepsilon_{LCT}}{2k_B M_1^3 M_2^3 z}\left[\begin{array}{l}
8\phi_1^5\left(-3M_2M_1(M_1 - M_2) + M_1^3 - M_2^3\right) \\
+4\phi_1^4\left\{\begin{array}{l} +4M_2^3 - 6M_1^3 \\ +M_1M_2\left[\begin{array}{l} (M_1 - M_2)(M_1v_2 - M_2v_1) \\ -(M_1 - M_2)^2 + 16M_1 - 14M_2 \end{array}\right] \end{array}\right\} \\
+2\phi_1^3\left\{\begin{array}{l} -4M_2^3 + 12M_1^3 \\ +M_1M_2\left(\begin{array}{l} -2(M_2M_1 + 4M_1 - 4M_2)(M_1v_2 - M_2v_1) \\ +M_1M_2\left(\begin{array}{l} (z+3)M_1M_2(M_1 - M_2) \\ +4(M_1 - M_2)^2 - 28M_1 + 20M_2 \end{array}\right) \end{array}\right) \end{array}\right\} \\
+\phi_1^2 M_1\left(\begin{array}{l} 2\left(M_1M_2\left(-M_1M_2^3 - 3M_1M_2 + 6M_2^2 - 2M_1 + 4M_2\right) - 2M_2^3\right) \\ +z^2M_1^2M_2^3 - 2zM_1M_2^2(M_1M_2 + M_1 - 2M_2) - 8(M_1 - M_2)^2 \\ +8M_2^2v_1(M_2 - M_1M_2 - M_1) \\ +4M_1M_2v_2(M_1M_2 - 2M_2 + 2M_1) \end{array}\right) \\
+\phi_1 M_1^2 M_2^2\left(-2zM_2(1 - M_1) - 6M_2 + 2M_1M_2 + 4M_2v_1 - M_1M_2z^2\right)
\end{array}\right] \tag{22}$$

$$\delta_2 = -\frac{\varepsilon_{LCT}^2\phi_1^2}{4k_B^2M_1^2M_2^2}\begin{bmatrix} 12\phi_1^4\left(M_1-M_2\right)^2 \\ +2\phi_1^3\left(-16\left(M_1-M_2\right)^2-4M_2^2+M_1M_2\begin{pmatrix}4-5\left(M_1-M_2\right)\\+2M_1v_2-2M_2v_1\end{pmatrix}\right) \\ +\phi_1^2\begin{pmatrix}+M_1^2M_2^2\left(z+6\right)+2M_2M_1^2\left(9-4v_2\right)-4M_1M_2^2\left(8-3v_1\right)\\+48M_2^2-72M_2M_1+28M_1^2\end{pmatrix} \\ +\phi_1\begin{pmatrix}-24M_2^2-8M_1^2+2M_2M_1\left(12-3M_1+17M_2-M_2M_1\left(6+z\right)\right)\\+4M_1M_2\left(M_1v_2-3M_2v_1\right)\end{pmatrix} \\ +M_1M_1\left(M_1M_2\left(6+z\right)+4M_2\left(v_1-3\right)-2M_1\right)+4M_2^2\end{bmatrix} \tag{23}$$

The quadratic temperature dependence in Eq. 19 ensures that all principle types of phase behavior in a binary system, namely a) only UCST, b) only LCST, c) closed loop miscibility gaps and d) UCST + LCST for the same system (hour-glass phase diagrams), can be calculated. However; in the case where $\delta_2 = 0$ only one miscibility gap with UCST or LCST behavior can be modeled, where the sign of δ_1 determine, if a UCST or a LCST phase behavior is present.

Let us discuss two special cases. The first case A should by characterized by

$$M = M_1 = M_2 \qquad b_{i,j} = 0 \tag{24}$$

The conditions in Eq. 24 imply that the two components having the same molecular weight and no branching points occur in the molecules. Eqs. 21 - 23 in combination with Eq. 24 lead to:

$$\delta_{0,\mathrm{A}} = \frac{\phi_1\ln\left(\phi_1\right)+\left(1-\phi_1\right)\ln\left(1-\phi_1\right)}{M} \tag{25}$$

$$\delta_{1,\mathrm{A}} = \frac{\varepsilon_{LCT}\phi_1\left(1-\phi_1\right)\left(6+2z+M\left(z^2-2z-2\right)\right)}{2k_BMz} \tag{26}$$

Assuming for the coordination number $z >= 6$ then Eq. 26 tell us that only the sign of the interaction energy of the considered mixture in the limit of Eq. 24 determines if LCST or UCST behavior can be found.

$$\delta_{2,A} = -\frac{\varepsilon_{LCT}^2\phi_1^2\left(1-\phi_1\right)^2\left(M^2\left(z+6\right)-14M+4\right)}{4k_B^2M^2} \tag{27}$$

According Eq. 27 the quantity $\delta_{2,A}$ can only become zero if the second bracket in the numerator of Eq. 27 vanished. The coordination number z can have value between 6 and 12. In this range the root of the bracket leads to segment number smaller then 1 or the complex numbers without any physical sense. These findings mean, that the LCT can describe all types of phase diagrams. This is in contrast to the classical FH-theory, where LCST-behavior or closed miscibility gaps or hour-glass phase diagrams can only be modeled, if the classical χ is assumed to be function of temperature. The solution of the critical condition (Eq. 18) in combination with Eq. 24 provides the critical temperatures:

$$T_{C1,2,A} = \pm\sqrt{3}\frac{\varepsilon_{LCT}}{k_B M}(1-\phi_1)\phi_1\sqrt{M\left((z+6)M^2-14M+4\right)} \tag{28}$$

The last equation demonstrate that one of the critical temperatures is located at negative temperatures and hence without any physical sense.

The next special case B which we can discuss reads:

$$M = M_1 = M_2 \qquad v_1 > 0 \quad v_2 = 0 \tag{29}$$

Hence, the mixture consists of two molecules having the same molecular weight, but component 1 have a branching structure and component 2 is linear. In this case the critical temperatures are given by:

$$T_{c1,2,B} = \frac{(1-\phi_1)\phi_1\varepsilon_{LCT}}{Mz(1-2\phi_1)k_B}\left[\begin{array}{l}-6Mv_1\phi_1(1-\phi_1) \\ \pm\sqrt{\begin{array}{l}3Mz^2\left(4+M(6v_1-14)+M^2(6+z)\right) \\ -12\phi_1 Mz^2\left(4+M(9v_1-14)+M^2(6+z)\right) \\ +12\phi_1^2 M\left(z^2\left(4+M(17v_1-14)+M^2(6+z)\right)+3Mv_1^2\right) \\ -24\phi_1^3 M^2 v_1\left(5z^2+3v_1\right)+36\phi_1^4 M^2 v_1^2\end{array}}\end{array}\right] \tag{30}$$

According Eq. 30 only one positive critical temperature for $\varepsilon_{LCT} > 0$ and $\phi_1 < 0.5$ can be calculated. The presence of two positive critical temperatures require a critical segment fraction above 0.5 for $\varepsilon_{LCT} > 0$. The introduction of the architecture in the theoretical framework can lead not only to a quantitative change in the phase diagram, but also to qualitative difference in comparison with the FH-theory.

The architecture and the size asymmetry also affect the range of T over which mean-field and Ising-type critical behaviors are observed [215, 228]. These temperature range are expressed in terms of the Ginzburg number Gi, which provide a rough estimate of the magnitude of the reduced temperature $\tau \equiv (T-T_C)/T$ at which the crossover from mean-field to Ising-type behavior occurs. Mean-field theory holds for $\tau \geq 10\,Gi$, while the Ising critical behavior corresponds to $\tau \leq Gi/10$. The range $Gi/10 < \tau < 10\,Gi$ describes a crossover domain with $\tau \approx Gi$ [228].

Important contribution emerge also from the so-called "equation of state effects", the variation of polymer properties with thermodynamic state and, in particular, with pressure [229]. This polymer fluid compressibility is introduced into the lattice model by allowing lattice sites to be empty. The void volume fractions were determined from the appropriate equation of state for a given system [229].

The LCT is extended to study all the possible nematic orderings of rectangular mesogens, namely, rodlike, discotic, and biaxial nematics for oblong mesogens, and rodlike and discotic nematics for square mesogens [230]. The model employed so far treats each polymer as a completely flexible entity, subject only to the excluded volume constraints against multiple occupancy of a lattice site by any two united atom groups. However; bond angle constraints and steric interaction of, e.g., hydrogen atoms and side groups to local chain stiffness and the semiflexibility of polymer molecules. For example, torsional motions about single C-C backbone bonds in many olefins are limited to a trans and a pair of gauche conformations with differing energies. The generalization of the LCT to include explicit trans-gauche energy differences was applied to study the combined influences of chain stiffness disparities, monomer molecular structures, energetic asymmetries, and nonrandom mixing on the miscibilities of binary polymer blends [217, 231, 232]. The LCT was also used to determine the essential microscopic parameters that influence the phase separation in binary blends of linear semiflexible lattice chains with equal polymerization indices [231, 233, 234]. A generalized entropy theory of glass formation was developed by merging the LCT for the thermodynamics of semiflexible polymer melts at constant pressure with the Adam-Gibbs relation between the structural relaxation time and the configurationally entropy [235, 236, 237, 238]. Characteristic temperatures and structural relaxation times for different classes of glass-forming polymer liquids are computed using a revised entropy theory of glass formation that permits the chain backbone and the side groups to have different rigidities [239]. The theory is applied to glass formation at constant pressure or constant temperature, where the calculations provide new insights into physical factors influencing the breadth of the glass transition and the associated growth of relaxation times [239].

The LCT was used to study the microscopic molecular factors affecting the miscibilities of A_xB_{1-x}/C binary mixtures (where the homopolymer C is either different or identical to the A_xB_{1-x} random copolymer species) [240, 241, 242, 243, 244, 245, 246]. The LCT was also used to study the influence of monomer structure, i.e., short chain branching, on the miscibility of binary polymer blends [216, 234] , where compressible blend model with a single interaction energy for all united atom groups were applied.

A further extension of the LCT theory is the development of a thermodynamic model able to describe the influence of monomer structure and local correlations on the free energy of strongly interacting and self-assembling polymer systems [247, 248]. This extension combines a systematic high dimension (1/d) and high temperature expansion (that is appropriate for weakly interacting systems) with a direct treatment of strong interactions. The general theory is illustrated for a binary polymer blend whose two components contain "sticky" donor and acceptor groups, respectively. The free energy is determined as an explicit function of the donor-acceptor contact probabilities that depend, in turn, on the local structure and both the strong and weak interactions.

THERMODYNAMIC MODEL – WERTHEIM THEORY

Another characteristic feature of the hyperbranched polymers is the large number of functional groups, which can have special interactions, like association sites. These association sites allow the formation of cross-associates as well as self-associate. The LCT can be combined with a model for association, like the Wertheim-perturbation theory [188, 226] or the extended chemical association model [227].

In Wertheim's approach [249, 250, 251, 252], the overall density of the particles in the fluid is split in two parts. For a fluid consisting of particles with one attractive spot the formalism is in terms of two densities: the overall particle density ρ is divided in a density of non-bonded particles ρ_0 and in a density of particles ρ_1 that did form an attractive bond with another particle [249]:

$$\rho = \rho_0 + \rho_1 \tag{31}$$

In a fluid of particles with only one attractive spot we can thus recognize non-bonded and bonded particles. The total 2-particle distribution function $G(1,2)$ is constituted from contributions arising from the correlations between two particles which have not formed an attraction bond $g_{00}(1,2)$, two particles of which one has formed an attraction bond $g_{10}(1,2)$ and $g_{01}(1,2)$, and two particles that both have formed an attraction bond $g_{11}(1,2)$. The coordinates of particles 1 and 2 are given by (r_1, Ω_1) and (r_2, Ω_2), where r_i denotes the position and Ω_i the orientation of particle i. To get a multiple density approach, the pair interaction potential $u(1,2)$ depending on the orientation of particle 1 and 2 must be split in two parts, an isotropic repulsive part $u_R(r_1, r_2)$ and a directional attractive part $u_{att}(1,2)$. The Mayer function $f(1,2)$ can be divided in an attractive $F(1,2)$ and a purely repulsive part $f_R(r_1, r_2)$ [249]:

$$f(1,2) = f_R(r_1, r_2) + F(1,2) \tag{32}$$

with

$$f_R(r_1, r_2) = \exp\left(\frac{-u_R(r_1, r_2)}{k_B T}\right) - 1 \tag{33}$$

and

$$F(1,2) = \exp\left(-\frac{u_R(r_1,r_2)}{k_B T}\right)\left(\exp\left(-\frac{u_{att}(1,2)}{k_B T}\right)-1\right) \tag{34}$$

It has been proven [249, 250] that this division of the Mayer function allows a diagrammatic expansion of ρ in terms of the activity z, f_R - and F -bonds similar to LCT. According the suggested expansion, the overall density ρ can be split into densities of bonded and non-bonded particles. The ρ_0 and ρ_1 -values, are then both classified by a different part of the set of diagrams that constitutes ρ. Starting from the grand canonical partition function and using these expansions of ρ_0 and ρ_1, Wertheim [249, 250] derived an exact diagrammatic expansions of the structural correlations $g_{00}(1,2)$, $g_{01}(1,2)$, $g_{10}(1,2)$ and $g_{11}(1,2)$ in terms of ρ_0 and ρ_1, f_R - and F -bonds. He then defined, along the same lines as the direct correlation function is defined, for fluids consisting of hard spheres, partial direct correlation functions $c_{00}(1,2)$, $c_{01}(1,2)=c_{10}(1,2)$ and $c_{11}(1,2)$ and derived their diagrammatical expansions in terms of ρ_0, ρ_1 and $f_R(1,2)$ and $F(1,2)$. The partial correlation could then be related to an Orstein-Zernicke matrix equation. This procedure bears strongly resemblance to the derivation of the Orstein-Zernicke equation for hard spheres [253, 254]. To obtain the fluid structure, the Orstein-Zernicke matrix equation has to be combined with an appropriate radial distribution function $g(r_1,r_2)$ as a closure equation and a self consistency relation based on Eq. 31. The self consistency relation is a mass balance equation determining division in bonded and non-bonded. For fluids without directional forces, the Orstein-Zernicke matrix equation and a closure equation determine the fluid structure in terms of particle density, ρ. For a fluid with directional forces, the Orstein-Zernicke matrix equation and a closure equation generate the correlations for $g_{00}(1,2)$, $g_{10}(1,2)$ and $g_{11}(1,2)$ in terms of ρ_0 and ρ_1. These correlations define the distribution of the particles over bonded and non-bonded particles. For fluids consisting of hard spheres this distribution is not possible, because there is only one type of particles. For fluids with directional attractive forces, the ρ_0 and ρ_1 functions assign $g_{00}(1,2)$, $g_{10}(1,2)$ and $g_{11}(1,2)$ with help of Orstein-Zernicke matrix equation and on the other hand the $g_{00}(1,2)$, $g_{10}(1,2)$ and $g_{11}(1,2)$ determine the values of ρ_0 and ρ_1. This last step is necessary for internal consistency and is provided by the self consistency relation. The Wertheim theory has three constituent parts, an Orstein-Zernicke matrix equation, closure equation and a self consistence mass balance equation. Wertheim [251, 252] extended this formalism to a multi-component mixture with different interaction sites.

Caused by the incompressibility assumed in the LCT the Wertheim theory needs to be modified. One possibility is a lattice Wertheim theory suggested by Nies et al. [255, 256].

The Wertheim theory was transferred on a fully occupied lattice, so an incompressible fluid is regarded [255, 256]. The following expression for the Helmholtz energy of association was derived [255, 256]:

$$\frac{f_{asso}}{RT} = \sum_i \phi_i \left[\sum_{A_i} \left[\ln\left(X_{A_i}\right) - \frac{X_{A_i}}{2} \right] + \frac{1}{2} M_i^A \right] \tag{35}$$

where X_{A_i} is the segment molar fraction of the non-bonded polymer segments and M_i^A is the number of association sites at one molecule. The notion "Non-bonded polymer segments" means in this case that the segments do not contribute in association. The value X_{A_i} is calculated as follows:

$$X_{A_i} = \left[1 + \sum_j \sum_{B_j} \phi_j X_{A_j} \Delta^{A_i B_j} \right]^{-1} \tag{36}$$

where the summation over B_j runs over all molecules and association sites. The association strength $\Delta^{A_i B_j}$ is given by:

$$\Delta^{A_i B_j} = K_i \left(\exp\left(\frac{\varepsilon_{ij}^{Asso}}{k_B T} \right) - 1 \right) \tag{37}$$

with K_i being the ratio of the nearest-neighbour positions with the proper orientation to all possible orientations of the component i and ε_{ij}^{Asso} is the association energy. The difference between the lattice Wertheim association model and the association model used in the SAFT equation of state lies in the association volume, K_i, as in the SAFT model the density is used to calculate the association volume. Using this version of association theory, the association sites have to be distributed on a molecule. For example a water molecule has four association sites, one at each proton of the hydrogen and one at each lone electron pair of the oxygen. After the localization of the association sites, the possible interactions of the association sites have to be defined; for instance that only a proton and a lone electron pair can interact with each other and not two proton sites or two electron pair sites. This leads for multi-component cross associating systems to a non linear system of equations of type Eq. 36, which has to be solved numerically to calculate the Helmholtz energy or the Gibbs energy, respectively. In opposite to this, in the case of only one associating molecule, the association can be determined analytically. To consider the cross-association between different molecules mixing rules for the association volume and the association energy are applied to calculate the parameters for the cross-association [188]:

$$K_{ij} = \frac{K_i + K_j}{2} \tag{38}$$

and:

$$\varepsilon_{ij}^{Asso} = \sqrt{\varepsilon_{ii}^{Asso}\varepsilon_{jj}^{Asso}} \tag{39}$$

By combination of the LCT with the lattice Wertheim approach both important effects occurring in systems built up from hyperbranched polymers can be take into account for the calculations of thermodynamic properties and phase equilibria.

Phase Equilibria Calculation – Examples

The general equation for the Gibbs energy of mixing (Eq. 10) in combination with Eqs. 11 to 16 provided in the last section allows the calculation of phase diagrams of various binary mixtures. The application to a special type of mixture can be realized with the help of Eqs. 7 and 8. In the following section chapters the use of these equations are discussed for polymer solutions as well as for polymer blends.

Hyperbranched Polymer + Solvent

The thermodynamic description of the LLE of hyperbranched polymer solutions presents a significant challenge to any polymer theory concerned with modeling the molecular factors controlling the mixing behavior. This analysis is designed to understand factors influencing the phase behavior and to aid in sorting out some of the puzzling data related to LLE of hyperbranched polymer solutions.

The incompressibility assumption implies that the LCT for binary systems (Eq. 10) made from hyperbranched polymer (component 1) and a solvent (component 2) contains only one adjustable parameter, the interaction energy, ε_{LCT}, defined in Eq. 9. All remaining quantities of the LCT are determined from the architecture of the polymer and the solvent. The first step in the LLE-modeling is the determination of the parameters (M_i, b_{3i}, b_{4i}) for the molecules present in the mixture.

Hyperbranched macromolecules generally exhibit a much less perfect structure than dendrimers, due to the presence of not only dendritic and terminal units, but also a large fraction of linear segments (Figure 2). This can be characterized by the degree of branching. A first definition of the degree of branching (DB) was introduced by Frechet et al. [25]. This DB-value is given by:

$$DB = \frac{D+T}{D+T+L} \tag{40}$$

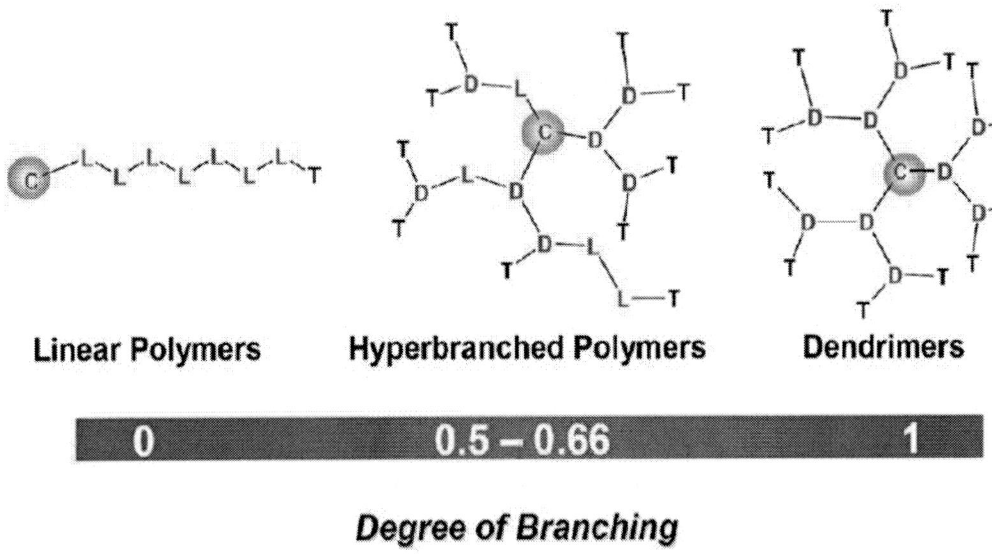

Figure 2. Schematic representation of the degree of branching for linear and hyperbranched polymers compared to dendrimers [54].

where D, L, T represent the relative abundances of the dendritic, linear and terminal units in the polymer, respectively. In Eq. 40 the linear segments are also counted as branching points, which leads to an overestimation of the DB in the case of low molecular weight polymers. Wurm and Frey [54] suggested the following expression:

$$DB = \frac{2D}{2D+L} \tag{41}$$

that gives a more accurate value for DB, even for the low molecular weight regime. It is important to point out that a dendrimer is limited to distinct generations that are less accessible in high generations, while with hyperbranched polymers it is possible to access a broad range of molecular weights and degree of branching [54]. However; in order to apply the LCT to hyperbranched polymers three characteristic parameters are necessary. The quantities given in Eqs. 7 and 8 reflect the internal architecture and size of the molecules, subject to the limitations imposed by the lattice morphology. The structure is chosen, whenever possible, as most closely corresponding to a united atom model in which united atom groups, such as CH_n reside at single lattice sites. In this chapter solutions of three hyperbranched polyesters with the generation numbers $g = 2$ (Boltorn H20), $g = 3$ (Boltorn H30) and $g = 4$ (Boltorn H40) are studied.

A detailed experimental characterization of these hyperbranched polymers can be found in the literature [30, 31]. All these molecules possess the same core: $O[CH_2C(C_2H_5)(CH_2O\text{–})_2]_2$. Depending on the generation number g the molecules additionally include a different number of groups A: $COC(CH_3)(CH_2O-)_2$ and B: $COC(CH_3)(CH_2OH)_2$. The general formulae of the polymers are for $g = 2$:

$(core)A_4B_8$, for $g = 3$: $(core)A_{12}B_{16}$ and for $g = 4$: $(core)A_{28}B_{32}$. According to these formulae (neglecting polydispersity) the molar masses take the values 1642 g/mol ($g = 2$), 3498 g/mol ($g = 3$) and 7210g/mol ($g = 4$). The number of OH- end groups equals the number of B-units. Figure 3 shows schematically a hyperbranched polyester with the generation number, $g = 3$.

To describe the architecture of a hyperbranched polymer two specification in addition to the generation number g will be used. The separator length n is the number of segments between two branching points. It denotes the number of segments of an A unit or a B unit. The number of core segments is given by the quantity n_0. It is assumed that one water molecule occupies one lattice site, so the number of core segments and the separator length can be determined by dividing the core, A and B units in groups that have a molar mass comparable to a water molecule. The solvents (water, propan-1-ol) were taken as linear molecules, where water occupies one lattice place and propan-1-ol three of them. The values of the separator length, number of core segments and generation number are collected in Table 1.

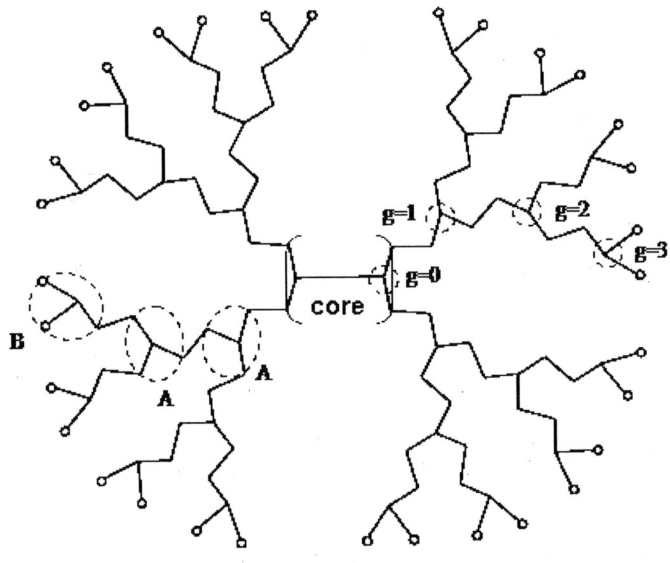

Figure 3. Schematic presentation of a hyperbranched polymer of generation number g=3 [187].

Table 1. Architectural parameters describing hyperbranched polyester

Separator length n	4
Number of core segments n_0	7
Generation number g	$g = 2$ (Boltorn H20) $g = 3$ (Boltorn H30) $g = 4$ (Boltorn H40)

Using these architectural parameters, the topological coefficients of the LCT can be calculated, where the number of segments of a hyperbranched polyester molecule can be calculated as follows [223]:

$$M = 4\left(2^g - 1\right)n + n_0 \tag{42}$$

Each A-unit and each B-unit possesses one branching point of degree 3. The core contains two such branching points. Branching points of degree 4 and higher do not exist in these polymer molecules. So the number of branching points is calculated as follows [223]:

$$b_3 = 4\left(2^g - 1\right) + 2 \tag{43}$$

In contrast to the experimental procedure, where the concentration is usually given in mass fraction, w_i, in the theoretical framework the concentrations of the components are measured by the segment fraction, ϕ_i. Both quantities are connected via:

$$\phi_1 = \frac{M_1 N_1}{M_1 N_1 + M_2 N_2} = \frac{\dfrac{M_P}{M_{M,1}} w_P}{\dfrac{M_1}{M_{M,1}} w_1 + \dfrac{M_2}{M_{M,2}} w_2} \tag{44}$$

where the indices 1 and 2 indicate that the specific value ($M_{M,i}$ is the molar mass of the component i) belonging to the polymer or to the solvent.

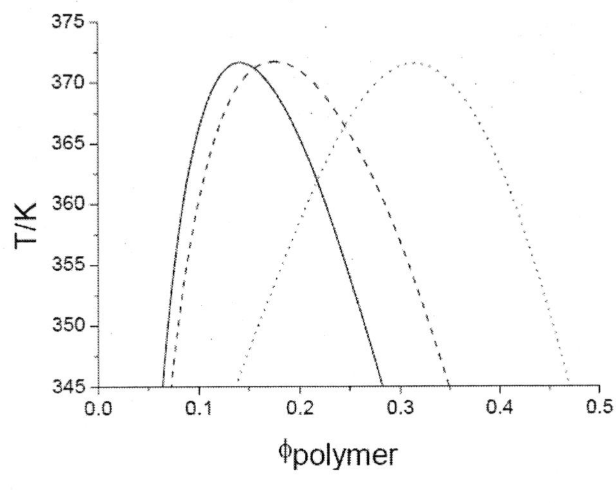

Figure 4. Influence of the coordination number (dotted line: $z = 6$, $\varepsilon_{LCT} / k_B = 124K$, dashed line: $z = 8$, $\varepsilon_{LCT} / k_B = 79.8K$, solid line: $z = 12$, $\varepsilon_{LCT} / k_B = 46.84K$) on the phase equilibria on hyperbranched polyester $\left(g = 2, n_0 = 7, n = 4\right)$ [223].

In order to study the influence of the association phenomena on the phase equilibrium the self-association of the solvents, the self-association of the polymer and the cross-association between solvent and polymer are modeled using the Wertheim lattice perturbation theory. This Wertheim theory needs a location of the association sites on each component in the mixture. Water is modeled as a 4C-modell, meaning that one association place is located at each lone electron pair of the oxygen and one association site is at each hydrogen atom. To consider the association phenomena of solutions containing hyperbranched polyester, the OH-groups are modeled by an association site on one lone electron pair of the oxygen and one association site on the hydrogen. The other lone electron pair of the oxygen is assumed to be hidden, because of steric considerations. Propan-1-ol is modeled using the same assumptions.

First, some model calculations for the system composed of hyperbranched polymer and a solvent occupying only on lattice site will be performed. Figure 4 demonstrated different binodale lines obtained with different numbers for the coordination number, z. The energy parameter ε_{LCT} can be fitted to critical temperature. Within the FH-theory only the well-known χ-parameter occur. However; within the LCT this χ-parameter contains the coordination number z as well as the interaction energy, ε_{LCT}. Hence, both parameters are connected. If one chose the coordination number, z, then the interaction energy ε_{LCT} must also be changed in order to represent the same critical temperature.

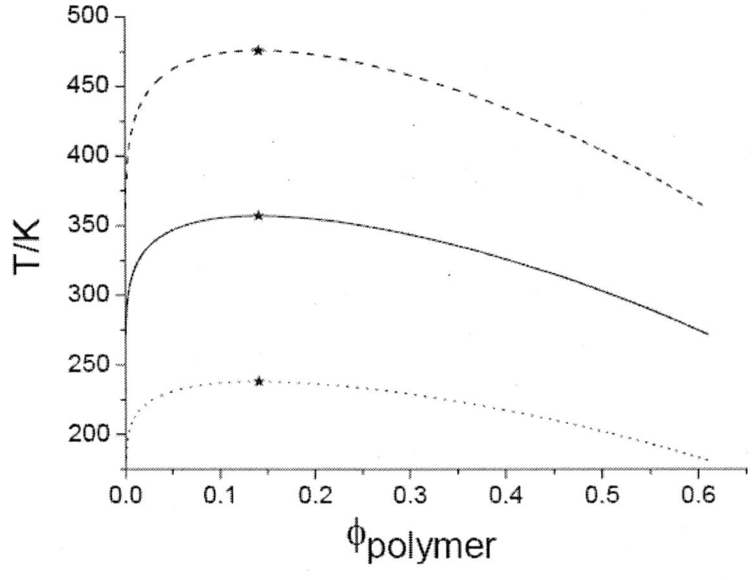

Figure 5. Dependence of the liquid-liquid equilibria (stars: critical points, dashed line: $\varepsilon_{LCT} / k_B = 60K$, solid line: $\varepsilon_{LCT} / k_B = 45K$, dotted line: $\varepsilon_{LCT} / k_B = 30K$) on the interaction energy for the hyperbranched polyester $(g = 2, n = 4, n_0 = 7)$ [223].

Figure 4 shows the influence of the coordination number on the liquid-liquid phase split of hyperbranched polyester solutions calculated by the LCT using different interaction energies to align the critical temperatures because the shape of the three cloud point curves should be compared. The cloud-point curve calculated with $z = 6$ shows a sharp decline of the high concentrated polymer branch and a slow rise on the diluted side of the curve. This behavior is not typical for polymer solutions. Using higher coordination numbers leads to more realistic cloud-point curves. Therefore, in the following calculations of LLE the coordination number $z = 12$ is used. Moreover, the LCT is a series expansion of the partition function in the inverse coordination number and the interaction energy, $\varepsilon_{LCT} / k_B T$. Using higher coordination numbers leads to a faster convergence of both series as also the interaction energy decline for larger coordination numbers.

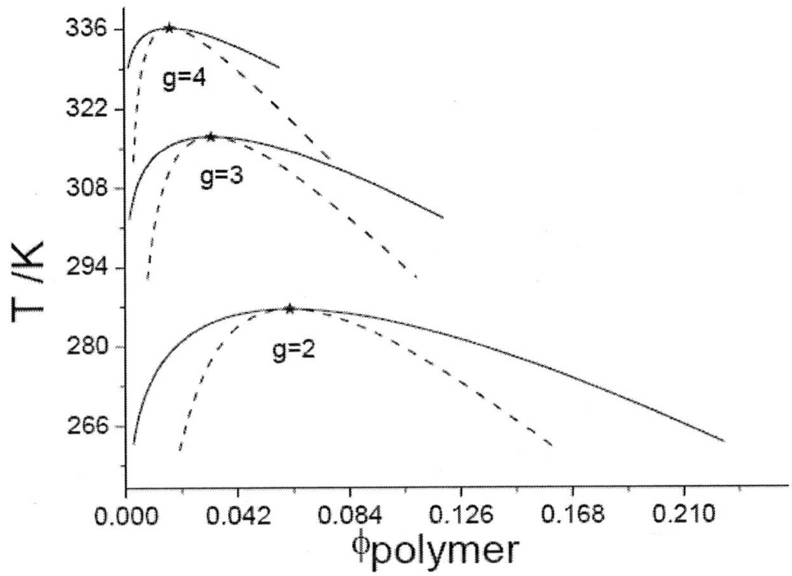

Figure 6. Influence of the generation number, g, on the phase equilibria (solid line: binodal, broken line: spinodal, star: critical point) on hyperbranched polyester solutions $(\varepsilon_{LCT} / k_B = 30K, n_0 = 2, n = 3, z = 12)$.

Comparing the three cloud point curves calculated with $z = 12$, but different values for ε_{LCT}, leads to the conclusion, that the interaction parameter ε_{LCT} has no impact on the critical composition or on the shape of the cloud point curve (Figure 5). By increasing or decreasing the interaction parameter ε_{LCT} the critical temperature is shifted to higher or lower values. That means, that the critical composition and the shape of the cloud point curve is determined by coefficients such as the separator length or the number of core segments, which are given by the architecture of the polymer. By growing separator length, n, the critical point is shifted to higher critical temperatures and lower critical segment fractions [223]. Increasing the number of core segments, n_0, leads to higher critical temperatures, lower critical composition and a narrower cloud point curve [223]. This effect can also be traced back to the fact that an increase of the number of core segment also increases the number of segments. But this influence is not as strong as the influence of the separator length, because the increase of the number of segments is equal to the increase of the number of core segments. Increasing the generation number of hyperbranched polymers, g, leads to higher critical temperatures, lower critical segment fractions and to narrower LLE [223]

(Figure 6). This effect can be explained by the fact that the number of segments increases in power of the generation number, g. The critical temperature of the hyperbranched polyester solution is lower compared to their linear analogous and the critical composition of hyperbranched polyester solutions is higher than the critical temperatures of the linear analogous [223].

Beside the architecture of a hyperbranched polymer also the number as well as the kind of functional groups has an important impact on the liquid-liquid equilibria of hyperbranched polymer solutions.

Applying the Wertheim association theory two parameters, namely the association volume, K_i, and the association energy, ε_i^{Asso}, must be specified. This theory uses two parameters, which have to be fitted on the LLE. For water these parameters can be obtained using the vapor pressure curve resulting in $\varepsilon_{water}^{Asso}/k_B = 1800K$ and $K_{water} = 0.01$. However; for the polymer no independent physical quantities are available. Therefore, these parameters must be fitted to the LLE.

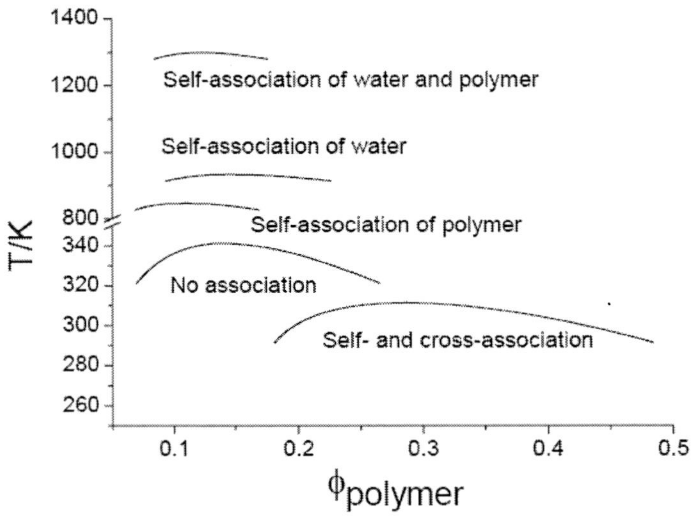

Figure 7. Impact of different kinds of association (self- and cross-association) on the spinodal curve of aqueous solutions of hyperbranched polyester $(g = 2)$ and the association parameter ($K_1 = 0.0673$, $\varepsilon_1^{Asso}/k_B = 1100K, K_2 = 0.01, \varepsilon_2^{Asso}/k_B = 1800K$) [188, 223].

To analyze the influence of association, the spinodal curves of the system hyperbranched polyester ($g = 2$) and water are calculated with different kinds of association (Figure 7). The parameters used for this model calculation are the same parameters used to calculate the liquid-liquid equilibrium of Boltorn H20. If only self-association of the polymer and water is considered, the critical point is shifted to unrealistic high critical values. Just the consideration of the self- and cross-association together leads to realistic critical temperatures. With help of Figure 7 the influence of cross- and self-association compared to the non-polar polymer in a non-polar solvent is investigated. By considering the self-and cross-association the critical point is shifted to higher segment fraction of the polymer and towards lower critical temperatures. Regarding these results it has to be pointed out that if the association phenomena are considered in the calculation both the self- and cross-association have to be included. These findings are contradictory to the results of Jang and Bae [257], who concluded that the self-association of water has the greatest influence on the liquid-liquid equilibrium of aqueous hyperbranched polyester solutions.

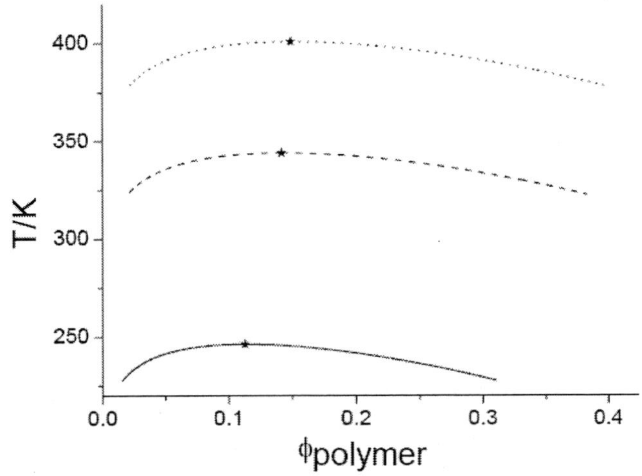

Figure 8. Influence of the association energy on the binodal curves (dotted line: $\varepsilon_1^{Asso}/k_B = 1100K$, $\varepsilon_2^{Asso}/k_B = 2000K$, dashed line: $\varepsilon_1^{Asso}/k_B = 1100K$, $\varepsilon_2^{Asso}/k_B = 1800K$, solid line: $\varepsilon_1^{Asso}/k_B = 1500K$, $\varepsilon_2^{Asso}/k_B = 1800K$) and the critical point (stars) of aqueous hyperbranched polyester solution using LCT ($\varepsilon_{LCT}/k_B = 30K$) in combination with Wertheim theory ($K_1 = 0.045$, $K_2 = 0.01$) [223].

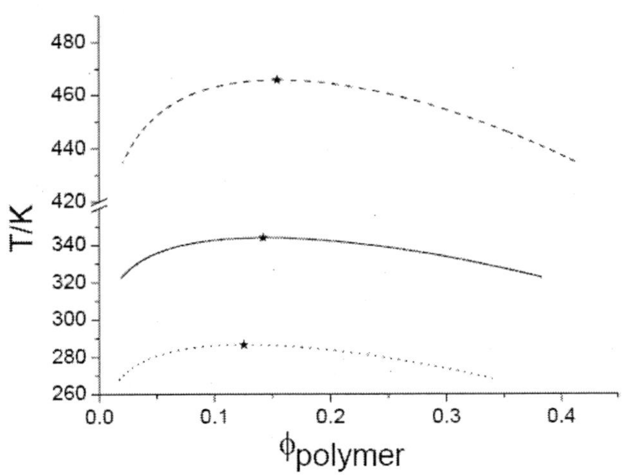

Figure 9. Influence of the association volume on the LLE (stars: critical point, solid line: $K_1 = 0.045$, $K_2 = 0.01$, dashed line: $K_1 = 0.045$, $K_2 = 0.02$, dotted line: $K_1 = 0.06$, $K_2 = 0.01$) of aqueous hyperbranched polyester solution using LCT ($\varepsilon_{LCT}/k_B = 30K$) in combination with Wertheim theory ($\varepsilon_1^{Asso}/k_B = 1500K$, $\varepsilon_2^{Asso}/k_B = 1800K$) [223].

The influence of the association energy on the liquid-liquid phase equilibria of aqueous hyperbranched polyester solutions is shown in Figure 8. By increasing the association energy of water the critical temperature and the critical segment fraction of the polymer are shifted to higher values. On the other hand, the critical temperature and the critical polymer segment fraction gets smaller by increasing the association energy of the polymer end groups. Another effect of changing the association energy is the broadening of the miscibility gap by increasing the association energy of water and the narrowing of the cloud point curve by rising up the association energy of the polymer. This behavior of the cloud point curve by increasing the association energy of water can be explained by the fact that water tends to association cluster of water molecules. Due to these reasons the miscibility gap broadens. By increasing the association energy of the polymer molecules, water tends to build up association cluster with polymer molecules and the polymer molecules are hindered to build large clusters with each other because of steric effects. So the miscibility gap gets closer and the critical temperature becomes lower.

Figure 9 shows the influence of the association volume on the phase equilibria of hyperbranched polyester + water system. Increasing the association volume of water or the association volume of the polymer has a similar effect on the miscibility gap as changing the association energy of water and the polymer. By increasing the association volume of water the critical temperature and the critical segment fraction is shifted to higher values. On the other hand, an increase of the association volume of the polymer leads towards lower critical temperatures and lower critical composition. The broadening of the cloud point curve by increasing the association volume of water can be explained in the same way as by increasing the association energy because the association strength is proportional to the association volume. Moreover, there is also a narrowing of the immiscibility gap by increasing the association volume of the polymer.

For the binary system composed of Boltorn H20 and water two data sets related to the LLE could be found in the literature [187, 188] (Figure 10). However; both data sets differ strongly from each other, where the data given in Ref. [188] are located at much higher temperatures. The demixing temperatures were estimated during heating and during cooling. Both temperatures were very close together, within maximal 2 K [188]. The data in Ref. [188] do not show the shoulder found by Yang and Bae [187]. Having in mind the melting temperature of Boltorn H20 is above 323K maybe the data found by Yang and Bae [187] are rather connected to the solid-liquid equilibria instead the liquid-liquid equilibria.

The branch of the cloud-point curve for the diluted polymer solution could by descript in excellent agreement with the experimental data, if the LCT either alone or in combination with the Wertheim theory is utilized (Figure 9). However; deviations between the experimental and theoretical cloud-point curve in the concentrated range occur. Taking the association forces in terms of self- and cross association into account leads to a slightly improvement of the calculated results by shifting the polymer volume fraction in the concentrated phases to larger values and hence the application of the Wertheim approach improves the calculation results in the right direction. Having in mind that the increased number of the used adjustable parameter it can be concluded the higher effort is not warrantable. Including only self association and not cross association in the theoretical calculations leads to extreme high demixing temperatures (Figure 7), were the polymer is hardly stable. Only if both effects (self- and cross association) are taken into account the calculation results agree with the experimental findings. This statement is contrary to the results found by Yang and Bae [257]. Using another version of the LCT in combination with another association model for the self association of water Yang and Bae [257] concluded the dominant impact is the self association of water on the LLE. However; like discussed in the literature [224] the reason for this conclusion can be errors in the applied version of the LCT. Similar results could be obtained for the system Boltorn H20 + propan-1-ol [188, 223]. These binary phase equilibrium investigations form the bases for the study of the ternary system made from Boltorn H20 + water + propan-1-ol [226].

Other hyperbranched polymer from the Boltorn family is Boltorn U3000. Domańska et al. [183] measured phase equilibria of polymer solutions in propan-1-ol and butan-1-ol. Boltorn U3000 is fatty acid modified polyester of generation number 3. The molecule consists also of the core $C(CH_2O-)_4$, of 12 separator groups A: $COC(CH_3)(CH_2O-)_2$ and 16 groups B: $COC(CH_3)(CH_2OH)(CH_2OR)$ with the end groups OH and OR. Here, R origins from the 16-carbon-long alkyl acid (R: $CH_3-(CH_2)_{14}CO-$). Again, the core is divided into 5 segments and the separator group A into 3 segments. Considering the 16-carbon-long group R the group B has 20 segments. All together a Boltorn U3000 molecule may be described by $M_1 = 373$ segments, where molar mass is $7192\ g/mol$. Furthermore, there are 28 branching points of degree 3 ($b_{3,1} = 28$) and one branching point of degree 4 ($b_{4,1} = 1$).

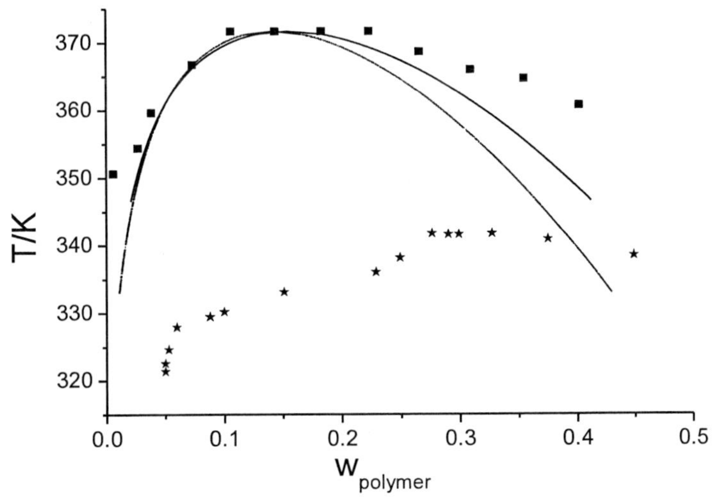

Figure 10. Experimental (squares [188], stars [187]) and calculated (solid line: LCT + Wertheim theory, broken line: LCT) LLE for the system Boltorn H20 + water.

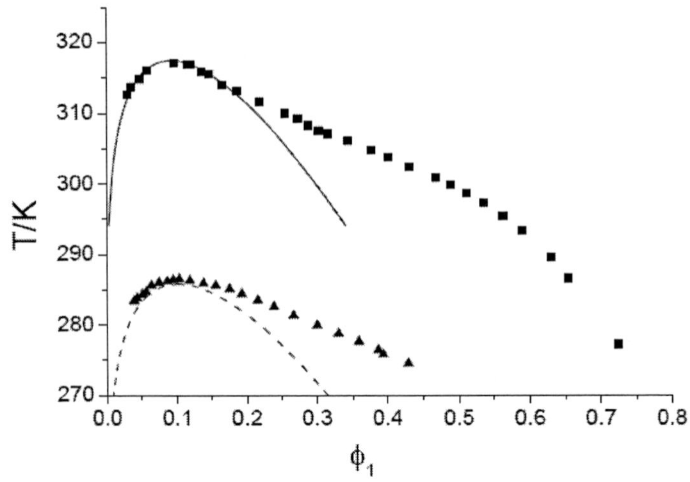

Figure 11. Experimental (squares: Boltorn U3000 + propan-1-ol [183], triangles: Boltorn U3000 + butan-1-ol) and calculated (solid line: Boltorn U3000 + propan-1-ol, broken line: Boltorn U3000 + butan-1-ol) LLE.

The segment numbers of the linear solvents propan-1-ol and butan-1-ol were assumed to be $M_2 = 3$ and $M_2 = 4$. Caused by 16 OH groups of Boltorn U3000, similar to the solution containing Boltorn H20 self- and cross-association occurs. The association phenomena can be treated with a chemical association model [227] or with the Wertheim perturbation theory. In this contribution we focus our attention to the combination of the LCT with the Wertheim perturbation theory. The relevant parameters within the LCT are: $g = 3, n = 3, n_0 = 5$. In order to minimize the number of adjustable parameter within the Wertheim perturbation theory we take to parameter describing propan-1-ol from the above discussed system propan-1-ol + Boltorn H20 [188], namely $K_2 = 0.11$ and $\varepsilon_2^{Asso} / k_B = 1745K$.

Table 2. Model parameters for solution of Boltorn U3000 in propan-1-ol or in butan-1-ol

Parameter	Boltorn U3000 + propan-1-ol	Boltorn U3000 + butan-1-ol
ε_{LCT} / k_B	10.9 K	8.35 K
K_1	0.023	0.023
$\varepsilon_1^{Asso} / k_B$	1200 K	1200 K
K_2	0.011	0.01
$\varepsilon_2^{Asso} / k_B$	1745 K	1710 K

Using the experimental data from the literature [183] the three remaining parameters ($\varepsilon_{LCT}, K_1, \varepsilon_1^{Asso}$) must be fitted. The phase split for the polymer solution Boltorn U3000 + butan-1-ol were calculated keeping the parameter K_1, ε_1^{Asso} from the system Boltorn U3000 + propan-1-ol. For this system the parameter $\varepsilon_{LCT}, K_2, \varepsilon_2^{Asso}$ must be adjusted to the experimental data given in the literature [183]. The calculated LLE together with the experimental data [183] are depicted in Figure 11. The adjusted parameters are listed in Table 2.

Similar to the results obtained for the system Boltorn H20 in water (Figure 10) or in propan-1-ol [188], the LCT in combination with the Wertheim approach describe the branched of the cloud point curve related to the diluted polymer solution very close to the experimental data. However; the branch of the cloud point curve describing the composition of the polymer-rich phase some derivation occur. For the system Boltorn U3000 + propan-1-ol the experimental cloud point curve shows a characteristic shoulder. Our theoretical framework was not able to describe this shoulder.

POLYMER BLENDS

Polymer blends are formulated by mixing polymers with different chemical structures to create new materials with beneficial properties of the individual components. While FH-theory explains some basic trends in blend miscibility, the theory completely neglects the dissimilarity in monomer structures that is central to the fabrication of real blends. The application of the LCT for mixtures made from two linear polymers or copolymers was intensively studied by Freed et al. [211-215, 221]. An excellent review could be found in the literature [215], where the influence of the monomer structure on the compatibility of the blend is discussed in great detail. Several illustrative applications of the LCT provide new molecular-scale interpretations for many nontrivial phenomena occurring in polymer systems [215]. The applications also demonstrate the predictive ability of the theory and its usefulness in analyzing thermodynamic data over a wide variety of polymer mixtures, ranging from binary homopolymer blends of various types of copolymer systems [215].

Table 3. Studied PBD/PS blends

blend	$M_{M,PBD}[g/mol]$	$M_{M,PS}[g/mol]$	M_{PBD}	M_{PS}	$\varepsilon_{LCT}/k_B\,[K]$
1	1100	1340	10	13	29.95
2	1100	1670	10	16	31.1
3	1100	4370	10	42	21.5
4	2350	1900	22	18	23.0
5	2350	3300	22	32	21.6

Recently [258], several new experimental and theoretical phase equilibrium data of different polymer blends were published. The theoretical calculations were performed using two different equations of state (Sanchez-Lacombe EOS and Perturbed Chain-Statistical Associated Fluid Theory - EOS [PC-SAFT]), where a nice agreement between the experimental data and the modeling result could be achieved [258]. Both equations of state require three pure component parameters and one binary interaction parameter, if no association is modeled.

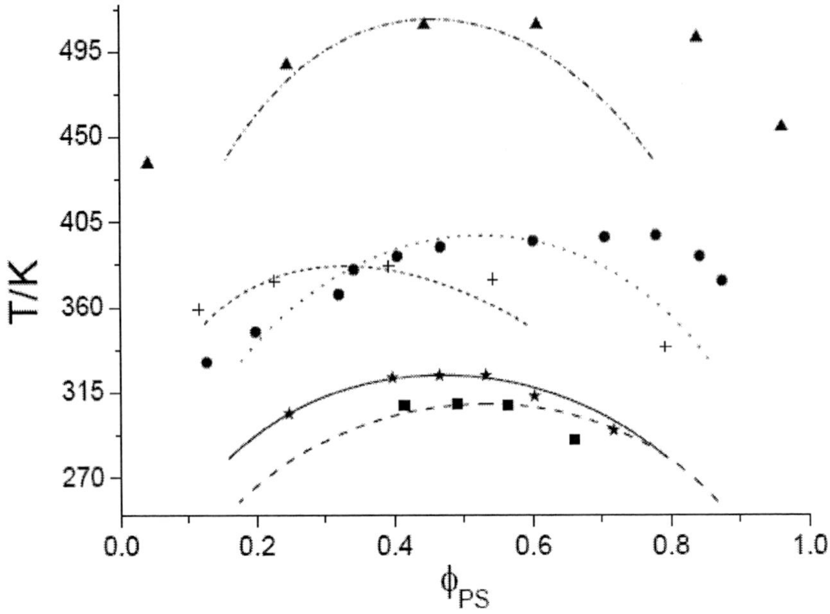

Figure 12. Experimental [258] (squares: blend 1, stars: blend 2, crosses: blend 3, circles: blend 4, triangles: blend 5) calculated LLE using the parameters given in Table 3.

Even, if polar groups able to form associates are present, no associations were taken into account [258]. Now we would like to investigate, if the LCT is also able to model the experimental phase diagrams. For comparison we use only the LCT and neglect also the association forces. This approach means that only one parameter must be fitted to experimental data, and no pure-component physical properties, like density data, are involved. For this purpose we select two mixtures, namely polybutadiene (PBD) + polystyrene (PS) and polyethylene glycol mono -methyl ether (PEGE) + 2,6-dimethyl phenylene oxide (PPO).

The first mixture, composed from PDB und PS, five different blends, differing in the molecular weight of the polymers (Table 3) experimental cloud point curves are available. Within the LCT we assume every polystyrene monomer occupy one lattice site and every butadiene monomer two lattice sites.

The calculated phase diagrams together with the experimental data [258] are plotted in Figure 12. First, from this figure it can be concluded that the LCT does a good job in modeling the experimental phase diagram. The results obtained with the LCT are very similar then the results obtained with the Sanchez-Lacombe EOS [258]. However; one must have in mind the different number of adjustable parameters. The calculations performed with the PC-SAFT-EOS are closer to the experimental data [258]. Maybe the LCT calculation can be improved by the incorporation of empty lattice sites.

The second example deals with blend made from PEGE and from PPG, where the monomers of both polymers carry polar groups. In the first approach the association will be neglected. We assume the PPG monomer occupy four lattice sites and the PEGE three of them. The parameters are collected in Table 4 and the phase diagrams are demonstrated in Figure 13.

Table 4. Studied PEGE/PPG blends

blend	$M_{M,PEGE}[g/mol]$	$M_{M,PPG}[g/mol]$	M_{PEGE}	M_{PPG}	$\varepsilon_{LCT}/k_B[K]$
1	750	2000	33	78	8.55
2	550	2000	24	78	9.35

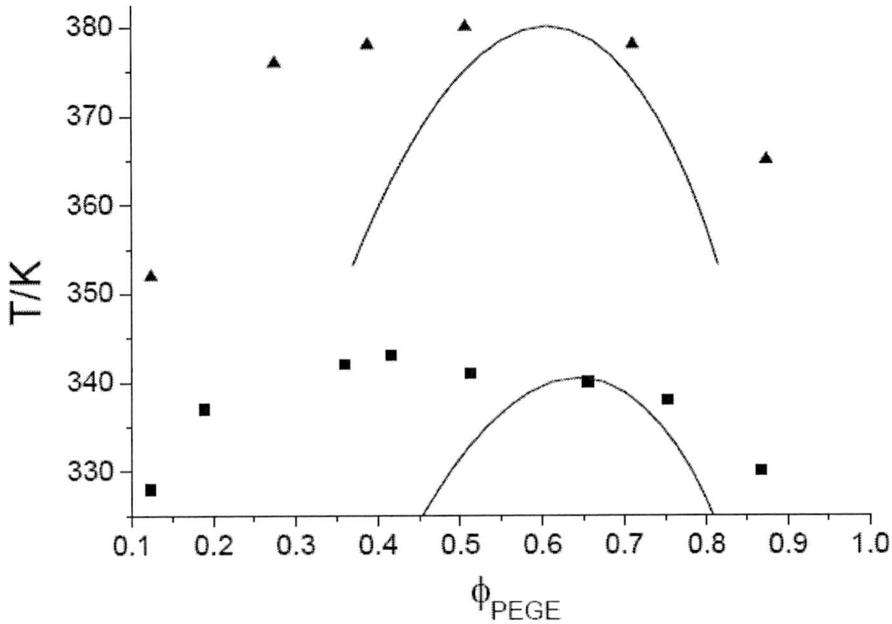

Figure 13. Experimental [258] (triangles: blend 1, squares: blend 2) calculated LLE using the parameters given in Table 4.

Table 5. Studied blends made from Boltorn H20 (1) + linear polymer (2)

blend	M_1	$\varepsilon_{LCT}/k_B[K]$
1	1	61
2	10	10
3	50	3.42
4	250	1.88
5	700	1.4

The theoretical results depict in Figure 13 show clearly the limits of the applied theoretical framework. The calculated miscibility gaps are too small in comparison with the experimental findings. One possible reason for this situation can be the polar interaction. However; the equations of state (SL-EOS and PC-SAFT) were able to model the experimental phase diagram, without to take the polar interaction into account. PPG can also show typical surfactant properties, like micelle formation. However; micelle formation is not covered by the used theory.

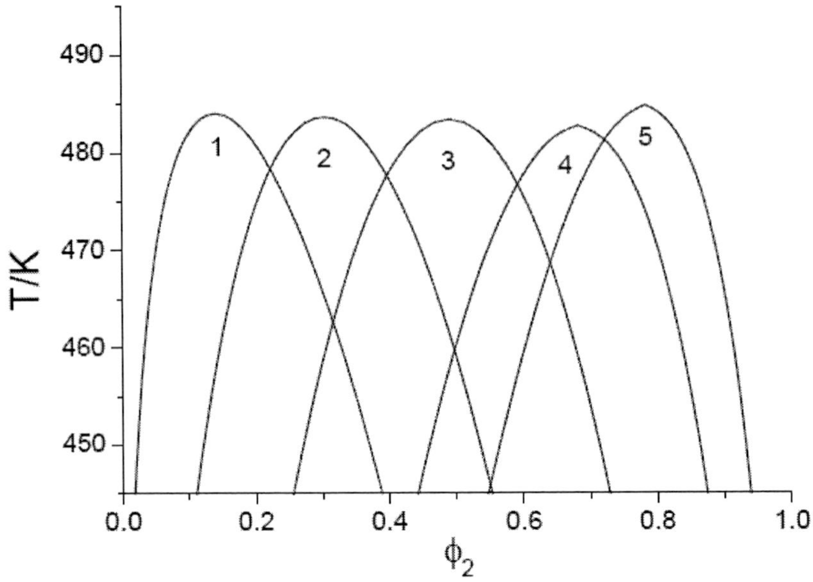

Figure 14. Calculated demixing curves for mixtures made from linear polymers with different molecular weights (1) and hyperbranched polymer Boltorn H20 (2) using the parameters given in Table 5. The numbers indicate the number of blend.

Unfortunately for blends built up from linear polymers and hyperbranched polymers no experimental data could be found in the public literature. Therefore, we are limited to model calculations. Let us study the phase behavior of a mixture of a linear polymer having different molecular weights with a hyperbranched polymer, for example Boltorn H20. The architectural parameters of the hyperbranched polymer are $g = 2, n = 4$ and $n_0 = 7$. The assumed segment-number and the energy parameter are listed in Table 5, where the energy parameters were chosen to ensure a miscibility gap with the identical critical temperature. The calculated cloud-point curves are plotted in Figure 14.

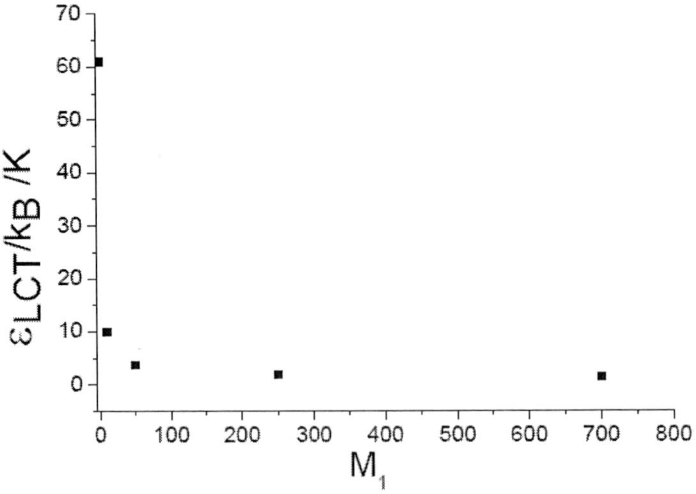

Figure 15. Interaction energy ε_{LCT} as function of the segment number for the system linear polymers with different molecular weights (1) and hyperbranched polymer Boltorn H20 (2).

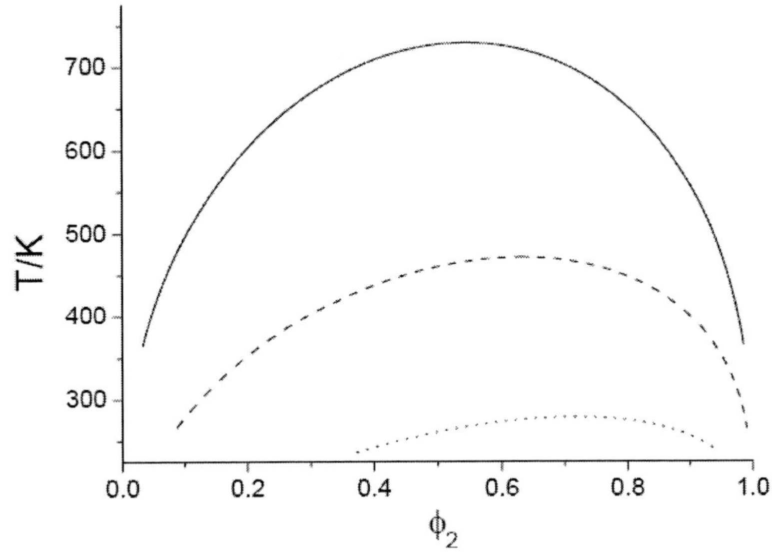

Figure 16. Calculated LLE for a mixture of Boltorn U3000 (1) with different Boltorn H-polymers (2) (solid line: Boltorn H40, dashed line: Boltorn H30, dotted line: Boltorn H20) using LCT with $\varepsilon_{LCT} / k_B = 1K$.

With increasing molecular weight of the linear polymer the critical concentration shifts to lower values of ϕ_2 (Figure 14). Additionally, with increasing molecular weight of the linear polymer a smaller value of ε_{LCT} is required to generate a miscibility gap. The dependency of ε_{LCT} on the molecular weight is plotted in Figure 15. In the limit of high molecular weights only very small interaction energy is sufficient to create a demixing behavior. This finding is very similar to the calculation results of the FH theory for the mixtures of two linear polymers [1, 5]. In order to calculate the phase composition the improved description of the entropy of mixing by the LCT in comparison with FH theory plays an important role, because the entropic effect prevail the enthalpic effect.

Next, we would like to study the phase behavior for the mixture of two hyperbranched polymers. Unfortunately, no experimental data could be found in the public literature. Therefore, we can perform only model calculations without comparison with experimental data. For this purpose we chose the mixture of Boltorn U3000 and Boltorn H20 or Boltorn H30 or Boltorn H40. The architecture of both hyperbranched polymers were explained in the chapter above. We assume a very low value for the interaction energy ε_{LCT} in order to see if a miscibility gap can be created by the different molecular mass and the different polymer architecture. For the mixture composed of Boltorn H20 + Boltorn U3000 a miscibility gap exist (Figure 16), but the corresponding demixing temperatures are below the melting temperature of the polymers. For the practical application of the polymer blend a high compatibility can be expected. This is in contrast to the mixture Boltorn H30 + Boltorn U3000, where the miscibility gap is located in the temperature range suitable for several applications. At constant temperature (i.e. 300K) up to 10% Boltorn H30 can be blended with Boltorn U3000 without phase separation. This can be useful to control several physical properties. On the other hand, only a very limited amount of Boltorn U3000 can be added to

Boltorn H30. A mixture made from Boltorn H40 and Boltorn U3000 is only thermodynamic stabile at small concentrations. The calculated demixing temperature can hardly be found in the experiment, because the polymer shows some degradations at temperatures over 400K.

If one assumed that the interaction energy, ε_{LCT}, do not depend on the architecture of the considered hyperbranched polymer, then $\varepsilon_{LCT} = 0$ for a mixture made from Boltorn H20 and Boltorn H40 is valid [224]. In this case our model predicts a homogeneous mixture of the whole temperature range. However; if a slight dependence of the interaction energy on the polymer architecture, maybe on the segment number or the connected quantity the generation number, is permissible then a miscibility gap can be predicted. For instance, if $\varepsilon_{LCT} / k_B = 1K$ a critical temperature of $T_C = 247K$ can be calculated using the LCT approach. This critical temperature is well below the melting temperature of both polymers. However; if the interaction energy is slightly increased, maybe to $\varepsilon_{LCT} / k_B = 1.5K$, a critical temperature of $T_C = 371\,K$ and a critical segment fraction of $\phi_C = 0.32$ can be calculated. This result can be interpreted, that also hyperbranched polymers of the same chemistry, but different architecture can show phase separation behavior. Using FH-theory for linear polymers the well-known interaction parameter χ can also be a function of the molecular weight. This was often found for oligomers [1, 2, 5].

In comparing lattice cluster theory predictions and experimental data, it must be borne in mind that the theory is based upon an admittedly oversimplified lattice model. Nevertheless, the theory provides a description of the gross influences on thermodynamic properties of polymer architecture. The improved theories might be constructed by further developments of the LCT to incorporate more aspects of reality, or they could require use of much more sophisticated atomistic non-lattice theories. The later approach would necessitate very heavy numerical computations, especially for the coexistence curves, and they are to be contrasted with the almost trivial computational requirements for applications of the analytical formulas of the LCT. Beside the architecture of the hyperbranched polymer covered with the LCT the number and the type of the terminal functional groups must be taken into account in the thermodynamic equations. These specific interactions can be modeled by self- and cross association phenomena with the help of the Wertheim perturbation theory.

Thermodynamics of Interfacial Properties

One of the major applications of hyperbranched polymers is their use in polymer blends or as additives in linear polymers. The effectively depends strongly on the surface properties, especially, on the tendency of the hyperbranched polymer to migrate to the surface to form a lubricating layer. The surface energy plays a key role in polymer processing and blending. For the application of hyperbranched polymer for encapsulation purposes of different guest molecules in medicine [96, 259] or dyeing [81] the surface energy plays also a dominant role. The surface energy can be controlled by the type of functional terminal groups [260] or charges. The behavior of hyperbranched polymers on the surface is substantially dependent on the formation of micellar-like structures [261]. For this reason several experimental investigations were performed in order to characterize the surface properties.

The surface properties of neat and modified Boltorn® hyperbranched polyesters were studied by water-contact-angle measurements and the micro-Wilhelmy wetting technique to determine the solid and melt surface tensions, respectively [262, 263]. The surface tensions of the melts are extremely high for hydroxyl-terminated hyperbranched polyesters and approach that of water, indicating the preferential surface segregation of hydroxyl terminal groups [262]. The substitution of hydroxyl groups with alkyl chains dramatically reduces the melt surface tension to the point where a large degree of substitution produces the surface tension equivalent of pure alkanes [262]. The molecular adsorption of third (HBP3) and fourth (HBP4) generations of hyperbranched polyesters with 32 and 64 hydroxyl-terminal groups on a bare silicon surface was studied with scanning probe microscopy and was described in the terms of the Langmuir isotherm [264]. The shape of hyperbranched polymers with 32 hydroxyl-terminal groups within an adsorbed layer evolved from a pancake with a thickness less than 1 nm for very low surface coverage to densely packed wormlike bilayer structures with a thickness of about 3 nm for the highest surface coverage [264]. The molecules of the fourth generation hold a stable, close-to-spherical shape with a diameter of 2.5 nm throughout the entire range of surface coverage including both dense monolayer's and isolated molecules [264]. High intramolecular flexibility of HBP3 molecules as compared with constrained mobility of bulkier branches of HBP4 is considered to be responsible for different surface behavior [264].

Mikhaylova et al. [265] applied a combination of different surface-sensitive techniques to obtain better understanding of adsorption processes of model proteins on hydrophilic surfaces of hydroxyl (phenolic group) terminated hyperbranched aromatic polyesters (HBP-OH) in comparison to hydrophobic polystyrene (PS) surface. Surprisingly, the adsorbed amount of lysozyme and human serum albumin on the hydrophilic surface of HBP-OH having weak negative surface charge was larger than on the hydrophobic PS surface. The adsorption kinetics of both proteins on the two types of surfaces showed very fast adsorption process [265]. Axisymmetric drop shape analysis was applied to measure in situ the changes in the contact angle and interfacial tensions as a result of the adsorption of protein molecules from a sessile solution droplet [265]. The observed effects are very complex since not only the presence of proteins, including the arrangement and conformation of the adsorbed molecules, but also the properties of the polymer surfaces play a role [265].

Peleshanko et al. [261] reported the synthesis of novel hyperbranched amphiphilic poly(ethylene oxide)-polystyrene $(PEO\text{-}PS)_n$ copolymers obtained by controlled radical polymerizations. Langmuir monolayers displayed reversible amphiphilic behavior at the air-water interface [261]. The random, mixed character of short hydrophilic and hydrophobic fragments results in peculiar surface behavior: unlike regular linear and star block copolymers, the amphiphilic hyperbranched macromolecules with higher PEO content are spread at the air-water interface and short PEO fragments are not submerged into the water subphase even at high compression [261].

The nature of the solid-liquid interface of hydroxyl-terminated hyperbranched aromatic polyester (HPB-OH) thin films was studied by Mikhaylova et al. [266]. Contact angle measurements, atomic force microscopy, zeta potential measurements and spectroscopic ellipsometry were used to characterize the physical and chemical structure of the HBP-OH thin films as well as the adsorption-relevant surface properties [266]. All surface-sensitive methods used indicate marked swelling behavior of HBP layers in aqueous phosphate buffer saline [266]. Moreover, the degree of swelling, zeta potential as well as surface free energy, strongly depend on the time the films had been annealed above their glass transition temperature [266, 267].

The influence of hyperbranched polymer grafted polypropylene (PP-HBP) on the interfacial adhesion between fusion bonded bilayers of polypropylene (PP) and polyamide 6 (PA6) and on the properties of PP/PA6 blends was investigated [268]. The interfacial adhesion between PP-HBP compatibilized bilayers was ten times higher compared to maleic anhydride grafted PP (PP-MAH) compatibilized bilayers [268]. This is attributed to the higher diffusivity and functionality of PP-HBP leading to the formation of more PP-PA6 copolymers at the interface [268]. The high diffusivity of PP-HBP is an asset for multilayer film extrusion, while for blends; the high functionality permits the use of less compatibilizer for similar property improvements [268].

Recently, dendritic core-shell architectures which are based on hyperbranched polyglycerol for the solubilisation of hydrophobic drugs have been synthesized and characterized [81]. The core of hyperbranched polyglycerol has been modified with hydrophobic biphenyl groups or perfluorinated chains to increase the core hydrophobicity of the macromolecules. These amphiphilic core-shell type architectures were then used to solubilize pyrene, nile red, and a perfluoro tagged diazo dye, as well as the drug nimodipine in water. Surface tension measurements and scanning electron microscope were used to reveal the aggregation properties of these complexes. The formation of supramolecular aggregates with diameters of $20\ nm$ and critical aggregate concentrations of $2 \cdot 10^{-6}\ mol / L$ have been observed [81].

Besides the thermodynamics of the surface properties, the kinetics, which is also strongly depended on the polymer architecture [269], is an essential feature for several applications. Díez-Pascual et al. [269] reported a comparative study between Langmuir and Gibbs monolayers of a hyperbranched polyol, poly(propylene glycol) homopolymers, and poly(propylene glycol)-poly(ethylene glycol) copolymers with different structure and molecular weight. With the help of dynamic surface tension and surface pressure measurements the adsorption kinetics were studied, where the kinetics were consistent with a rapid diffusion stage followed by a slow reorganization at the air-water interface [269].

Recently, the surface tension of the pure hyperbranched polymer, Boltorn U3000, and binary mixtures of Boltorn U3000 with alcohols (butan-1-ol, hexan-1-ol) have been measured at atmospheric pressure in the range of temperatures from $298.15\ K$ to $328.15\ K$, or of Boltorn U3000 with methyl-tert-butylether (MTBE) in the range of temperatures from $298.15\ K$ to $318.15\ K$ [270]. The influence of hyperbranched polyesters with different functional end groups on the surface tension of mixtures with an oligo(ester diol) was investigated [271] using Wilhelmy balance technique. The results indicate that the surface tension of the pure hyperbranched polyesters strongly depends on the functionality of the end groups, where the functionalization of the hydroxyl end groups by short alkyl chains (methyl, tert-butyl) reduced the surface tension depending on the degree of substitution [271]. The surface tension of the mixtures with the hydroxyl-terminated hyperbranched polyester was slightly increased at higher concentrations of the hyperbranched polymer compared to the surface tension of the pure ester diol [271]. On the other hand, the surface tension of mixtures could be considerably decreased using 1% of hyperbranched polyester polyols partially substituted with short alkyl chains [271], whereas the modified hyperbranched polyesters act as surface active agents. Peleshanko et al. [272] presented a review focused on the recent developments in the field of the highly branched molecules (excluding dendrimers) with emphasis on their surface behavior, microstructure, surface morphology, and properties.

Unfortunately, no experimental data related to the miscibility of system containing hyperbranched polymers could be found in the literature.

Theories of polymer interfacial properties have been developed along two fundamentally different lines [273], the self consistent field techniques and free energy functional methods, which have been designed originally for describing the strong and weak segregation limits, respectively. However; only a limit amount of work dealing with hyperbranched polymers can be found in the literature. For example, Qian et al. [274] used self-consistent (SCF) mean-field lattice simulations to study entropy-driven segregation of various branched and hyperbranched polymeric additives in chemically similar linear polymer hosts. The simulations account for the effect of molecular architecture on local configurationally entropy in the blends, but ignore the effect of architecture on local density and blend compressibility. Star, dendrimer, and comb-like additives are all found to be enriched at the surface of chemically identical linear host polymers [274]. The magnitude of their surface excess increases with increased number of chain ends and decreases with increased segmental crowding near the branch point [274]. Provided the number of arms and molecular weight of the branched additives are maintained constant, Qian et al. [274] found that the simplest branched architecture, the symmetric star, exhibits the strongest preference for the surface of binary polymer blends.

In this contribution we focus our attention to a special version of free energy functional methods, namely the modified density gradient theory, originally developed by van der Waals [275] and Lord Rayleigh [276] and later rediscovered by Cahn-Hilliard [277]. The theory of van der Waals [275] and Lord Rayleigh [276] leads to a description of the state of matter within the vapor-liquid interface. The goal of the theory was to elucidate the density profiles and to evaluate the associate surface tension. Van der Waals [275] used in his description a local free energy density consisting of a uniform term, evaluated at the local density, and a kind of gradient energy density term which is proportional to the square of the density gradient. The interfacial density profiles can be obtained by minimizing this free energy density under the constraint that the total number of molecules is constant, and resulting in an expression that allows evaluation of surface tensions. In version provides by Cahn-Hilliard [277] the density gradient approach leads to a general expression of Helmholtz energy of an inhomogeneous system, in which the homogeneous portion of the free energy can be taken from any equation of state or activity coefficient model, while the inhomogeneous contribution is truncated at the leading square gradient approximation. Several examples using this approach for the calculation of surface properties, like interfacial tension, density profiles within the interface and selective enrichment in the interface can be found in the literature [278, 279, 280, 281, 282, 283, 284, 285, 286].

In 1986, Wang et al. [287] presented a continuum model for polymer chains near an asymmetric (A-B) liquid-liquid interface where each side of the interface can have different polymer-surface interactions. For example, one side can attract the macromolecule, while the other can repel it. The model contains different monomer free energies and different excluded volume interactions for the macromolecule in the two solvents [287].

Later [288, 289, 290] the de Gennes modified square gradient approximation as arising from the entropic contributions associated with chain connectivity were used. McMullen et al. [288] employed density functional methods to derive the free energy and grand potential functionals appropriate to homopolymers and blends based on the FH theory. Later on, this theoretical framework was extended to block copolymers [289] permitting the treatment of compressible systems and therefore of more strongly first-order microphase separations. Freed and coworkers [291, 292] discussed the influence of the concentration dependent interaction parameter, χ , within the FH limit on the interfacial properties, especially on the interfacial width and tension. In order to study the surface tension of dilute polymer solutions a renormalization group approach was applied [293]. An interpolation function extends the calculations to intermediate values of the polymer-surface interaction strength and provides the surface pressure as a function of both polymer-polymer and polymer-surface interactions [293].

Using a compressible version of FH theory within the density functional treatment for the liquid-liquid interface of phase separated binary polymer blends [294] involved several numerical problems, because the coupled, nonlinear Euler-Lagrange equations for the interfacial profile and tension are shown to become numerically unstable and permit analytic solution in the asymptotic profile wings. This analysis was limited to symmetric binary blends because this restriction permits to compute numerically the central portion of the interfacial profile [294].

Freed [295] provide a bridge between the density functional and self-consistent-field formulations for inhomogeneous polymer systems by deriving the self-consistent-field equations from a density functional approach. The final density functional emerges in the form of a Landau-type expansion about an analytically tractable zeroth-order inhomogeneous reference system, and the important presence of chain connectivity contributions is quite evident [295]. In 1996, Freed [296] represented an analytical theory for the competing influences of polymer-surface and polymer-polymer interactions, density and composition variations, and blend asymmetries on the surface profiles of a multi-component polymer blend near an interacting, impenetrable interface. The theory is explicitly applied in the limit of small continuum model polymer-surface interaction parameters, a limit which still enables treating all qualitative behaviors of polymers that individually tend either to aggregate toward or to segregate from the surface [296]. The formulation is based on an analytic combined self-consistent field-density functional theory for inhomogeneous polymer systems, where the compressible polymer system was described with a generic Gaussian chain-random mixing type model, and the bulk phase with a Sanchez-Lacombe-type model [296].

The theory for the interfacial properties of compressible binary polymer systems [294] was generalized to treat phase separated asymmetric binary blends. Lifschitz and Freed [297] used the Sanchez-Lacombe approximation for the homogeneous system free energy and a Cahn-Hilliard-de Gennes expression for the square gradient contribution. These two ingredients enable the same theory to describe both the compositions of the coexisting phases and the interfacial properties [297]. It was found that the coupled equations for the two concentration profiles are numerically unstable [297]. In order to solve this problem a perturbation-variational method for symmetric blends was introduced and extended to treat blends with asymmetry due to differing polymerization indices and/or interaction parameters [297]. The zeroth order approximation, which uses a hyperbolic tangent interpolation between the generally different densities in the coexisting phases, is found to be extremely accurate in predicting the interfacial tensions and widths [297].

Dudowicz et al. [298] investigated a minimal equilibrium polymerization model for the competition between self-assembly on a boundary and in solution that arises when an assembling system is in the presence of an adsorbing interface. Adsorption generally occurs upon cooling, but assembly (equilibrium polymerization) may arise either upon cooling or heating. Both cases are shown to exhibit a coupling between adsorption and self-assembly [298]. When both assembly and adsorption proceed upon cooling, a change in the ratio of the enthalpy of adsorption to the enthalpy of assembly in solution can switch the system between a predominance of self-assembly in solution to assembly on the substrate [298]. If assembly is promoted by heating and adsorption by cooling, as in many self-assembling proteins in aqueous solution, then a self-assembly analog of a closed loop phase boundary is found [298].

This contribution is focused on interfacial properties of binary systems in liquid-liquid equilibrium, which are of considerable practical importance in the description of liquid extraction processes or in all other applications mentioned above. First the interfacial properties will be considered at atmospheric pressure. Keeping the pressure constant, the density gradient theory (DGT) will be blended with an activity coefficient model [278]. The Cahn-Hilliard theory [277] describes the thermodynamic properties of a system where an interface exists between two equilibrium phases. In the liquid-liquid equilibrium of binary systems, consisting of components A and B, at atmospheric pressure p, the temperature T and the compositions of the two bulk phases (I and II) are given by the thermodynamic phase equilibrium conditions. It was assumed that the system is at constant temperature and pressure and to contain a fixed number of moles of each component. If there is a volume change upon mixing, then the total volume V will be a variable. In many practical situations V can be regarded as constant, so that the Helmholtz and Gibbs free energies are equivalent. In the interface between two phases, only the concentration, expressed in segment fraction, ϕ_B, varies continuously from the bulk concentration of phase I ϕ_B^I to the bulk concentration of phase II ϕ_B^{II}. The phase diagram can be calculated using a free energy model related to the LCT and the Wertheim perturbation theory representing the association forces [188, 223].

Following the method suggested in the literature [188, 223, 278] the interfacial tension σ for a planar interface in a binary system at equilibrium can be computed by:

$$\sigma = 2N \int_{\phi_B^I}^{\phi_B^{II}} \sqrt{\beta \, \Delta g(\phi_B)} \, d\phi_B \tag{45}$$

where β is the so-called influence parameter and can be fitted to one experimental interfacial tension at one temperature. $\Delta g(\phi_B)$ is the grand thermodynamic potential. The composition profile can be determined by [188, 223, 278]:

$$z = z_0 + \int_{\phi_{0B}}^{\phi_B} \sqrt{\frac{\beta}{\Delta g(\phi_B)}} \, d\phi_B \tag{46}$$

where z_0 and ϕ_{0B} represent an arbitrarily chosen origin and composition. The z-origin is arbitrarily located at a composition of $\left(\phi_B^I + \phi_B^I\right)/2$. This means $z_0 = 0$ and $\phi_{0B} = \left(\phi_B^I + \phi_B^I\right)/2$. A distance z may be determined for any ϕ_B lying between the bulk concentrations by evaluating the integral numerically.

Knowledge of β and $\Delta g\left(\phi_B\right)$ allows to calculate the interfacial tension with help of Eq. 45 and the composition profile using Eq. 46. In principle, $\Delta g\left(\phi_B\right)$ can be evaluated using any model free energy function, and hence also the LCT or the LCT in combination with the Wertheim theory. Usually, β is fitted using one interfacial tension at one temperature [278]. Unfortunately, no experimental values of interfacial tension of systems containing hyperbranched polymers are available in the literature. For this reason we discuss not the interfacial tension itself, but the reduced interfacial tension, which is set to $\sigma_{red} = \sigma / \sqrt{\beta}$. The procedure has the advantage, if β is known, for example from one experimental data point; all other interfacial tensions can be calculated. As consequence also the interfacial profiles are given in reduced quantities.

INTERFACIAL PROPERTIES - CALCULATION EXAMPLES

Beside the phase equilibrium the DGT allows also to investigate the interfacial properties between the coexisting phases. One important quantity related to the interface is the interfacial tension. Using Eq. 45 in combination with LCT or in combination with LCT and Wertheim theory permits the calculation of the reduced interfacial tension of the system Boltorn H20 + water. The calculated reduced interfacial tensions as function of the distance from the critical point are plotted in Figure 17. At the critical point the interfacial tension has to be zero. With increasing distance from the critical point the interfacial tension increases too. Similar to the results obtained for the phase-equilibrium calculation the association forces have an impact on the interfacial tension, but not a dominating influence. The association forces lead to a higher interfacial tension. With decreasing temperature this effect is more pronounced. The same situation could be found for the system propan-1-ol + Boltorn H20 and propan-1-ol + Boltorn U3000 (Figure 18). However; the impact of the association forces is more important in the aqueous polymer solution compared with the solution, where the polymer is dissolved in propan-1-ol. The possible reason for this finding can be the different association strength. At constant distance from the critical point the interfacial tension of the aqueous polymer solution is much higher than in the system propan-1-ol + Boltorn H20 or propan-1-ol + Boltorn U3000. Using only the LCT the predicted interfacial tension for both systems at a given distance from the critical point is very similar (i.e. $T_C - T = 10\text{K}$, water + Boltorn H20 $\sigma_{red} = 0.15$, propan-1-ol + Boltorn H20 $\sigma_{red} = 0.12$, propan-1-ol + Boltorn U3000 $\sigma_{red} = 0.061$). Taking the association forces into account the solutions differ in the interfacial tension (i.e. $T_C - T = 10\text{K}$, water + Boltorn H20 $\sigma_{red} = 0.22$, propan-1-ol +

Boltorn H20 $\sigma_{red} = 0.17$, propan-1-ol + Boltorn U3000 $\sigma_{red} = 0.11$), where the impact of association force is for the aqueous solution more pronounced. The difference of interfacial tension for both polymers (Boltorn H20 and Boltorn U3000) in the same solvent, namely propan-1-ol, can be explained by the different miscibility in terms of length of the tie line (Figure 10 and Figure 11). If experimental data are available, the reduced quantities can be recalculated in interfacial tensions.

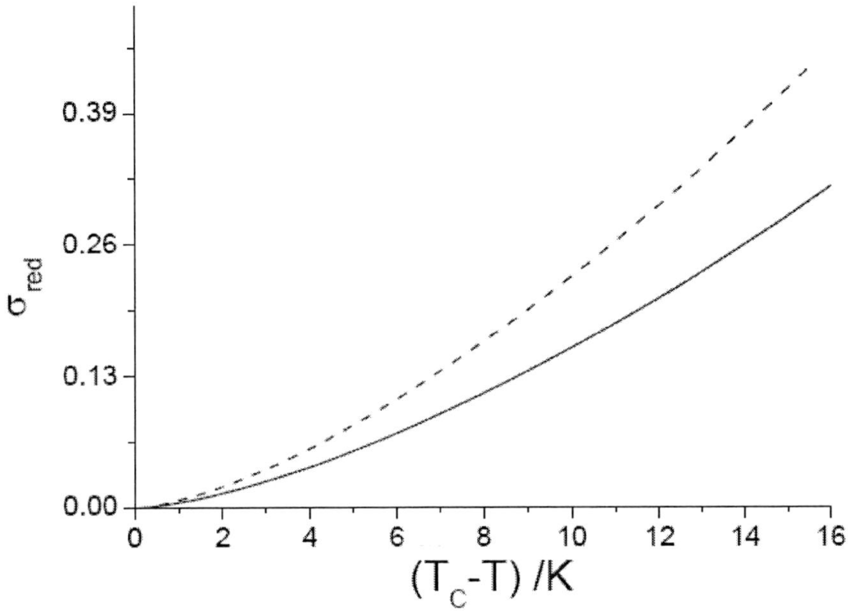

Figure 17. Predicted interfacial tension using LCT (solid line) and LCT + Wertheim theory (broken line) for the system water + Boltorn H20 [188, 223].

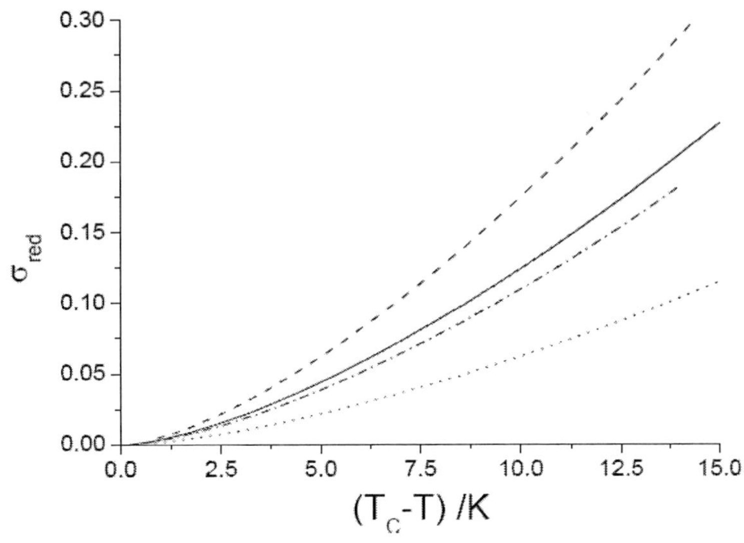

Figure 18. Predicted interfacial tension using LCT (solid line: Boltorn H20 + propan-1-ol [188, 223], dashed line: Boltorn U3000 + propan-1-ol) and LCT + Wertheim theory (dotted line: Boltorn H20 + propan-1-ol [188, 223], dotted-dashed line: Boltorn U3000 + propan-1-ol).

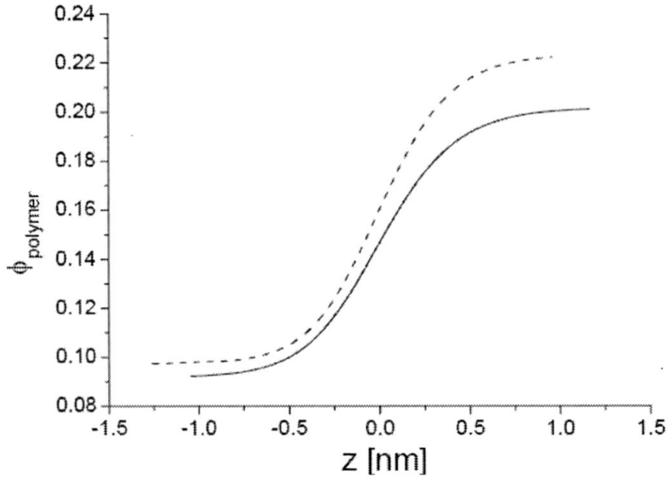

Figure 19. Predicted concentration profile using LCT (solid line) or LCT and Wertheim theory (broken line) for the system water + Boltorn H20 at T=369.2K [188, 223].

Not only the interfacial tension itself, but also the concentration profiles in the interface depend slightly from the intermolecular forces (Figure 19 and Figure 20). The concentration profile of the liquid-liquid interface varies only in the direction normal to the interface, which we take as the z direction. The interfacial profiles are calculated with the help of Eq. 46 in combination with LCT or LCT and Wertheim theory for the system water + Boltorn H20 (Figure 19) and for the system propan-1-ol + Boltorn H20 (Figure 20). Analyzing the interfacial profiles given in Figure 19 and Figure 20 show again the impact of the association forces. The incorporation of the Wertheim theory leads to a sharper concentration profiles in comparison of the theoretical results, if only the LCT is used. The sharper concentration profiles are strictly connected with a smaller interfacial thickness. No extreme in the concentration profiles could be detected, and hence no relative enrichment in the interface is predicted by the suggested theoretical framework.

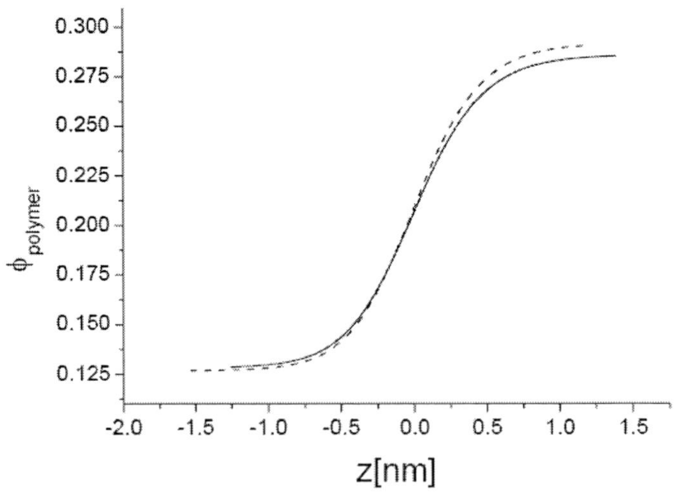

Figure 20. Predicted concentration profile using LCT (solid line) or LCT and Wertheim theory (broken line) for the system propan-1-ol + Boltorn H20 at T=366.9 K [188, 223].

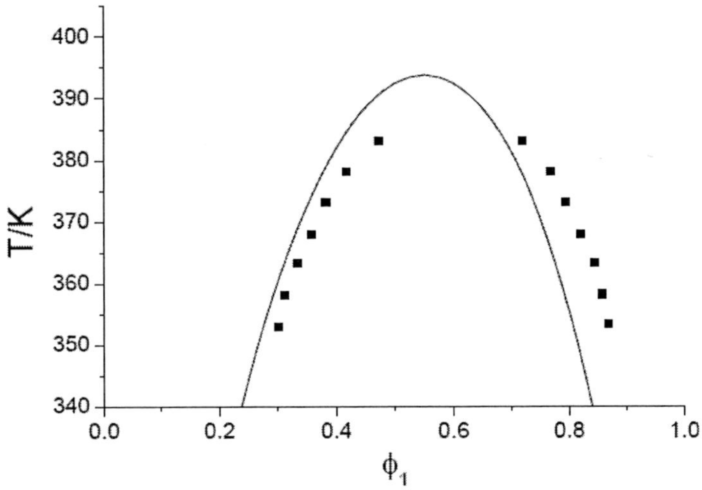

Figure 21. Experimental [299] (squares) and calculated (LCT, line) cloud point curve for the binary blend PDMS (1) + PTMDSE (2).

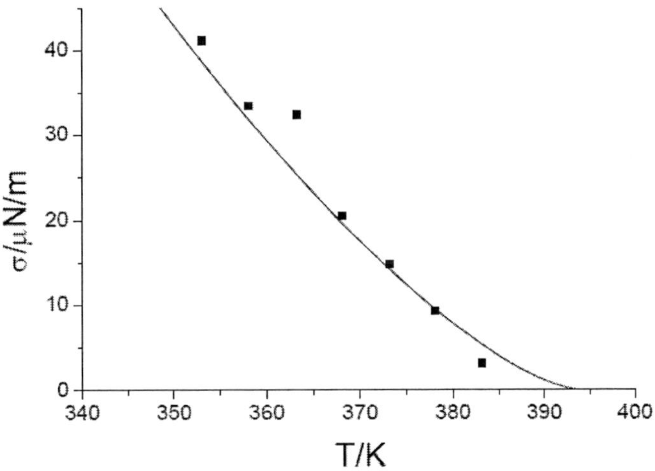

Figure 22. Experimental [299] (squares) and calculated (LCT, line) interfacial tension for the binary blend PDMS (1) + PTMDSE (2).

Sakane et al. [299] measured the cloud point curve and the interfacial tension of phase-separated mixtures of two linear polymers, namely poly(dimethylsiloxane) (PDMS) (1) and poly(1,1,3,3-tetramethyldisiloxanylethylene) (PTMDSE) (2), using the sessile-drop method. Additionally, they adopted also the square-gradient in order to describe the interfacial tension of phase-separated polymer mixtures [299]. In contrast to this contribution Sakane et al. [299], combined the Cahn-Hilliard theory with the classical Flory-Huggins theory. Assuming the monomer of PDMS occupies one lattice site and the other monomer two of them leads to the following characteristic parameters within the LCT:

$$M_1 = 145, \quad M_2 = 218, \quad \varepsilon_{LCT} / k_B = 0.91K \tag{47}$$

where the interaction energy, ε_{LCT}, was fitted to the experimental cloud-point curve taken from the literature [299. The calculated phase diagram together with the experimental data points are plotted in Figure 21. The comparison shows that the selected theoretical framework leads to a satisfactory agreement with the experimental data, especially if one had in mind. that only one adjustable parameter, ε_{LCT}, is used. The phase diagram forms the basic for the calculation of the interfacial tension with the help of Eq. 45. To obtain explicit numerical results for the interfacial tension, we need to fix the value of β. The parameter β was fitted to one experimental interfacial temperature at one temperature. This allows the prediction of the temperature dependency of the interfacial tension.

Figure 22 shows a comparison between the experimental data [299] and the calculated interfacial tension for the binary blend PDMS + PTMDSE. From this result it can be concluded that the combination of the LCT with the density gradient theory can be applied to predict the temperature dependency of the interfacial tension of phase separated polymer blends with a high accuracy. This can be very useful, because the experiments are very challenging due to the high viscosity.

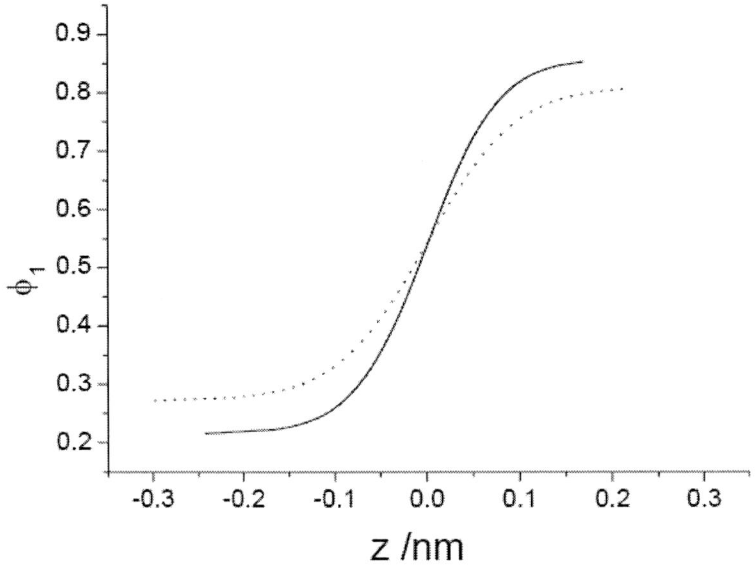

Figure 23. Predicted concentration profile using LCT at two temperatures (solid line: T=330K, dotted line 351K) for the system PDMS (1) + PTMDSE (2).

In Figure 23 two concentration profiles of PDMS within the interface at two different temperatures are plotted versus the coordinate z perpendicular to the interface. This system to feature an UCST behavior (Figure 21) and hence the critical temperature is located at higher temperatures. At the critical temperature the interfacial profile must diverge, because the interfacial thickness must become infinity. Therefore, the interfacial profile at higher temperature ($T = 351K$ in Figure 23) is broader than at lower temperatures ($T = 330K$ in Figure 23). No extremes could by found in the concentration profile meaning that no selective enrichment of one component in the interface occurs.

CONCLUSION

The increasing attention given to hyperbranched polymer for several new applications in different fields ranching from medicine over energy technology until chemical engineering is the driving force of their research. Among others, the development of thermodynamic models is one important point in this direction. The major hurdle in the thermodynamics of hyperbranched polymers is the complexity of these polymers caused by the architecture and the large number of functional groups responsible for specific interaction between the polymer molecules itself and between the polymer and the surrounding solvent. By means of an elaborate lattice cluster theory, allowing describing the polymer architecture, in combination with Wertheim perturbation approach, allowing modeling association forces, both effects could be taken into account in the thermodynamic model. Additionally, the square gradient theory can be utilized for the investigation of inhomogeneous system having an interface. The burden of this theoretical framework is to compute the interfacial properties, like interfacial tension and interfacial profiles, in phase-separated systems.

The endeavor to demonstrate the full potential of this approach we compare the theoretical results in terms of phase equilibria and interfacial tension with experimental data.

Overall, the results indicate how polymer architecture can exert a tremendous influence on the physical and solution properties of a material. A significant benefit of the LCT lies in the algebraic nature of the expressions for thermodynamic properties that apply for all compositions, molecular weights, interaction energies, monomer structure, and temperature. The Gibbs energy of mixing can be calculated using a simple polynomial, which is derived in this contribution. The LCT excess free energy for a binary system in the incompressible limit depends on the interaction energies only through the single dimensionless interaction energy, $\varepsilon_{LCT} = \varepsilon_{11} + \varepsilon_{22} - 2\varepsilon_{12}$.

In the light of this idea, we carried out calculations of phase diagram for polymer solutions containing hyperbranched polymers or polymer blends of linear or hyperbranched polymers.

ACKNOWLEDGMENTS

Authors gratefully acknowledge the financial support of this research by the German Science Foundation (DFG En 291/7-1) and the Max-Buchner Foundation (Grant Nr. 2864), as well Perstorp, Sweden for the polymers.

REFERENCE

[1] R. Koningsveld, W.H. Stockmayer, E. Nies, *Polymer Phase Diagrams*. Oxford University Press, 2001.

[2] K. Kamide, *Thermodynamics of Polymer Solutions – Phase Equilibria and Critical Phenomena*. in A.D. Jenkins (Ed.) Polymer Science Library 9, Elsevier, 1990.

[3] J.M. Prausnitz, Molecular thermodynamics for some applications in biotechnology. *J. Chem. Thermodynamics* 35 (2003) 21-39.

[4] P.A. Albertsson, *Partition of Cell Particles and Macromolecules*, 3rd edition. John Wiley and Sons, New York, 1986.
[5] P.J. Flory, *Principles of Polymer Chemistry*, Cornell University Press, Ithaca, 1953.
[6] J.M.J. Frechet, D.A. Tomalia, *Dendrimers and Other Dendritic Polymers*. Wiley Series in Polymer Science, John Wiley and Sons, Ltd. 2001.
[7] D.A. Tomalia, The dendritic state. *Materials Today* 8 (2005) 34-46.
[8] P.J. Flory, Molecular Size Distribution in Three Dimensional Polymers. VI. Branched Polymers Containing A—R—B_{f-1} Type Units. *J. Am. Chem. Soc.* 74 (1952) 2718-2723.
[9] P. A. Gunatilake, G. Odian, D.A. Tomalia, Thermal polymerization of a 2-(carboxyalkyl)-2-oxazoline, *Macromolecules* 21 (1988) 1556-1562.
[10] Y.H. Kim, O.W. Webster, Water-Soluble Hyperbranched Polyphenylene: A Unimolecular Micelle. *J. Am. Chem. Soc.* 112 (1990) 4592-4593.
[11] H.R. Kricheldorf, T. Stukenbrock, New polymer syntheses. XC. A-B-A triblock copolymers with hyperbranched polyester A-blocks. *J. Polym. Sci., Part A: Polym. Chem.* 36 (1998) 31-38.
[12] A. Hult, M. Johansson, E. Malmström, Hyperbranched Polymers, *Adv. Poly. Sci.* 143 (1999) 1-34.
[13] A.M. Fischer, H. Frey, Soluble hyperbranched poly(glycolide) copolymers. *Macromolecules* 43 (2010) 8539-8548.
[14] J. Su, L. Zhang, Q. Wu, Amphiphilic dendritic polymers. *Prog. Chem.* 20 (2008) 1980-1986.
[15] M. Jikei, M.A. Kakimoto, Hyperbranched polymers: A promising new class of materials. *Prog. Polym. Sci.* (Oxford) 26 (2001) 1233-1285.
[16] D. Yan, J. Hou, X. Zhu, J.J. Kosman, H.S. Wu, A new approach to control crystallinity of resulting polymers: Self-condensing ring opening polymerization. *Macrom. Rapid Comm.* 21 (2000) 557-561.
[17] T. Huber, F. Böhme, H. Komber, J. Kronek, J. Luston, D. Voigt, B. Voit, New hyperbranched poly(ether amide)s via nucleophilic ring opening of 2-oxazoline-containing monomers. *Macrom. Chem. Phys.* 200 (1999) 126-133.
[18] M. Trollså s, B. Atthoff, H. Claesson, J.L. Hedrick, Hyperbranched poly(ε-caprolactone)s. *Macromolecules* 31 (1998) 3439-3445.
[19] F. Wurm, H. Schüle, H. Frey, Amphiphilic Linear-Hyperbranched block copolymers with linear poly(ethylene oxide) and hyperbranched poly(carbosilane) block. *Macromolecules* 41 (2008) 9602-9611.
[20] R.V. Srinivasa, A.B. Samui, Effect of molecular architecture and size of mesogen on phase behavior and photoactive properties of photoactive liquid crystalline hyperbranched polyester epoxies containing benzylidene moiety. *J. Polym. Sci., Part A: Polym. Chem.* 46 (2008) 552-563.
[21] V.I. Irzhak, A structural characteristic of hyperbranched polymers. *Polym. Sci.* – Ser. B 51 (2009) 143-148.
[22] D.M.A. Buzza, Power law polydispersity and fractal structure of hyperbranched polymers. *Eur. Phys. J.* E 13 (2004) 79-86.
[23] S.D.M.A. Buzza, T.C.B. McLeish, D. Parker, A.J. Keeney, W.J. Feast, Rheology and molecular weight distribution of hyperbranched polymers. *Macromolecules* 35 (2002) 9605-9612.

[24] A.H.E. Müller, D. Yan, M. Wulkow, Molecular parameters of hyperbranched polymers made by self-condensing vinyl polymerization. 1. Molecular weight distribution. *Macromolecules* 30 (1997) 7015-7023.

[25] J.J.J. Frechet, M. Henmi, I. Gitsov, S. Aoshima, M.R. Leduc, R.B. Grubbs, Self-Condensing Vinyl Polymerization: An Approach to Dendritic Materials. *Science* 269 (1995) 1080-1083.

[26] H.T. Chang, J.M.J. Frechet, Proton-Transfer Polymerization: A New Approach to Hyperbranched Polymers. *J. Am. Chem. Soc.* 121 (1999) 2313-2314.

[27] H.C. Kolb, M.G. Finn, K.B. Sharples, Click Chemistry: Diverse Chemical Function from a Few Good Reactions. *Angew. Chem. Int. Ed.* 40 (2001) 2004-2021.

[28] P. Wu, A.K. Feldman, A.K. Nugent, C.J. Hawker, A. Schell, B. Voit, J. Pyun, J.M.J. Frechet, K.B. Sharpless, V.V. Fokin, M.J. Joralemon, R.K. O' Reilly, J.B. Matson, K.L. Wooley, Dendrimers Clicked Together Divergently. *Macromolecules* 38 (2005) 5436-5443.

[29] A. Qin, J.W.Y. Lam, B.Z. Tang, Click polymerization: Progresses, challenges, and opportunities. *Macromolecules* 43 (2010) 8693-8702.

[30] E. Zagar, M. Zigon, Characterization of a Commercial Hyperbranched Aliphatic Polyester Based on 2,2,-Bis(methylol) propionic Acid. *Macromolecules* 35 (2002) 9913-9925.

[31] E. Zagar, M. Zidon, S. Podzimek, Characterization of commercial aliphatic hyperbranched polyesters. *Polymer* 47 (2006) 166-175.

[32] N.K. Jain, U. Gupta, Application of dendrimer-drug complexation in the enhancement of drug solubility and bioavailability. *Expert Opinion on Drug Metabolism and Toxicology* 4 (2008) 1035-1052.

[33] G. Kreutzer, C. Ternat, T.Q. Nguyen, C.J.G. Plummer, J.A.E. Mason, V. Castelletto, I.W. Hamley, F. Sun, S.S. Sheiko, A. Herrmann, L. Ouali, H. Sommer, W. Fieber, M.I. Velazco, H.A. Klok, Water-soluble, unimolecular containers based on amphiphilic multiarm star block copolymers. *Macromolecules* 39 (2006) 4507-4516.

[34] T. Satoh, Unimolecular micelles based on hyperbranched polycarbohydrate cores. *Soft Matter* 5 (2009) 1972-1982.

[35] T. Satoh, Y. Kitajyo, R. Sakai, A. Narumi, T. Takahashi, H. Kaga, N. Kaneko, T. Kakuchi, Size-selective encapsulation property of unimolecular reverse micelle consisting of hyperbranched D-glucan core and L-leucine ethyl ether shell. *Macrom. Symp.* 279 (2009) 145-150.

[36] E. Burakowska, R. Haag, Dendritic polyglycerol core-double-shell architectures: Synthesis and transport properties, *Macromolecules* 42 (2009) 5545-5550.

[37] K.R. Kumar, D.E. Brooks, Comparison of hyperbranched and linear polyglycidol unimolecular reverse micelles as nanoreactors and nanocapsules. *Macrom. Rapid Comm.* 26 (2005) 155-159.

[38] T. Satoh, M. Tamaki, Y. Kitajyo, T. Maeda, H. Ishihara, T. Imai, H. Kaga, T. Kakuchi, Synthesis of unimolecular reversed micelle consisting of a poly(L-lactide) shell and hyperbranched D-mannan core. *J. Polym. Sci., Part A: Polym. Chem.* 44 (2006) 406-413.

[39] Y. Liang, W. Decheng, C. Xiangyu, J. Ming, P. Hongting, Unimolecular micelle derived from hyperbranched polyethylenimine with well-defined hybrid shell of

polyethylene oxide and polystyrene: A versatile nanocapsule. *J. Polym. Sci., Part A: Polym. Chem.* 48 (2010) 681-691.

[40] V. Percec, P. Chu, M. Kawasumi, Toward "willowlike" thermotropic dendrimers, *Macromolecules* 27 (1994) 4441-4453.

[41] V. Percec, P. Chu, G. Ungar, J. Zhou, Rational design of the first nonspherical dendrimer which displays calamitic nematic and smectic thermotropic liquid crystalline phases. *J. Am. Chem. Soc.* 117 (1995) 11441-11454.

[42] Y.H. Kim, Lyotropic liquid crystalline hyperbranched aromatic polyamides. *J. Am. Chem. Soc.* 114 (1992) 4947-4948.

[43] K.L. Wooley, J.M.J. Fréchet, C.J. Hawker, Influence of shape on the reactivity and properties of dendritic, hyperbranched and linear aromatic polyesters. *Polymer* 35 (1994) 4489-4495.

[44] K. Matyjaszweski, J.T. Shigemonto, J.M.J. Frechet, M.C. Leduc, Controlled/"Living" Radical Polymerization with Dendrimers Containing Stable Radicals. *Macromolecules* 29 (1995) 4167-4171.

[45] E. Malmström, A. Hult, Hyperbranched polymers: A review. *J. Macrom. Sci. – Rev. Macrom. Chem. Phys.* 37 (1997) 555-579.

[46] O.A. Matthews, A.N. Shipway, J.F. Stoddart, Dendrimers - Branching out from curiosities into new technologies. *Prog. Polym. Sci.* (Oxford) 23 (1998) 1-56.

[47] M. Fischer, F. Vögtle, Dendrimere: From Design to Application – A Progress Report. *Angew. Chem. Int. Ed.* 38 (1999) 884-905.

[48] B. Voit, Hyperbranched polymers: A chance and a challenge. *Comptes Rendus Chimie* 6 (2003) 821-832.

[49] C. Gao, D. Yan, Hyperbranched polymers: From synthesis to applications, *Prog. Polym. Sci.* (Oxford) 29 (2004) 183-275.

[50] H.J. Huang, S. Ramaswamy, U.W. Tschirner, B.V. Ramarao, A review of separation technologies in current and future biorefineries, *Sep. Purif. Techn.* 62 (2008) 1-21.

[51] M. Irfan, M. Seiler, Encapsulation using hyperbranched polymers: From research and technologies to emerging applications. *Ind. Eng. Chem. Res.* 49 (2010) 1169-1196.

[52] D. Astruc, E. Boisselier, C. Ornelas, Dendrimers designed for functions: From physical, photophysical, and supramolecular properties to applications in sensing, catalysis, molecular electronics, photonics, and nanomedicine. *Chem. Rev.* 110 (2010) 1857-1959.

[53] D. Wilms, S.E. Stiriba, H. Frey, Hyperbranched polyglycerols: From the controlled synthesis of biocompatible polyether polyols to multipurpose applications. *Acc. Chem. Res.* 43 (2010) 129-141.

[54] F. Wurm, H. Frey, Linear-dendritic block copolymers: The state of the art and exciting perspectives. *Prog. Polym. Sci.* (Oxford) 36 (2011) 1-52.

[55] J. Khandare, R. Haag, Pharmaceutically used polymers: Principles, structures, and applications of pharmaceutical delivery systems. *Handbook of Experimental Pharmacology* 197 (2010) 221-250.

[56] R.H. Müller, K. Mäder, S. Dohla, Solid lipid nanoparticles (SLN) for controlled drug delivery - a review of state of the art. *Eur. J. Pharm. Biopharm.* 50 (2000) 161-77.

[57] C.M. Paleos, D. Tsiourvas, Z. Sideratou, L.A. Tziveleka, Drug delivery using multifunctional dendrimers and hyperbranched polymers. *Expert Opinion on Drug Delivery* 7 (2010) 1387-1398.

[58] J. Liu, X. Zhu, D. Yan, Self-assembly of hyperbranched polymers and its biomedical applications. *Adv. Mat.* 22 (2010) 4567-4590.

[59] L.A. Tziveleka, C. Kontoyianni, Z. Sideratou, D. Tsiourvas, C.M. Paleos, Novel functional hyperbranched polyether polyols as prospective drug delivery systems. *Macrom. Biosci.* 6 (2006) 161-169.

[60] P. Kolhe, E. Misra, R.M. Kannan, S. Kannan, M. Lieh-Lai, Drug complexation, in vitro release and cellular entry of dendrimers and hyperbranched polymers. *Intern. J. Pharm.* 259 (2003) 143-160.

[61] P. Kolhe, J. Khandare, O. Pillai, S. Kannan, M. Lieh-Lai, R. Kannan, Hyperbranched polymer-drug conjugates with high drug payload for enhanced cellular delivery. *Pharm. Res.* 21 (2004) 2185-2195.

[62] S. Kannan, P. Kolhe, V. Raykova, M Glibatec, R.M. Kannan, M. Lieh-Lai, D. Bassett, Dynamics of cellular entry and drug delivery by dendritic polymers into human lung epithelial carcinoma cells. *J. Biomat. Sci., Polym. Ed.* 15 (2004) 311-330.

[63] C. Yiyun, X. Tongwen, Dendrimers as Potential Drug Carriers. Part I. Solubilization of Non-Steroidal Anti-Inflammatory Drugs in the Presence of Polyamidoamine Dendrimers. *Eur. J. Med.* Chem. 40 (2005) 1188-1192.

[64] O. Perumal, J. Khandare, P. Kolhe, S. Kannan, M. Lieh-Lai, R.M. Kannan, Effects of branching architecture and linker on the activity of hyperbranched polymer-drug conjugates. *Bioconj. Chem.* 20 (2009) 842-846.

[65] S. Suttiruengwong, J. Rolker, I. Smirnova, W. Arlt, M. Seiler, L. Lüderitz, Y. Pérez De Diego, P.J. Jansens, Hyperbranched polymers as drug carriers: Microencapsulation and release kinetics. *Pharm. Dev. Techn.* 11 (2006) 55-70.

[66] E. Güç, G. Günüdz, U. Gündüz, Fatty acid based hyperbranched polymeric nanoparticles for hydrophobic drug delivery. *Drug Dev. Ind. Pharm.* 36 (2010) 1139-1148.

[67] R.K. Kainthan, C. Mugabe, H.M. Burt, D.E. Brooks, Unimolecular micelles based on hydrophobically derivatized hyperbranched polyglycerols: Ligand binding properties. *Biomacromolecules* 9 (2008) 886-895.

[68] C. Mugabe, B.A. Hadaschik, R.K. Kainthan, D.E. Brooks, A.I. So, M.E. Gleave, H.M. Burt, Paclitaxel incorporated in hydrophobically derivatized hyperbranched polyglycerols for intravesical bladder cancer therapy. *BJU International* 103 (2009) 978-986.

[69] C. Kontoyianni, Z. Sideratou, T. Theodossiou, L.A. Tziveleka, D. Tsiourvas, C.M. Paleos, A novel micellar PEGylated hyperbranched polyester as a prospective drug delivery system for Paclitaxel. *Macrom. Biosci.* 8 (2008) 871-881.

[70] T. Boulikas, I. Tsogas, Microtubule-targeted antitumor drugs: Chemistry, mechanisms and nanoparticle formulations. *Gene Therapy Mol. Biol.* 12 (2008) 313-358.

[71] X. Zhang, X. Zhang, Z. Wu, X. Gao, C. Cheng, Z. Wang, C. Li, A hydrotropic β-cyclodextrin grafted hyperbranched polyglycerol co-polymer for hydrophobic drug delivery. *Acta Biomat.* 7 (2011) 585-592.

[72] L. Zhang, C.H. Hu, S.X. Cheng, R.X. Zhuo, Hyperbranched amphiphilic polymer with folate mediated targeting property. *Coll. Surf. B: Biointerf.* 79 (2010) 427-433.

[73] C. Mugabe, R.T. Liggins, D. Guan, I. Manisali, I. Chafeeva, D.E. Brooks, M. Heller, J.K. Kackson, H.M. Burt, Development and in vitro characterization of paclitaxel and

docetaxel loaded into hydrophobically derivatized hyperbranched polyglycerols. *Intern. J. Pharm.* 404 (2011) 238-249.

[74] S. Reven, J. Grdadolnik, J. Kristl, E. Žagar, Hyperbranched poly(esteramides) as solubility enhancers for poorly water-soluble drug glimepiride. *Intern. J. Pharm.* 396 (2010) 119-126.

[75] M.A. Quadir, M.R. Radowski, F. Kratz, K. Licha, P. Hauff, R. Haag, Dendritic multishell architectures for drug and dye transport. *J. Contr. Release* 132 (2008) 289-294.

[76] T. Wang, M. Li, H. Gao, Y. Wu, Nanoparticle carriers based on copolymers of poly(ε-caprolactone) and hyperbranched polymers for drug delivery. *J. Coll. Interf. Sci.* 353 (2011) 107-115.

[77] K.J. Haxton, H.M. Burt, Hyperbranched polymers for controlled release of cisplatin. *Dalton Transactions* 43 (2008) 5872-5875.

[78] J. Liu, W. Huang, Y. Pang, X. Zhu, Y. Zhou, D. Yan, The in vitro biocompatibility of self-assembled hyperbranched copolyphosphate nanocarriers. *Biomaterials* 31 (2010) 5643-5651.

[79] S. Santra, C. Kaittanis, J.M. Perez, Cytochrome c encapsulating theranostic nanoparticles: A novel bifunctional system for targeted delivery of therapeutic membrane-impermeable proteins to tumors and imaging of cancer therapy. *Mol. Pharm.* 7 (2010) 1209-1222.

[80] T. Wang, Y. Wu, M. Li, Novel hyperbranched poly (amine-ester)-poly (lactide-co-glycolide) polymeric micelles as amphotericin B carriers. *Polym. Bull.* 65 (2010) 425-442.

[81] I.N. Kurniasih, H. Liang, J.P. Rabe, R. Haag, Supramolecular aggregates of water soluble dendritic polyglycerol architectures for the solubilization of hydrophobic compounds. *Macrom. Rapid Comm.* 31 (2010) 1516-1520.

[82] C. Xia, X. Ding, Y. Sun, H. Liu, Y. Li, Hyperbranched-upon-dendritic macromolecules as unimolecular hosts for controlled release. *J. Polym. Sci., Part A: Polym. Chem.* 48 (2010) 4013-4019.

[83] M. Prabaharan, J.J. Grailer, S. Pilla, D.A. Steeber, S. Gong, Amphiphilic multi-arm-block copolymer conjugated with doxorubicin via pH-sensitive hydrazone bond for tumor-targeted drug delivery. *Biomaterials* 30 (2009) 5757-5766.

[84] A. Breitenbach, Y.X. Ti, T. Kissel, Branched biodegradable polyesters for parenteral drug delivery systems. *J. Contr. Release* 64 (2000) 167-178.

[85] J. Rojo, R. Delgado, Glycodendritic structures: Promising new antiviral drugs. *J. Antimicrob. Chemotherapy* 54 (2004) 579-581.

[86] C.M. Paleos, L.A. Tziveleka, Z. Sideratou, D. Tsiourvas, Gene delivery using functional dendritic polymers. *Expert Opinion on Drug Delivery* 6 (2009) 27-38.

[87] C.M. Paleos, D. Tsiourvas, Z. Sideratou, Molecular engineering of dendritic polymers and their application as drug and gene delivery systems. *Mol. Pharm.* 4 (2007) 169-188.

[88] H. Tian, X. Chen, H. Lin, C. Deng, P. Zhang, Y. Wei, X. Jing, Micellization and reversible pH-sensitive phase transfer of the hyperbranched multiarm PEI-PBLG copolymer. *Chemistry - A Eur. J.* 12 (2006) 4305-4312.

[89] V. Russ, H. Elfberg, C. Thoma, J. Kloeckner, M. Ogris, E. Wagner, Novel degradable oligoethylenimine acrylate ester-based pseudodendrimers for in vitro and in vivo gene transfer. *Gene Therapy* 15 (2008) 18-29.

[90] M. Krämer, J.F. Stumbé, G. Grimm, B. Kaufmann, U. Krüger, M. Weber, R. Haag, Dendritic polyamines: Simple access to new materials with defined treelike structures for application in nonviral gene delivery. *ChemBioChem* 5 (2004) 1081-1087.

[91] P. Banerjee, W. Reichardt, R. Weissleder, A. Bogdanov Jr., Novel hyperbranched dendron for gene transfer in vitro and in vivo. *Bioconj. Chem.* 15 (2004) 960-968.

[92] X. Shuai, T. Merdan, F. Unger, M. Wittmar, T. Kissel, Novel Biodegradable Ternary Copolymers hy-PEI-g-PCL-b-PEG: Synthesis, Characterization, and Potential as Efficient Nonviral Gene Delivery Vectors. *Macromolecules* 36 (2003) 5751-5759.

[93] H. Tian, W. Xiong, J. Wei, Y. Wang, X. Chen, X. Jing, Q. Zhu, Gene transfection of hyperbranched PEI grafted by hydrophobic amino acid segment PBLG. *Biomaterials* 28 (2007) 2899-2907.

[94] J.A. Khan, R.K. Kainthan, M. Ganguli, J.N. Kizhakkedathu, Y. Singh, S. Maiti, Water soluble nanoparticles from PEG-based cationic hyperbranched polymer and RNA that protect RNA from enzymatic degradation. *Biomacromolecules* 7 (2006) 1386-1388.

[95] L.A. Tziveleka, A.M.G. Psarra, D. Tsiourvas, C.M. Paleos, Synthesis and evaluation of functional hyperbranched polyether polyols as prospected gene carriers. *Intern. J. Pharm.* 356 (2008) 314-324.

[96] W.J. Kim, A.C. Bonoiu, T. Hayakawa, C. Xia, M.A. Kakimoto, H.E. Pudavar, K.S. Lee, P.N. Prasad, Hyperbranched polysiloxysilane nanoparticles: Surface charge control of nonviral gene delivery vectors and nanoprobes. *Intern. J. Pharm.* 376 (2009) 141-152.

[97] R. Reul, J. Nguyen, T. Kissel, Amine-modified hyperbranched polyesters as non-toxic, biodegradable gene delivery systems. *Biomaterials* 30 (2009) 5815-5824.

[98] J. Chen, C. Wu, D. Oupický, Bioreducible hyperbranched poly(amido amine)s for gene delivery. *Biomacromolecules* 10 (2009) 2921-2927.

[99] X. Wang, Y. He, J. Wu, C. Gao, Y. Xu, Synthesis and evaluation of phenylalanine-modified hyperbranched poly(amido amine)s as promising gene carriers. *Biomacromolecules* 11 (2010) 245-251.

[100] Y.Z. You, Z.Q. Yu, M.M. Cui, C.Y. Hong, Preparation of photoluminescent nanorings with controllable bioreducibility and stimuli-responsiveness. *Angew. Chem. Int. Ed.* 49 (2010) 1099-1102.

[101] R. Reul, J. Nguyen, T. Kissel, Amine-modified hyperbranched polyesters as non-toxic, biodegradable gene delivery systems. *Biomaterials* 30 (2009) 1-10.

[102] Z. Zhong, Y. Song, J.F.J. Engbersen, M.C. Lok, W.E. Hennink, J. Feijen, A versatile family of degradable non-viral gene carriers based on hyperbranched poly(ester amine)s. *J. Contr. Release* 109 (2005) 317-329.

[103] R.K. Kainthan, D.E. Brooks, In vivo biological evaluation of high molecular weight hyperbranched polyglycerols. *Biomaterials* 28 (2007) 4779-4787.

[104] J. Blacklock, Y.Z. You, Q.H. Zhou, G. Mao, D. Oupický, Gene delivery in vitro and in vivo from bioreducible multilayered polyelectrolyte films of plasmid DNA. *Biomaterials* 30 (2009) 939-950.

[105] O.G. Brown, C. Bergquist, P. Ferm, K.L. Wooley, Unusual, promoted release of guests from amphiphilic cross-linked polymer networks. *J. Am. Chem. Soc.* 127 (2005) 11238-11239.

[106] C. Wiener, M.W. Brechbiel, H. Brothers, R.L. Magin, O.A. Gansow, D.A. Tomalia, P.C. Lauterbur, Dendrimer-based metal chelates: A new class of magnetic resonance imaging contrast agents. *Magn. Reson. Med.* 31 (1994) 1-8.

[107] M.H.M. Oudshoorn, R. Rissmann, J.A. Bouwstra, W.E. Hennink, Synthesis and characterization of hyperbranched polyglycerol hydrogels. *Biomaterials* 27 (2006) 5471-5479.

[108] M.H.M. Oudshoorn, R. Penterman, R. Rissmann, J.A. Bouwstra, D.J. Broer, W.E. Hennink, Preparation and characterization of structured hydrogel microparticles based on cross-linked hyperbranched polyglycerol. *Langmuir* 23 (2007) 11819-11825.

[109] E.G.R. Fernandes, V. Zucolotto, A.A.A. De Queiroz, Electrospinning of hyperbranched poly-L-lysine/polyaniline nanofibers for application in cardiac tissue engineering. *J. Macrom. Sci., Part A: Pure and Applied Chem.* 47 (2010) 1203-1207.

[110] S. Luo, C. Ling, X. Hu, X. Liu, S. Chen, M. Han, J. Xia, Thermoresponsive unimolecular micelles with a hydrophobic dendritic core and a double hydrophilic block copolymer shell. *J. Coll. Interf. Sci.* 353 (2011) 76-82.

[111] S. Pedrón, C. Peinado, P. Bosch, K.S. Anseth, Synthesis and characterization of degradable bioconjugated hydrogels with hyperbranched multifunctional cross-linkers. *Acta Biomat.* 6 (2010) 4189-4198.

[112] Z. Sanping, X. Weilin, Cyclodextrin-containing supramolecular hydrogels. *Prog. Chem.* 22 (2010) 916-926.

[113] S. Pedrón, K. Anseth, J.A. Benton, P. Bosch, C. Peinado, Bioapplications of networks based on photo-cross-linked hyperbranched polymers. *Macrom. Symp.* 291-292 (2010) 307-313.

[114] J. Shepherd, P. Sarker, S. Rimmer, L. Swanson, S. MacNeil, I. Douglas, Hyperbranched poly(NIPAM) polymers modified with antibiotics for the reduction of bacterial burden in infected human tissue engineered skin. *Biomaterials* 32 (2011) 258-267.

[115] S. Xu, Y. Luo, R. Graeser, A. Warnecke, F. Kratz, P. Hauff, K. Licha, R. Haag, Development of pH-responsive core-shell nanocarriers for delivery of therapeutic and diagnostic agents. *Bioorg. Med. Chem. Letters* 19 (2009) 1030-1034.

[116] A. Castonguay, A.K. Kakkar, Dendrimer templated construction of silver nanoparticles. *Adv. Coll. Interf. Sci.* 160 (2010) 76-87.

[117] N. Karak, R. Konwarh, B. Voit, Catalytically active vegetable-oil-based thermoplastic hyperbranched polyurethane/ silver nanocomposites. *Macromol. Mat. Eng.* 295 (2010) 159-169.

[118] S.S. Mahapatra, N. Karak, Silver nanoparticle in hyperbranched polyamine: Synthesis, characterization and antibacterial activity. *Mat. Chem. Phys.* 112 (2008) 1114-1119.

[119] C. Aymonier, U. Schlotterbeck, L. Antonietti, P. Zacharias, R. Thomann, J.C. Tiller, S. Mecking, Hybrids of silver nanoparticles with amphiphilic hyperbranched macromolecules exhibiting antimicrobial properties. *Chem. Comm.* 8 (2002) 3018-3019.

[120] Y. Shen, M. Kuang, Z. Shen, J. Nieberle, H. Duan, H. Frey, Gold nanoparticles coated with a thermosensitive hyperbranched polyelectrolyte: Towards smart temperature and pH nanosensors. *Angew. Chem. Int. Ed.* 47 (2008) 2227-2230.

[121] D. Wan, Q. Fu, J. Huang, Synthesis of amphiphilic hyperbranched polyglycerol polymers and their application as template for size control of gold nanoparticles, *J. Appl. Polym. Sci.* 101 (2006) 509-514.

[122] Y. Zhang, H. Peng, W. Huang, Y. Zhou, D. Yan, Facile preparation and characterization of highly antimicrobial colloid Ag or Au nanoparticles. *J. Coll. Interf. Sci.* 325 (2008) 371-376.

[123] J. Keilitz, M.R. Radowski, J.D. Marty, R. Haag, F. Gauffre, C. Mingotaud, Dendritic polymers with a core - Multishell architecture: A versatile tool for the stabilization of nanoparticles. *Chem. Mat.* 20 (2008) 2423-2425.

[124] J.D. Marty, E. Martinez-Aripe, A.F. Mingotaud, C. Mingotaud. Hyperbranched polyamidoamine as stabilizer for catalytically active nanoparticles in water. *J. Coll. Interf. Sci.* 326 (2008) 51-54.

[125] S.R. Puniredd, C.M. Yin, Y.S. Hooi, P.S. Lee, M.P. Srinivasan, Dendrimer-encapsulated Pt nanoparticles in supercritical medium: Synthesis, characterization, and application to device fabrication. *J. Coll. Interf. Sci.* 332 (2009) 505-510.

[126] X.Y. Liu, F. Cheng, Y. Liu, H.J. Liu, Y. Chen, Preparation and characterization of novel thermoresponsive gold nanoparticles and their responsive catalysis properties. *J. Mat. Chem.* 20 (2010) 360-368.

[127] Q. Tang, F. Cheng, X.L. Lou, H.J. Liu, Y. Chen, Comparative study of thiol-free amphiphilic hyperbranched and linear polymers for the stabilization of large gold nanoparticles in organic solvent. *J. Coll. Interf. Sci.* 337 (2009) 485-491.

[128] X. Liu, K. Matsumura, Y. Tomita, K. Yasui, K. Kojima, K. Chikama, Nonlinear optical responses of nanoparticle-polymer composites incorporating organic (hyperbranched polymer)-metallic nanoparticle complex. *J. Appl. Phys.* 108 (2010) 073102.

[129] H. Li, L. Zhang, J.K. Jo, C.S. Ha, Y.A. Shchipunov, I. Kim, Facile synthesis of titania/hyperbranched polyglycidol nanohybrids with controllable morphologies: from solid spheres, capsules to tubes. *J. Nanopart. Res.* 13 (2011) 1-12.

[130] S. Wei, Y. Zhu, Y. Zhang, J. Xu, Preparation and characterization of hyperbranched aromatic polyamides/Fe_3O_4 magnetic nanocomposite. *React. Funct. Polym.* 66 (2006) 1272-1277.

[131] X. Gao, X. Zhang, X. Zhang, C. Cheng, Z. Wang, C. Li, Encapsulation of BSA in polylactic acid-hyperbranched polyglycerol conjugate nanoparticles: Preparation, characterization, and release kinetics. *Polym. Bull.* 65 (2010) 787-805.

[132] M. Arkas, D. Tsiourvas, Organic/inorganic hybrid nanospheres based on hyperbranched poly(ethylene imine) encapsulated into silica for the sorption of toxic metal ions and polycyclic aromatic hydrocarbons from water. *J. Hazard. Mat.* 170 (2009) 35-42.

[133] E.G. Oosterom, J.N.H. Reek, P.C.J. Kamer, P.W.N.M. van Leeuwen, Transition metal catalysis using functionalized dendrimers. *Angew. Chem. Int. Ed.* 40 (2001) 1828-1849.

[134] L. Tuchbreiter, S. Mecking, Hydroformylation with dendritic-polymer-stabilized rhodium colloids as catalyst precursors. *Macrom. Chem. Phys.* 208 (2007) 1688-1693.

[135] G. Zhang, X. Fan, J. Kong, Y. Liu, Y. Xie, Hyperbranched organosilicon polymers. *Prog. Chem.* 20 (2008) 326-338.

[136] K. Tahara, H. Shimakoshi, A. Tanaka, Y. Hisaeda, Synthesis, characterization and catalytic function of a B 12-hyperbranched polymer. *Dalton Transactions* 39 (2010) 3035-3042.

[137] M. Gladitz, S. Reinemann, H.J. Radusch, Preparation of silver nanoparticle dispersions via a dendritic-Polymer template approach and their use for antibacterial surface treatment. *Macromol. Mat. Eng.* 294 (2009) 178-189.

[138] J. Fang, H. Kita, K.I. Okamoto, Gas permeation properties of hyperbranched polyimide membranes. *J. Membr. Sci.* 182 (2001) 245-256.

[139] J. Rolker, M. Seiler, L. Mokrushina, W. Arlt, Potential of Branched Polymers in the Filed of Gas Absorption: Experimental Gas Solubilities and Modeling. *Ind. Eng. Chem. Res.* 46 (2007) 6572-6583.

[140] H. Gao, D. Wang, W. Jiang, S. Guan, Z. Jiang, Gas permeability of fluorinated hyperbranched polyimide. *J. Appl. Polym. Sci.* 109 (2008) 2341-2346.

[141] Y. Yin, L. Yang, M. Yoshino, J. Fang, K. Tanaka, H. Kita, K.I. Okamoto, Synthesis and gas permeation properties of star-like poly(ethylene oxide)s using hyperbranched polyimide as central core. *Polymer J.* 36 (2004) 294-302.

[142] T. Suzuki, Y. Yamada, Physical and gas transport properties of novel hyperbranched polyimide - Silica hybrid membranes. *Polym. Bull.* 53 (2005) 139-146.

[143] T. Suzuki, Y. Yamada, K. Itahashi, 6FDA-TAPOB hyperbranched polyimide-silica hybrids for gas separation membranes. *J. Appl. Polym. Sci.* 109 (2008) 813-819.

[144] T. Suzuki, Y. Yamada, Y. Tsujita, Gas transport properties of 6FDA-TAPOB hyperbranched polyimide membrane. *Polymer* 45 (2004) 7167-7171.

[145] J. Fang, H. Kita, K.I. Okamoto, Gas Permeation Properties of hyperbranched polyimide membranes. *J. Membr. Sci.* 182 (2001) 245-56.

[146] D.M. Sterescu, D.F. Stamatialis, M. Wessling, Boltorn-modified polyimide gas separation membranes. *J. Membr. Sci.* 310 (2008) 512-521.

[147] D.M. Sterescu, D.F. Stamatialis, E. Mendes, J. Kruse, K. Rätzke, F. Faupel, M. Wessling, Boltom-modified poly(2,6-dimethyl-1,4-phenylene oxide) gas separation membranes, *Macromolecules* 40 (2007) 5400-5410.

[148] L. Boogh, B. Pettersson, J.A.E. Manson, Dendritic hyperbranched polymers as tougheners for epoxy resins, *Polymer* 40 (1999) 2249-2261.

[149] P. Ghosh, W.M. Lackowski, R.M. Crooks, Two new approaches for patterning polymer films using templates prepared by microcontact printing. *Macromolecules* 34 (2001) 1230-1236.

[150] Y. Hong, S.J. Coombs, J.J. Cooper-White, M.E. MacKay, C.J. Hawker, E. Malmström, N. Rehnberg, Film blowing of linear low-density polyethylene blended with a novel hyperbranched polymer processing aid. *Polymer* 41 (2000) 7705-7713.

[151] Y. Hong, J.J. Cooper-White, M.E. Mackay, C.J. Hawker, E. Malmström, N. Rehnberg, A novel processing aid for polymer extrusion: Rheology and processing of polyethylene and hyperbranched polymer blends. *J. Rheol.* 43 (1999) 781-793.

[152] S.M. Burkinshaw, P.E. Fröhling, M. Mignanelli, The effect of hyperbranched polymers on the dyeing of polypropylene fibres. *Dyes and Pigments* 53 (2002) 229-235.

[153] P.E. Fröhling, Dendrimers and dyes - A review. *Dyes and Pigments* 48 (2001) 187-195.

[154] R. Allabashi, M. Arkas, G. Hörmann, D. Tsiourvas, Removal of some organic pollutants in water employing ceramic membranes impregnated with cross-linked silylated dendritic and cyclodextrin polymers. *Water Res.* 41 (2007) 476-486.

[155] M. Arkas, D. Tsiourvas, C.M. Paleos, Functional dendritic polymers for the development of hybrid materials for water purification. *Macromol. Mat. Eng.* 295 (2010) 883-898.

[156] T. Itoh, M. Ikeda, N. Hirata, Y. Moriya, M. Kubo, O. Yamamoto, Ionic conductivity of the hyperbranched polymer-lithium metal salt systems. *J. Power Sources* 81-82 (1999) 824-829.

[157] E. Barriau, H. Frey, A. Kiry, M. Stamm, F. Grö hn, Negatively charged hyperbranched polyether-based polyelectrolytes. *Coll. Polym. Sci.* 284 (2006) 1293-1301.

[158] T. Itoh, N. Hirata, Z. Wen, M. Kubo, O. Yamamoto, Polymer electrolytes based on hyperbranched polymers. *J. Power Sources* 97-98 (2001) 637-640.

[159] Z. Wang, M. Ikeda, N. Hirata, M. Kubo, T. Itoh, O. Yamamoto, Thermal, electrochemical, and spectroscopic characterizations of hyperbranched polymer electrolyte. *J. Electrochem. Soc.* 146 (1999) 2209-2215.

[160] T. Itoh, S. Horii, S. Hashimoto, T. Uno, M. Kubo, Polymer electrolytes plasticized with hyperbranched polymer for lithium polymer batteries. *Ionics* 10 (2004) 450-457.

[161] L. Rakesh, A. Müller, P. Chhetri, Development of a hyperbranched fuel cell membrane material for improved proton conductivity. *Fluid Dynam. Mat. Proc.* 6 (2010) 179-202.

[162] T. Itoh, K. Hirai, M. Tamura, T. Uno, M. Kubo, Y. Aihara, Synthesis and characteristics of hyperbranched polymer with phosphonic acid groups for high-temperature fuel cells. *J. Solid State Electrochem.* 14 (2010) 2179-2189.

[163] T. Itoh, T. Sakakibara, Y. Takagi, M. Tamura, T. No, M. Kubo, Y. Aihara, Proton-conducting electrolyte membranes based on hyperbranched polymer with a sulfonic acid group for high-temperature fuel cells. *Electrochim. Acta* 55 (2010) 1419-1424.

[164] D.M. Tigelaar, M.A.B. Meador, W.R. Bennett, Composite electrolytes for lithium batteries: Ionic liquids in APTES cross-linked polymers. *Macromolecules* 40 (2007) 4159-4164.

[165] F.M. Wang, H.C. Wu, C.S. Cheng, C.L. Huang, C.R. Yang, High ionic transfer of a hyperbranched-network gel copolymer electrolyte for potential electric vehicle (EV) application. *Electrochim. Acta* 54 (2009) 3788-3793.

[166] F. Wu, T. Feng, C. Wu, Y. Bai, L. Ye, J. Chen, Thermally stable hyperbranched polyether-based polymer electrolyte for lithium-ion batteries. *J. Phys. D: Appl. Phys.* 43 (2010) 035501.

[167] I. Gur, N.A. Fromer, C.P. Chen, A.G. Kanaras, A.P. Alivisatos, Hybrid solar cells with prescribed nanoscale morphologies based on hyperbranched semiconductor nanocrystals. *Nano Letters* 7 (2007) 409-414.

[168] J. Qiao, C. Yang, Q. He, F. Bai, Y. Li, Hyperbranched conjugated polymers for photovoltaic applications. *J. Appl. Polym. Sci.* 92 (2004) 1459-1466.

[169] Y. Tang, P. Shen, T. Ding, H. Huang, B. Zhao, S. Tan, Hyperbranched conjugated polymers with donor-π-acceptor architecture as organic sensitizers for dye-sensitized solar cells. *Eur. Polym. J.* 46 (2010) 2033-2041.

[170] Q. Cao, P. Liu, Hyperbranched polyurethane as novel solid–solid phase change material for thermal energy storage. *Eur. Polym. J.* 42 (2006) 2931-2939.

[171] Q. Cao, L. Liao, H. Xu, Study on the influence of thermal characteristics of hyperbranched polyurethane phase change materials for energy storage. *J. Appl. Polym. Sci.* 115 (2010) 2228-2235.

[172] L. Liao, Q. Cao, H. Liao, Investigation of a hyperbranched polyurethane as a solid-state phase change material. *J. Mat. Sci.* 45 (2010) 2436-2441.

[173] Y.H. Kim, Hyperbranched Polymers 10 Years After. *J. Polym. Sci. Part A: Polym. Chem.* 36 (1998) 1685-1698.

[174] L.A. Kleintjens, H.M. Schoffeleers, R. Koningsveld, Entmischung in hochmolekularen Vielstoffsystemen, XVII. Eine Wechselwirkungsfunktion für Lösungen verzweigter Polymerer. *Ber. Bunsenges. Phys. Chem.* 81 (1977) 980-985.

[175] T.W. de Loos, W. Poot, R.N. Lichtenthaler, The Influence of Branching on High-Pressure Vapor-Liquid Equilibria in Systems of Ethylene and Polyethylene. *J. Supercrit. Fluids* 5 (1992) 282-286.

[176] A. Striolo, J.M. Prausnitz, Osmotic second virial coefficient for linear and star poly(ethylene oxide). *Polymer* 42 (2001) 4773-4775.

[177] J. W. Kennedy, M. Gordon and R. Koningsveld, Generalization of the Flory-Huggins treatment of polymer solutions. *J. Polym. Sci., Part C: Polym. Symp.* 39 (1972) 43–69.

[178] L.A. Kleintjens, R. Koningsveld, Liquid-Liquid Phase Separation in Multicomponent Polymer Systems. 18. Effect of Short-Chain Branching. *Macromolecules* 13 (1980) 303-310.

[179] S. Chen, M. Banaszak, M. Radosz, Phase Behavior of Poly(ethylene -1 butene) in Subcritical and Supercritical Propane: Ethyl Branches Reduce Segment Energy and Enhance Miscibility. *Macromolecules* 28 (1995) 1812-1817.

[180] F.J. Blas, L.F. Vega, Thermodynamic properties and phase equilibria of branched chain fluids using first- and second-order Wertheim's thermodynamic perturbation theory. *J. Chem. Phys.* 115 (2001) 3906-3915.

[181] M.K. Kozlowska, J.F. Bas, C.S. Schacht, J. Gross, T.W. de Loos, Phase Behavior of Hyperbranched Polymer Systems: Experiments and Application of the Perturbed-Chain Polar SAFT Equation of State. *J. Phys. Chem.* B 113 (2009) 1022-1029.

[182] U. Domańska, K. Paduszyński, Z. Ołek-Tryznowska, (Liquid + liquid) equilibria of binary systems containing hyperbranched polymer Boltorn® H2004 - Experimental study and modelling in terms of lattice-cluster theory. *J. Chem. Thermodynamics* 43 (2011) 167-171.

[183] U. Domańska, K. Paduszyński, Z. ZoŁek-Tryznowska, Liquid-liquid phase equilibria of binary systems containing hyperbranched polymer B-U3000: Experimental study and modeling in terms of lattice cluster theory. *J. Chem. Eng. Data* 55 (2010) 3842-3846.

[184] M. Seiler, J. Rolker, W. Arlt, Phase Behavior and Thermodynamic Phenomena of Hyperbranched Polymer Solutions. *Macromolecules* 36 (2003) 2085-2092.

[185] V. Munoz Portela, E.J.M. Straver, T.W. de Loos, High-Pressure Phase Behavior of the System Propane - Boltorn H320. *J. Chem. Eng. Data* 54 (2009) 2593-2598.

[186] M. Seiler, W. Arlt, H. Kautz, H. Frey, Experimental data and theoretical considerations on vapor-liquid and liquid-liquid equilibria of hyperbranched polyglycerol and PVA solutions. *Fluid Phase Equilibria* 201 (2002) 359-379.

[187] J.G. Jang, Y.C. Bae, Phase behaviors of hyperbranched polymer solutions. *Polymer* 40 (1999) 6761-6768.

[188] T. Zeiner, P. Schrader, S. Enders, D. Browarzik, Phase- and Interfacial Behavior of Hyperbranched Polymer Solutions. *Fluid Phase Equilibria* 302 (2011) 321-330.

[189] U. Domańska, Z. Zołek-Tryznowska, M.M. Tshibangu, D. Ramjugernath, T.M. Letcher, Separation of an alcohol and a tetrahydrofuran, methyl tert-butyl ether, or ethyl tert-butyl ether by solvent extraction with a hyperbranched polymer at T = 298.15 K. *J. Chem. Eng. Data* 55 (2010) 2879-2885.

[190] U. Domanska, Z. Zolek-Tryznowska, A. Pobudkowskas, Separation of Hexane/Ethanol Mixtures. LLE of Ternary Systems (Ionic Liquid or Hyperbranched Polymer + Ethanol + Hexane) at T=298.15. *J. Chem. Eng. Data* 54 (2009) 972-976.

[191] C.S. Schacht, S. Bahramali, D. Wilms, H. Frey, J. Gross, T.W. de Loos, Phase Behavior of the system hyperbranched polyglycerol + methanol + carbon dioxide. *Fluid Phase Equilibria* 299 (2010) 252-258.

[192] J.G. Lieu, M. Liu, J.M.J. Frechet, J.M. Prausnitz, Vapor-Liquid Equilibria for Dendritic-Polymer Solutions. *J. Chem. Eng. Data* 44 (1999) 613-620.

[193] C. Mio, S. Kiritsov, Y. Thio, R. Brafman, J. Prausnitz, C. Hawker, E.E. Malmstrom, Vapor-Liquid Equilibria for Solutions of Dendritic Polymers. *J. Chem. Eng. Data* 43 (1998) 541-550.

[194] U. Domanska, Z. Zolek-Tryznowska, Solubility of hyperbranched polymer, Boltorn W-3000, in alcohols, ethers and hydrocarbons. *J. Chem. Thermodynamics* 42 (2010) 1304-1309.

[195] S.J. Ko, Y.C. Bae, S.H. Kong, S.O. Ryu, Phase Behaviors of Hyperbranched Solid Polymer Electrolyte/Salt Systems: The Structure Effect. *J. Polym. Sci., Part B: Polym. Phys.* 42 (2004) 2254-2262.

[196] A. Polese, C. Mio, A. Bertucco, Infinite-dilution activity coefficients of polar and nonpolar solvents in solutions of hyperbranched polymers. *J. Chem. Eng. Data* 44 (1999) 839-845.

[197] G.S. Dritsas, K. Karatasos, C. Panayiotou, Investigation of Thermodynamic Properties of Hyperbranched Poly(ester amide) by Inverse Gas Chromatography. *J. Polym. Sci., Part B: Polym. Phys.* 46 (2008) 2166-2172.

[198] G.S. Dritsas, K. Karatasos, C. Panayiotou, Investigation of thermodynamic properties of hyperbranched aliphatic polyestern by inverse gas chromatography. *J. Chromatography* A 1216 (2009) 8979-8985.

[199] U. Domańska, Z. Zołek-Tryznowska, Thermodynamic properties of hyperbranched polymer, Boltorn U3000, using inverse gas chromatography. *J. Phys. Chem.* B 113 (2009) 15312-15321.

[200] U. Domanska, Z. Zolek-Tryznowska, Measurements of mass-fraction activity coefficient at infinite dilution of aliphatic and aromatic hydrocarbon, thiophene, alcohols, water, ethers, and ketones in hyperbranched polymer, Boltorn H2004, using inverse gas chromatography. *J. Chem. Thermodynamics* 42 (2010) 363-370.

[201] U. Domańska, Z. Zołek-Tryznowska, Mass-fraction activity coefficients at infinite dilution measurements for organic solutes and water in the hyperbranched polymer Boltorn W3000 using inverse gas chromatography. *Journal J. Chem. Eng. Data* 55 (2010) 1258-1265.

[202] M. Xu, X. Yan, R. Cheng, X. Yu, Investigation into the solution properties of hyperbranched polymer. *Polymer International* 50 (2001) 1338-1345.

[203] U. Domańska, Z. Zołek-Tryznowska, Measurement of the density and viscosity of binary mixtures of (hyper-branched polymer B-H2004 + 1-butanol, or 1-hexanol, or octanol, or methyl tert-butyl ether). *J. Chem. Thermodynamics* 42 (2010) 651-658.

[204] U. Domańska, Z. Zołek-Tryznowska, Temperature and composition dependence of the density and viscosity of binary mixtures of (hyperbranched polymer, B-U3000 + 1-alcohol, or ether). *J. Chem. Thermodynamics* 41 (2009) 821-828.

[205] I.B. Rietveld, D. Bedeaux, J.A.M. Smit, Osmotic compressibility of poly(propylene imine) dendrimers in deuterated methanol. *J. Coll. Interf. Sci.* 232 (2000) 317-325.

[206] H. Cheng, S. Xie, Y. Zhou, W. Huang, D. Yan, J. Yang, B. Ji, Effect of degree of branching on the thermoresponsive phase transition behaviors of hyperbranched multiarm copolymers: Comparison of systems with LCST transition based on coil-to-globule transition or hydrophilic-hydrophobic Balance. *J. Phys. Chem.* B 114 (2010) 6291-6299.

[207] X. Liu, F. Cheng, H. Liu, Y. Chen, Unusual salt effect on the lower critical solution temperature of hyperbranched thermoresponsive polymers. *Soft Matter* 4 (2008) 1991-1994.

[208] T.D. Martter, M.D. Foster, T. Yoo, S. Xu, G. Lizzaraga, R.P. Quirk, P.D. Butler, Nonuniversal Behavior of the Thermodynamic Interaction Parameter in Blends of Star and Linear Polybutadiene. *Macromolecules* 35 (2002) 9763-9772.

[209] D.J. Massa, K.A. Shriner, S.R. Turner, B.I. Voit, Novel Blends of Hyperbranched Polyesters and Linear Polymers. *Macromolecules* 28 (1995) 3214-3220.

[210] A.M. Nemirovski, M.G. Bawendi, K.F. Freed, Lattice models of polymer solutions: monomers occupying several lattice sites. *J. Chem. Phys.* 87 (1987) 7272-7284.

[211] K.F. Freed, M.G. Bawendi, Lattice theories of polymeric fluids. *J. Phys. Chem.* 93 (1989) 2194-2203.

[212] J. Dudowicz, K.F. Freed, Effect of Monomer Structure and Compressibility on the Properties of Multicomponent Polymer Blends and Solutions: 1. Lattice Cluster Theory of Compressible Systems. *Macromolecule*s 24 (1991) 5076-5095.

[213] J. Dudowicz, M.S. Freed, K.F. Freed, Effect of Monomer Structure and Compressibility on the Properties of Multicomponent Polymer Blends and Solutions: 2. Application to Binary Blends. *Macromolecules* 24 (1991) 5096-5111.

[214] J. Dudowicz, K.F. Freed, Effect of Monomer Structure and Compressibility on the Properties of Multicomponent Polymer Blends and Solutions: 3. Application to PS(D)/PVME Blends. *Macromolecules* 24 (1991) 5112-5123.

[215] K.F. Freed, J. Dudowicz, Influence of monomer molecular structure on the miscibility of polymer blends. *Adv. Poly. Sci.* 183 (2005) 63-126.

[216] K.W. Foreman, K.F. Freed, Generalizations of Huggins-Guggenheim-Miller-type theories to describe the architecture of branched lattice chains. *J. Chem. Phys.* 102 (1995) 4663-4672.

[217] K.W. Foreman, K.F. Freed, I.M. Ngola, Nonrandom mixing in polymer blends: Implications for phase behavior. *J. Chem. Phys.* 107 (1997) 4688-4704.

[218] J. Dudowicz, K.F. Freed, W.G. Madden, Role of molecular structure on the thermodynamic properties of melts, blends, and concentrated polymer solutions: comparison of Monte Carlo simulations with the cluster theory for the lattice model. *Macromolecules* 23 (1990) 4803-4819.

[219] D. Buta, K.F. Freed, I. Szleifer, Thermodynamic properties of lattice polymers: Monte Carlo simulations and mean-field theories. *J. Chem. Phys.* 112 (2000) 6040-6048.

[220] D. Buta, K.F. Freed, I. Szleifer, Monte Carlo test of the lattice cluster theory: Thermodynamic properties of binary polymer blends. *J. Chem. Phys.* 114 (2001) 1424-1431.

[221] D. Buta, K.F. Freed, Mixtures of lattice polymers with structured monomers. *J. Chem. Phys.* 120 (2004) 6288-6298.

[222] W.G. Madden, A.I. Pesci, K.F. Freed, Phase equilibria of lattice polymer and solvent: Tests of theories against simulations. *Macromolecules* 23 (1990) 1181-1191.

[223] T. Zeiner, *Phase- and Interfacial Behaviour of Hyperbranched Polymer Solutions*. Ph-D thesis, Technical University Berlin, 2010.

[224] T. Zeiner, D. Browarzik, S. Enders, Calculation of the liquid-liquid equilibrium of aqueous solutions of hyperbranched polymers. *Fluid Phase Equilibria* 286 (2009) 127-133.

[225] T. Zeiner, S. Enders, Flüssig-Flüssig Gleichgewichte von hyperverzweigten Polymeren in Lösungsmittelmischungen. CIT 82 (2010) 1378.

[226] T. Zeiner, S. Enders, Phase-Behavior of Hyperbranched Polymer Solutions in Mixed Solvents. *Chem. Eng. Sci.* submitted September 2010.

[227] T. Zeiner, C. Browarzik, D. Browarzik, S. Enders, Calculation of the liquid-liquid equilibrium of solutions of hyperbranched polymers with the lattice-cluster theory combined with an association model. *J. Chem. Thermodynamics* submitted 2011.

[228] in polymer blends? Lattice cluster theory computations for structured monomer, compressible systems. *J. Chem. Phys.* 99 (1993) 4804-4820.

[229] J. Dudowicz, K.F. Freed, Pressure dependence of polymer fluids: Application of the lattice cluster theory. *Macromolecules* 28 (1995) 6625-6641.

[230] W. Li, K.F. Freed, Lattice cluster theory for phase behavior of rectangular mesogens. II. Nearest-neighbor interactions, phase diagrams, and competitive nematic orderings. *J. Chem. Phys.* 103 (1995) 5693-5711.

[231] J. Dudowicz, K.F. Freed, J.F. Douglas, Lattice model of equilibrium polymerization. IV. Influence of activation, chemical initiation, chain scission and fusion, and chain stiffness on polymerization and phase separation. *J. Chem. Phys.* 119 (2003) 12645-12666.

[232] K.W. Foreman, K.F. Freed, Influence of stiffness, monomer structure, and energetic asymmetries on polymer blend miscibilities: Applications to polyolefins. *Macromolecules* 30 (1997) 7279-7295.

[233] K.W. Foreman, K.F. Freed, Microscopic parameters influencing the phase separation in compressible binary blends of linear semiflexible polymers. *J. Chem. Phys.* 106 (1997) 7422-7437.

[234] K.F. Freed, J. Dudowicz, Influence of short chain branching on the miscibility of binary polymer blends: Application to polyolefin mixtures. *Macromolecules* 29 (1996) 625-636.

[235] J. Dudowicz, K.F. Freed, J.F. Douglas, Entropy theory of polymer glass formation revisited. I. General formulation. *J. Chem. Phys.* 124 (2006) 064901.

[236] E.B. Stukalin, J.F. Douglas, K.F. Freed, Application of the entropy theory of glass formation to poly(α -olefins). *J. Chem. Phys.* 131 (2009) 114905.

[237] E.B. Stukalin, J.F. Douglas, K.F. Freed, Plasticization and antiplasticization of polymer melts diluted by low molar mass species. *J. Chem. Phys.* 132 (2010) 084504.

[238] J. Dudowicz, K.F. Freed, J.F. Douglas, Generalized entropy theory of polymer glass formation. *Adv. Chem. Phys.* 137 (2008) 125-222.

[239] J. Dudowicz, K.F. Freed, J.F. Douglas, Direct computation of characteristic temperatures and relaxation times for glass-forming polymer liquids. *J. Chem. Phys.* 123 (2005) 1-4.

[240] J. Dudowicz, K.F. Freed, Molecular influences on miscibility patterns in random copolymer/homopolymer binary blends. *Macromolecules* 31 (1998) 5094-5104.

[241] J. Dudowicz, K.F. Freed, Lattice cluster theory for copolymer blends: General theory in the incompressible system, long chain limit. *Polish Journal of Chemistry* 75 (2001) 527-545.

[242] J. Dudowicz, K.F. Freed, Lattice cluster theory for pedestrian. 2. Random copolymer systems. *Macromolecules* 33 (2000) 3467-3477.

[243] K.F. Freed, J. Dudowicz, Lattice cluster theory for pedestrians: Models for random copolymer blends. *Macrom. Symp.* 149 (2000) 11-16.

[244] J. Dudowicz, K.F. Freed, Explanation for the unusual phase behavior of polystyrene-b-poly(n-alkyl methacrylate) diblock copolymers: specific interactions. *Macromolecules* 33 (2000) 5292-5299.

[245] J. Dudowicz, K.F. Freed, Energetically driven asymmetries in random copolymer miscibilities and their pressure dependence. *Macromolecules* 30 (1997) 5506-5519.

[246] J. Dudowicz, K.F. Freed, Molecular origin of the free energy dependence on the monomer sequence in random copolymer systems. *Macromolecules* 29 (1996) 7826-7827.

[247] K.F. Freed. Extension of lattice cluster theory to strongly interacting, self-assembling polymeric systems. *J. Chem. Phys.* 130 (2009) 061103.

[248] J. Dudowicz, J.F. Douglas, K.F. Freed, Equilibrium polymerization models of re-entrant self-assembly. *J. Chem. Phys.* 130 (2009) 164905.

[249] M.S. Wertheim, Fluids with highly directional attractive forces. I. Statistical thermodynamics. *J. Stat. Phys.* 35 (1984) 19-34.

[250] M.S. Wertheim, Fluids with highly directional attractive forces. II. Thermodynamic perturbation theory and integral equations. *J. Stat. Phys.* 35 (1984) 35-47.

[251] M.S. Wertheim, Fluids with highly directional attractive forces. III. Multiple attraction sites. *J. Stat. Phys.* 42 (1986) 459-476.

[252] M. S. Wertheim, Fluids with highly directional attractive forces. IV. Equilibrium polymerization. *J. Stat. Phys.* 42 (1986) 477-492.

[253] T. Morita, K. Hiroike, A New Approach to the Theory of Classical Fluids. III. General Treatment of Classical Systems. *Progr. Theor. Phys.* 25 (1961) 537-578.

[254] G. Stell, *The Equilibrium Theory of Classical Fluids*; Benjamin; New York (1964).

[255] E. Nies, T. Li, H. Berghmans, R.K. Heenan, S.M. King, Upper Critical Solution Temperature Phase Behavior, Composition Fluctuations, and Complex Formation in Poly(Vinyl Methyl Ether)/DO Solutions: Small-Angle Neutron-Scattering Experiments and Wertheim Lattice Thermodynamics Perturbation Theory Predictions. *J. Phys. Chem.* B 110 (2006) 5321-5329.

[256] E. Nies, A. Ramzi, H. Berghmans, T. Li, R.K. Heenan, S. M. King, Composition Fluctuations, Phase Behavior, and Complex Formation in Poly(vinyl methyl ether)/DO Investigated by Small-Angle Neutron Scattering. *Macromolecules* 38 (2005) 915-924.

[257] J.G. Jang, Y. C. Bae, Phase behavior of hyperbranched polymer solutions with specific interactions. *J. Chem. Phys.* 114 (2001) 5034-5042.

[258] P.F. Arce, M. Aznar, Modeling of thermodynamic behavior of PVT properties and cloud point temperatures of polymer blends and polymer blend + carbon dioxide systems using non-cubic equations of state. *Fluid Phase Equilibria* 286 (2009) 17-27.

[259] C.S. Gudipati, J.A. Finlay, J.A. Callow, M.E. Callow, K.L. Wooley, The antifouling and fouling-release perfomance of hyperbranched fluoropolymer (HBFP)-poly(ethylene glycol) (PEG) composite coatings evaluated by adsorption of biomacromolecules and the green fouling alga ulva. *Langmuir* 21 (2005) 3044-3053.
[260] M.A. Elrehim, B. Voit, B. Bruchmann, K.J. Eichhorn, K. Grundke, C. Bellmann, Structural and end-group effects on bulk and surface properties of hyperbranched poly(urea urethane)s. *J. Polym. Sci., Part A: Polym. Chem.* 43 (2005) 3376-3393.
[261] S. Peleshanko, R. Gunawidjaja, S. Petrash, V.V. Tsukruk, Synthesis and interfacial behavior of amphiphilic hyperbranched polymers: Poly(ethylene oxide)-polystyrene hyperbranches. *Macromolecules* 39 (2006) 4756-4766.
[262] M.E. Mackay, G. Carmezini, B.B. Sauer, W. Kampert, On the surface properties of hyperbranched polymers. *Langmuir* 17 (2001) 1708-1712.
[263] E. Žagar, M. Žigon, Aliphatic hyperbranched polyesters based on 2,2-bis(methylol) propionic acid - Determination of structure, solution and bulk properties. *Prog. Polym. Sci.* (Oxford) 36 (2011) 53-88.
[264] A. Sidorenko, X.W. Zhai, S. Peleshanko, A. Greco, V.V. Shevchenko, V.V. Tsukruk, Hyperbranched polyesters on solid surfaces. *Langmuir* 17 (2001) 5924-5931.
[265] Y. Mikhaylova, V. Dutschk, M. Müller, K. Grundke, K.J. Eichhorn, B. Voit, Study of the solid-liquid interface of hydroxyl-terminated hyperbranched aromatic polyesters (HBP-OH) in aqueous media. II. Adsorption of model proteins *Coll. Surf. A: Physicochem. Eng. Aspects* 297 (2007) 19-29.
[266] Y. Mikhaylova, V. Dutschk, C. Bellmann, K. Grundke, K.J. Eichhorn, B. Voit, Study of the solid-liquid interface of hydroxyl-terminated hyperbranched aromatic polyesters (HBP-OH) in aqueous media. I. Characterisation of the properties and swelling behaviour of HBP-OH thin films. *Coll. Surf. A: Physicochem. Eng. Aspects* 279 (2006) 20-27.
[267] Y. Mikhaylova, E. Pigorsch, K. Grundke, K.J. Eichhorn, B. Voit, Surface properties and swelling behaviour of hyperbranched polyester films in aqueous media. *Macrom. Symp.* 210 (2004) 271-280.
[268] G. Jannerfeldt, L. Boogh, J.A.E. Manson, Tailored interfacial properties for immiscible polymers by hyperbranched polymers. *Polymer* 41 (2000) 7627-7630.
[269] A.M. Díez-Pascual, A. Compostizo, A. Crespo-Colín, R.G. Rubio, R. Miller, Adsorption of water-soluble polymers with surfactant character. Adsorption kinetics and equilibrium properties. *J. Coll. Interf. Sci.* 307 (2007) 398-404.
[270] U. Domańska, Z. Zolek-Tryznowska, Effect of temperature and composition on the surface tension and thermodynamic properties of binary mixtures of Boltorn U3000 with alcohols and ether. *J. Sol. Chem.* 39 (2010) 864-876.
[271] A. Ziemer, M. Azizi, D. Pleul, F. Simon, S. Michel, M. Kreitschmann, P. Kierkus, B. Voit, K. Grundke. Influence of hyperbranched polyesters on the surface tension of polyols. *Langmuir* 20 (2004) 8096-8102.
[272] S. Peleshanko, V.V. Tsukruk, The architectures and surface behavior of highly branched molecules. *Prog. Polym. Sci.* (Oxford) 33 (2008) 523-580.
[273] I.C. Sanchez, *Physics of Polymer Surface and Interface*, Butterworth-Heinemann: Boston, London, Oxford (1992).
[274] Z. Qian, V.S. Minnikanti, L.A. Archer, Surface segregation of highly branched polymer additives in linear hosts. *J. Polym. Sci., Part B: Polym. Phys.* 46 (2008) 1788-1801.

[275] J. D. van der Waals, *Over de Continuiteit van den Gasen Vloeistoftoestand*. University of Leiden, 1873 (thesis).
[276] W.S. Lord Rayleigh, On the Theory of Surface Forces. *Phil. Mag*. 30 (1890) 285-298.
[277] J.W. Cahn, J.E. Hilliard, Free Energy of a Nonuniform System. I. Interfacial Free Energy. *J. Chem. Phys*. 28 (1958) 258-267.
[278] S. Enders, K. Quitzsch, Calculation of Interfacial Properties of Demixed Fluids Using Density Gradient Theory. *Langmuir* 14 (1998) 4606-4614.
[279] H. Kahl, S. Enders, Calculation of Surface Properties of Pure Fluids Using Density Gradient Theory and SAFT-EOS. *Fluid Phase Equilibria* 172 (2000) 27-42.
[280] D. Dittmar, R. Eggers, H. Kahl, S. Enders, Measurement and Modelling of the Interfacial Tensions of Triglyceride Mixtures in Contact with Dense Gases. *Chem. Eng. Sci.* 57 (2002) 355-363.
[281] S. Enders, H. Kahl, Interfacial properties of binary mixtures. *Phys. Chem. Chem. Phys.* 4 (2002) 931-936.
[282] S. Enders, H. Kahl, J. Winkelmann, Interfacial Properties of Polystyrene in Contact with Carbon Dioxide. *Fluid Phase Equilibria* 228-229 (2005) 511-522.
[283] S. Enders, H. Kahl, Interfacial Properties of Water + Alcohol Mixtures. *Fluid Phase Equilibria* 263 (2008) 160-167.
[284] O.G. Nino-Amezquita, S. Enders, Ph. T. Jaeger, R. Eggers, Measurement and Prediction of Interfacial Tension of Binary Mixtures. *Ind. Eng. Chem. Res*. 49 (2010) 592-601.
[285] G. Nino-Amezquita, S. Enders, Prediction of Interfacial Tension of Binary Mixtures. *Computer-Aided Chem. Eng*. 28 (2010) 85-90.
[286] O.G. Niño Amézquita, S. Enders, P.T. Jaeger, R. Eggers, Interfacial properties of mixtures containing supercritical gases. *J. Supercrit. Fluids* 55 (2010) 724-34.
[287] Z.G. Wang, A.M. Nemirovsky, K.F. Freed, Interaction of a polymer chain with an asymmetric liquid-liquid interface. *J. Chem. Phys*. 85 (1986) 3068-3077.
[288] W.E. McMullen, K.F. Freed, A density functional theory of polymer phase transitions and interfaces. *J. Chem. Phys*. 92 (1990) 1413-1426.
[289] W.E. McMullen, K.F. Freed, A density functional theory of polymer phase transitions and interfaces. II. Block copolymers. *J. Chem. Phys*. 93 (1990) 9130-9144.
[290] M. Lifschitz, K.F. Freed, H. Tang, Hyperbolic tangent variational approximation for interfacial profiles of binary polymer blends. *J. Chem. Phys*. 103 (1995) 3767-3781.
[291] H. Tang, K.F. Freed, Interfacial studies of incompressible binary blends. *J. Chem. Phys.* 94 (1991) 6307-6322.
[292] X.C. Zeng, D.W. Oxtoby, H. Tang, K.F. Freed, On square-gradient theories of polymer blend interfaces. *J. Chem. Phys*. 96 (1992) 4816-4817.
[293] K.R. Myers, A.M. Nemirovsky, K.F. Freed, Surface tension of dilute polymer solutions. I. A renormalization group approach. *J. Chem. Phys*. 97 (1992) 2790-2803.
[294] M. Lifschitz, K.F. Freed, Interfacial behavior of compressible polymer blends. *J. Chem. Phys*. 98 (1993) 8994-9013.
[295] K.F. Freed, Interrelation between density functional and self-consistent-field formulations for inhomogeneous polymer systems. *J. Chem. Phys*. 103 (1995) 3230-3239.
[296] K.F. Freed, Analytic theory of surface segregation in compressible polymer blends. *J. Chem. Phys*. 105 (1996) 10572-10582.

[297] M. Lifschitz, K.F. Freed, Interfacial behavior of phase separated asymmetric compressible binary polymer blends. *J. Chem. Phys.* 105 (1996) 1633-1645.

[298] J. Dudowicz, J.F. Douglas, K.F. Freed, Competition between self-assembly and surface adsorption. *J. Chem. Phys.* 130 (2009) 084903.

[299] Y. Sakane, K. Inomata, H. Morita, T. Kawakatsu, M. Doi, T. Nose, Effects of low-molecular-weight additives on interfacial tension of polymer blends: experiments for poly(dimethylsiloxane)/poly(tetramethyldisiloxanylethylene) 1 oligo(dimethylsiloxane), and comparison with mean-field calculations. *Polymer* 42 (2001) 3883-3891.

In: Polymer Phase Behavior
Editors: T. P. Ehlers and J. K. Wilhelm, pp. 75-121
ISBN: 978-1-61324-336-7
© 2011 Nova Science Publishers, Inc.

Chapter 2

RAMAN STUDY OF THE PRESSURE AND TEMPERATURE INDUCED TRANSFORMATIONS IN CRYSTALLINE POLYMERS OF C_{60}

***K. P. Meletov* [1,*] *and G. A. Kourouklis* [2,#]**
1. Institute of Solid State Physics of the Russian Academy of Sciences, Moscow region, Russia
2. Physics Division, School of Technology, Aristotle University of Thessaloniki, Thessaloniki, Greece

ABSTRACT

High hydrostatic pressure causes a number of effects in fullerene polymers; the most interesting of them being further pressure-induced polymerization and subsequent structural phase transitions in partially polymerized fullerenes. The behavior of the phonon modes of the polymeric one-dimensional orthorhombic 1D-O phase, the two-dimensional tetragonal 2D-T and rhombohedral 2D-R phases of C_{60} have been studied as a function of pressure, up to ~30 GPa, at room temperature. The 1D-O polymeric phase, even at small pressures, undergoes pressure-enhanced photo-induced transformation to a new polymeric phase characterized by twinned polymeric chains. The photo-transformed 1D-O polymer and two-dimensional 2D-R and 2D-T polymeric phases undergo irreversible structural transformations at different pressures to new cross-linked three-dimensional polymeric structures. The phonon spectra of the high-pressure phases provide strong indication that the fullerene molecular cage is preserved in the recovered phases. The decomposition of the 2D-R polymer of C_{60} during high temperature treatment leads to the initial face centered cubic structure of the fullerene C_{60} monomer.

* 142432 Chernogolovka, Moscow region, Russia.
# GR-540 06 Thessaloniki, Greece.

1. INTRODUCTION

Polymerization involving C_{60} molecules is a complicated process since carbon atoms are found in both sp^2 and sp^3 hybridized states and therefore many possibilities appear for the formation of bonds due to the existence of 30 unsaturated double C=C bonds in the fullerene molecular cage. It has been established that the polymerization process can be initiated by photons of visible or ultraviolet light [1], alkali metal doping [2,3] and high-pressure/high-temperature (HPHT) treatment of the pristine C_{60} [4,5]. The covalent polymeric bonds are usually formed by the so-called [2+2] cyclo-addition reaction via the formation of four-membered rings between adjacent fullerene molecules. The intermolecular bonding leads to appreciable decrease of the intermolecular distance and deformation of the fullerene molecular cage, these effects result in the lowering of symmetry of the highly symmetric parent C_{60} molecule [1]. Photopolymerization is responsible for the existence of some fraction of polymeric material in any C_{60} specimen exposed to daylight illumination. The photopolymerized C_{60} material contains mainly its dimeric state, which is insoluble in commonly used solvents but reverts to the initial monomeric state by heating at ~500 K for a short time [1]. Pristine C_{60} material absorbs light very efficiently resulting in polymerization but this happens up to the radiation penetration depth, which is about 10 μm in thickness for visible light. Bulk quantities of C_{60} polymers have been available after the development of the HPHT polymerization technology, which give the opportunity to produce various types of crystalline polymers of C_{60} [4,5]. The HPHT polymers of C_{60} have attracted considerable attention because of their interesting structure and promising properties related to their hardness [6]. The crystalline structure and the dimensionality of HPHT polymers depends strongly on the pressure (P) and temperature (T) treatment conditions. Thus, the C_{60} molecules form linear polymeric chains having orthorhombic crystal structure (1D-O) and/or dimers and higher oligomers at lower P and T. At intermediate P and T two-dimensional polymeric layers are formed which have either a rhombohedral (2D-R) or a tetragonal (2D-T) crystal structure, while at higher P and T the face centered cubic structures, based on three-dimensionally (3D) cross-linked polymerization of the material, are formed [1,4-7]. In addition, the treatment of the pristine C_{60} material under high non-uniform pressure and high temperature leads to the creation of several disordered polymeric phases, the so-called ultrahard fullerite phases [8,9]. Detailed X-ray studies of these phases have revealed their 3D polymeric character [10,11].

The polymerization of C_{60} is effected by the destruction of a number of double C=C intramolecular bonds and the creation of intermolecular covalent bonds associated with sp^3-like fourfold coordinated carbon atoms in the fullerene molecular cage. The linear polymeric chains with 4 sp^3-like coordinated carbon atoms in the fullerene molecular cage are obtained for temperatures in the range 500-600 K and pressure higher than 1 GPa [5]. Parallel straight chains form two ordered orthorhombic structures that belong to pseudo-tetragonal *Immm* space group (pressure above 2-3 GPa) and orthorhombic *Pmnn* space group (pressure below 2 GPa) [12]. The planar polymeric layers, with 8 and 12 sp^3-like coordinated carbon atoms in the fullerene molecular cage, obtained at temperatures in the range 700-900 K and pressure 1.5-9 GPa, show tetragonal or trigonal unit cells [4,5]. These layers, stacked in a close-packed arrangement, form tetragonal and rhombohedral structures [4,5]. The tetragonal structure, usually observed at pressures below 5 GPa, can be stabilized in two types of stacking depending on the temperature/pressure path conditions; the *Immm* orthorhombic structure (actually tetragonal because the *a* and *b* axes are almost equal) and the $P\,4_2/mmc$ tetragonal structure [12]. The calculated lattice energies of these structures are very close; nevertheless, the energy of the $P\,4_2/mmc$ stacking is lower than that of the *Immm* stacking a fact that results in the growth of samples containing mainly the $P\,4_2/mmc$ structure with some inclusion of the *Immm* structure [13]. The rhombohedral structure, usually observed at pressures higher than 5 GPa, is formed by two types of layer stacking while both structures are described by the $R\bar{3}m$ space group [12].

The three-dimensional cross-linked polymeric structures can be obtained at pressures higher than 9 GPa and temperatures of about 600-700 K. The strong interest in these polymeric structures arises primarily from the expected extremely high hardness of these materials [14]. The rather disordered crystal structure of these polymers is face centered cubic; it becomes more and more disordered as the treatment temperature is raised [12]. Theoretical investigation [15] has predicted that the 3D-polymerized C_{60} might be formed by the application of uniaxial pressure perpendicular to the polymeric layers of the 2D-T phase of C_{60} belonging to pseudo-tetragonal *Immm* space group. According to the density-functional calculations, polymerization will take effect at a lattice constant of c=10.7 Å, which is attainable at a pressure of ~20 GPa, and results in the formation of a stable metallic phase having 24 sp^3-like and 36 sp^2-like hybridized carbon atoms per each C_{60} molecule [15]. Another theoretical study [16] predicted that uniaxial compression perpendicular to the chains in the 1D or to the polymeric layers in the 2D polymeric phases of C_{60} leads to the 3D polymerization with 52, 56 or even 60 sp^3-like coordinated carbon atoms per C_{60} molecular cage. These transformations are expected to take place at pressures lower than 14 GPa and the new phases are semiconducting with large bulk and shear moduli.

Our interest, over several years, has been focused on the experimental study of the pressure-induced transformations in crystalline polymeric phases of C_{60} and the structural stability of the new high-pressure phases by means of *in-situ* Raman spectroscopy using the diamond anvil cell (DAC) technique. The Raman scattering and infrared absorption spectra of various crystalline polymeric phases, prepared under carefully controlled conditions of HPHT treatment, have very rich and well defined structure. Their intensity distribution and peak positions differ significantly among the pristine C_{60}, 1D-O, 2D-R and 2D-T polymeric phases as has been shown by the detailed study of their optical spectra combined with their structural analysis [17]. In addition, the perturbations in the structure of the C_{60} cages, caused by external disturbances like pressure, temperature, chemical bond formation etc., are manifested in the phonon spectrum [18,19] giving important information on the environment of the molecular cage. Therefore, Raman spectroscopy can be used effectively for the identification of various polymeric phases of C_{60} as well as for the *in-situ* monitoring of high-pressure induced effects and phase transformations in the fullerene-based class of materials. In the present chapter, we are going to summarize the results of our perennial experimental study of the effect of pressure and temperature on the prototype linear and planar polymeric forms of C_{60}. Our aim is twofold. First, we are going to present our systematic work concerning the C_{60} materials, specifically the ones we may call template or prototype materials, like the linear and planar polymeric forms with the intention to document the conditions of their formation, namely, the chemical and thermodynamic parameters that control the way they are formed. With Raman spectroscopy, it is actually possible to probe the mechanism of the bond formation in the polymerization processes as well as the degree of polymerization. Second, we would like to stress the implications of this knowledge for the future, especially for the understanding of the newly emerged class of all carbon materials as well as for their applications. We believe that starting with the polymerization process of C_{60} additional interesting materials for applications are going to be produced which will exploit both carbon bond properties as well as their various forms of dimensionality. This becomes more appealing after the new insight on carbon materials provided by the graphene structure properties. We think that there may be in the future much more carbon materials in addition to graphene, which will explore and utilize the carbon bond properties, the symmetry as well as the dimensionality of these materials and use the polymerization of C_{60} in tailoring new materials for specific applications.

2. Research Method and Experimental Technique

The two-dimensional polymeric samples were prepared from sublimed 99.99% pure C_{60} powder pressurized in a piston/cylinder type high-pressure device and kept at high pressure/high temperature for a certain time. The synthesis of the 2D-T polymer of C_{60} was performed at pressures in the range 2.3-2.5 GPa and at a temperature of about 820 K similar to the procedure described in [20]. X-ray analysis of the samples, after the high-pressure/high-temperature treatment, confirmed that the crystal structure of the polymer is tetragonal ($a = b =$ 9.082 Å and $c =$ 14.990 Å). Detailed analysis has shown that it is described better in a space group *Immm* rather than $P\,4_2/mmc$ due to a possible high degree of stacking disorder in the direction perpendicular to the polymeric layers [12-13]. The 2D-R polymer of C_{60} was obtained by subjecting pristine C_{60} powdered material to a pressure of P≈5 GPa at a temperature of 773 K. The X-ray analysis, after the high-pressure/high temperature treatment, confirmed that the crystal structure of the polymeric sample is rhombohedral (space group: $R\bar{3}m$, $a =$ 9.22 Å and $c =$ 24.6 Å) [4]. The samples of the 1D-O polymer were prepared by pressurizing of pure C_{60} powder at 1.2 GPa and 573 K in a "toroid"-type high-pressure device. The X-ray diffraction analysis has confirmed that the samples have the orthorhombic packing of linear polymeric chains (space group *Pmnn: a* = 9.098 Å, $b =$ 9.831 Å, and $c =$ 14.72 Å) [21].

Raman spectra were recorded using a triple monochromator (DILOR XY-500) equipped with a CCD liquid-nitrogen cooled detector system. The spectra were taken in the back-scattering geometry by the use of a micro-Raman system comprising an OLYMPUS microscope equipped with objectives of 100× and 20× magnification and a spatial resolution of ~1 μm and ~3.5 μm, respectively. The spectral width of the system was ~3 cm^{-1}. The 514.5, 488 and 647.1 nm lines of the Ar^+ and Kr^+ lasers were used for excitation. The laser power kept lower than 2 mW, measured directly before the high-pressure cell, in order to avoid the polymer decomposition caused by laser heating effects. The photoluminescence (PL) spectra were recorded using a single monochromator JOBIN YVON THR-1000 equipped with a CCD liquid-nitrogen cooled detector system. The spectral width of the system was ~4 cm^{-1}. The 488 nm line of an Ar^+ laser was used for excitation of the luminescence spectra.

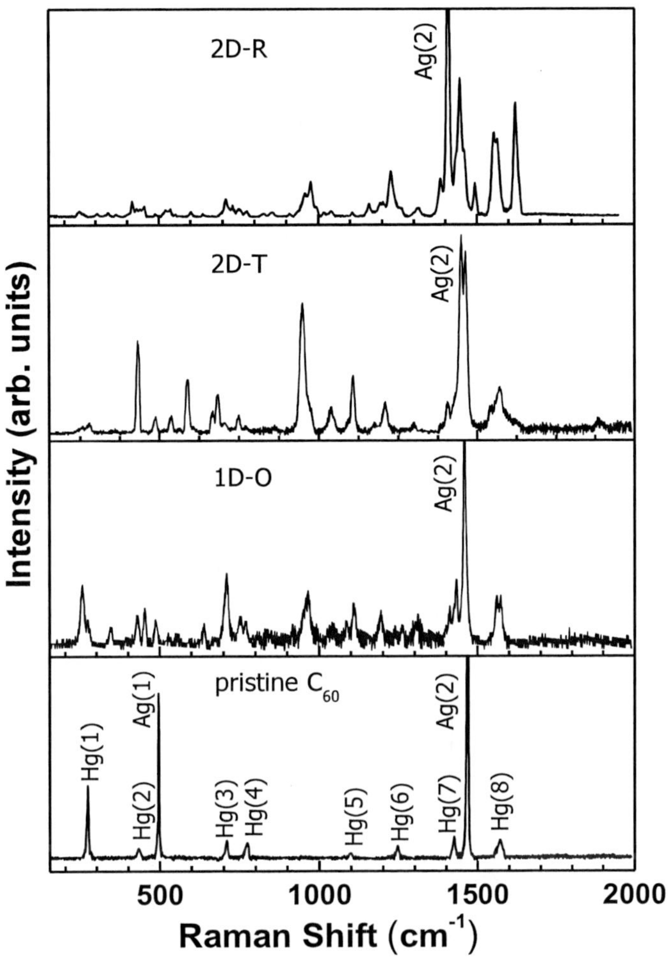

Figure 1. Raman spectra of various polymeric phases of C_{60} along with that of monomeric C_{60} recorded at normal conditions. The assignment given for the C_{60} refers to its molecular modes.

Measurements of the Raman and PL spectra at high pressures were carried out using the DAC of Mao-Bell type [22]. The 4:1 methanol-ethanol mixture was used as pressure transmitting medium and the ruby fluorescence technique was used for pressure calibration [23]. The Raman measurements at high temperature up to ~600 K were performed using a high temperature cell with a quartz window. The cell was equipped with a temperature controller unit that maintains temperatures up to 700 K with an accuracy of ±2 K. The heating and cooling rates in the temperature range of interest were less than ~15 K/min.

The polymeric samples with typical dimensions of ~100 μm were selected from the batch material by means of micro-Raman probing for their intense, clear and spatially uniform Raman response, characteristic for the 2D-T, 2D-R and 1D-O polymeric phases [17]. The Raman spectra of these polymers, in the frequency region 150-2000 cm^{-1} at normal conditions, are illustrated in Figure 1, along with that of the pristine C_{60}. The phonon frequencies, obtained by fitting the Raman data with Voigtian lines, are listed in Table 1.

Table 1. Phonon frequencies of the various polymeric phases of C_{60} and their assignment based on the C_{60} molecular modes

Mode	1D(O) ω (cm^{-1})	2D(T) ω (cm^{-1})	2D(R) ω (cm^{-1})	Mode	1D(O) ω (cm^{-1})	2D(T) ω (cm^{-1})	2D(R) ω (cm^{-1})
$H_g(1)$	-	-	245	$F_{1g}(2)$	967	972	977
$H_g(1)$	255	257	267	$F_{2u}(4)$	-	-	1016
$H_g(1)$	273	279	308	$F_{2u}(4)$	1044	1039	1037
$H_u(1)$	345	-	342	$H_g(5)$	-	-	1042
$F_{2u}(1)$	-	-	366	$H_g(5)$	1087	1090	1078
$G_u(1)$	-	-	406	$H_g(5)$	1110	1108	1109
$H_g(2)$	429	432	415	$G_g(3)$	-	1176	1158
$H_g(2)$	-	-	438	$G_g(3)$	1193	-	1195
$H_g(2)$	453	-	451	$F_{2g}(3)$	-	1208	1204
$A_g(1)$	488	487	492	$H_g(6)$	1239	-	1224
$F_{1u}(1)$	528	-	520	$H_g(6)$	-	-	1230
$F_{2g}(1)$	-	537	532	$H_g(6)$	1260	-	1260
$F_{1g}(1)$	555	562	558	$G_g(4)$	1307	1299	1314
$F_{1g}(1)$	-	589	-	$H_g(7)$	1398	-	1385
$H_u(2)$	-	-	579	$H_g(7)$	1411	1405	-
$H_u(2)$	-	610	596	$H_g(7)$	1426	1428	-
$H_u(3)$	639	-	640	$H_g(7)$	1433	-	-
$H_g(3)$	-	666	-	$H_g(7)$	1448	-	-
$H_g(3)$	-	684	695	$A_g(2)$	1458	1446	1410
$F_{2u}(2)$	-	-	709	$F_{1g}(3)$	1464	1464	1461
$H_g(3)$	710	706	712	$F_{1g}(3)$	-	-	1495
$H_g(3)$	-	-	731	$F_{2g}(4)$	-	1543	-
$H_g(4)$	748	753	749	$H_g(8)$	1562	-	1554
$F_{2g}(2)$	-	-	767	$H_g(8)$	1575	1571	1563
$H_g(4)$	771	772	776	$H_g(8)$	-	-	1569
$F_{2u}(3)$	-	-	827	$G_g(6)$	-	1599	1621
$H_u(4)$	843	863	856	$G_g(6)$	-	1624	1627
$H_u(4)$	917	-	868	$2\times G_g(2)$	-	1888	-
$G_g(2)$	955	950	958				

Polymerization results in the lowering of the molecular symmetry and therefore the Raman active fivefold-degenerated H_g modes are split and, in addition, new peaks may appear, originating from initially inactive modes of the C_{60} molecule. The assignment given in Table 1, is based on the icosahedral symmetry of the parent C_{60} molecule [17, 24, 25]. For simplicity, the irreducible representations of the C_{60} molecule are used to characterize the corresponding modes in the polymeric phases. The most important probe in the Raman study of the polymeric phases of C_{60} is the behavior of the $A_g(2)$ pentagonal pinch (PP) mode of the C_{60} molecule. This mode is related to the in-phase stretching vibration of the five double C=C bonds, which involves tangential displacements with a contraction of the pentagonal rings and expansion of the hexagonal rings. Its frequency is very sensitive to any perturbation on the molecular cage and particularly to the breakdown of some bonds involved in the polymerization process. The PP mode is downshifted in all polymeric phases of C_{60}, as the breakdown of double C=C bonds and the subsequent formation of the intermolecular covalent bonds leads to lower intramolecular average bond stiffness. In the case of the *fcc* crystal structure of the C_{60} monomer, the $A_g(2)$ mode is observed at 1468 cm^{-1}. In the 1D-O polymer this peak is observed at 1458 cm^{-1} in excellent agreement with [26]. In the 2D-T polymer, the downshift of the PP mode is expected to be larger than in the 1D-O polymer as additional inter-molecular bonds are formed. We observe indeed two peaks at 1446 and 1464 cm^{-1} in good agreement with [17]; the lower energy peak has been proposed to be the PP mode for the case of a 2D-T polymer, while the 1464 cm^{-1} peak has been attributed to the PP mode of C_{60} dimers [27]. However, from complementary analysis of Raman and infrared spectra of the 2D-T phase, it was concluded that the presence of an appreciable amount of dimers in the system is unlikely [6]. Therefore, this peak has another origin; a possibility may be the activation of an initially inactive mode in I_h symmetry $\{F_{1g}(3)\}$, which becomes active in the D_{2h} symmetry of 1D-O and 2D-T phases. Finally, in the case of the 2D-R phase the $A_g(2)$ mode exhibits the strongest softening and is observed at 1408 cm^{-1}. Note, that a weak peak at 1446 cm^{-1} is related most likely to small inclusion of the 2D-T polymer in the samples used for Raman measurements. Thus, the synthesis of 2D-R polymer requires carefully controlled pressure/temperature conditions to minimize the content of tetragonal phase in the samples.

3. Raman Spectra of the 2D-T Polymer and Pressure Induced Phase Transitions

The Raman spectra of the 2D-T polymer of C_{60} at various pressures up to 27.5 GPa and room temperature, in the frequency region 180-2050 cm^{-1}, are illustrated in Figure 2(a). The spectra were recorded upon pressure increase, while the spectral region around the strong triply degenerated T_{2g} mode of diamond, appearing at 1332 cm^{-1} at ambient pressure [28], is omitted. Apart from the pressure shift of the Raman bands, the first spectrum, recorded at 1.1 GPa, is similar to the typical Raman spectrum of the 2D-T polymeric phase at ambient conditions. The lowering of the cage symmetry, I_h of pristine C_{60} to D_{2h} in the 2D-T polymer, results in the splitting of the degenerate icosahedral intramolecular modes and to the activation of initially silent modes. Moreover, although the 2D-T phase retains the inversion center of the pristine C_{60} molecule, we cannot discard the possibility that imperfections in the crystal structure of the polymer and/or natural ^{13}C substitution may facilitate the appearance of some ungerade (u) modes in its Raman spectrum [29].

As the pressure increases, the Raman peaks shift to higher energies and their bandwidth gradually increases. The broadening of the Raman peaks is further enhanced for pressures higher than 10 GPa most probably due to the solidification of the pressure-transmitting medium. The pressure behavior of the Raman modes of the 2D-T polymer is fully reversible up to 12 GPa as it has been experimentally verified [29]. As can be clearly seen from Figure 2(a), the Raman peaks of the 2D-T polymer remain relatively narrow and well-resolved for pressures up to ~14 GPa, showing the stability of the used samples. For pressures above 14 GPa the Raman peak bandwidths increase considerably, while the peak intensities decrease. In addition, the peak broadening is accompanied by a gradual enhancement of the background (not shown in Figure 2, as the Raman spectra are presented after background subtraction).

The fluorescence from the 2D-T polymer of C_{60} is appearing in another energy region; therefore, this background is related, most likely, to the enhancement of strain and inhomogeneity within the sample induced at higher pressure.

Drastic changes in the Raman spectrum of the 2D-T polymer are first observed above 20 GPa; new distinct peaks appear in the spectrum while their intensities increase with increasing pressure. On the contrary, some of the initial Raman peaks disappear above this critical pressure. The Raman spectrum of the material for P≥20 GPa differs significantly from that at lower pressure and the observed changes can be attributed to a new high-pressure phase. From Figure 2(a) it is clear, even for an applied pressure as high as ~27.5 GPa, that the Raman spectrum of the high-pressure phase is well resolved with relatively narrow peaks. Moreover, the frequency positions of the majority of the peaks in the new phase can be tracked back to the peaks observed in the initial 2D-T polymeric phase of C_{60}. This means that the C_{60} molecular cages are preserved at pressures higher than ~20 GPa as the Raman peaks in the high-pressure phase have their origin in the intramolecular cage vibrations. Figure 2(b) shows the Raman spectra of the material upon pressure release. The decrease of pressure, from ~27.5 GPa down to ambient conditions, results in the gradual shift of the Raman peaks of the high-pressure phase to lower energies. The release of pressure does not lead to any observable changes in the Raman intensity distribution and the high-pressure phase remains stable and is recovered at ambient conditions. The bottom spectrum in Figure 2(b) was recorded at ~0.6 GPa, while the sample was recovered in air after disassembling the DAC and tested again by means of micro-Raman probing.

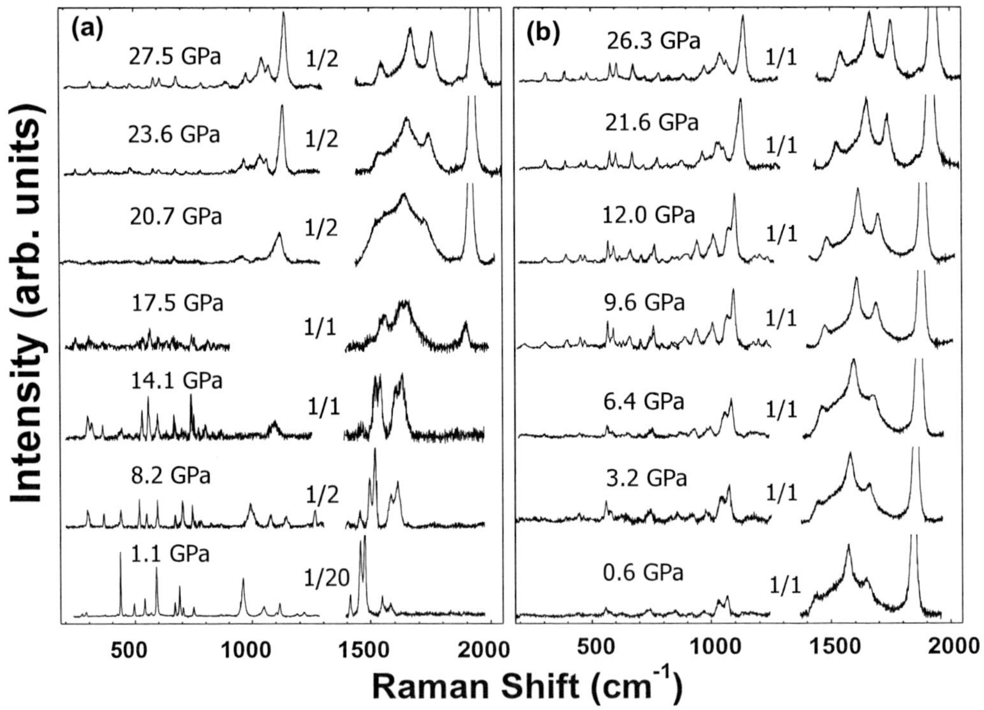

Figure 2. Raman spectra of the 2D-T polymer of C_{60} at 300 K and various pressures, recorded for (a) increasing and (b) decreasing pressure runs. The numbers 1/x indicate the relative scale of the spectra.

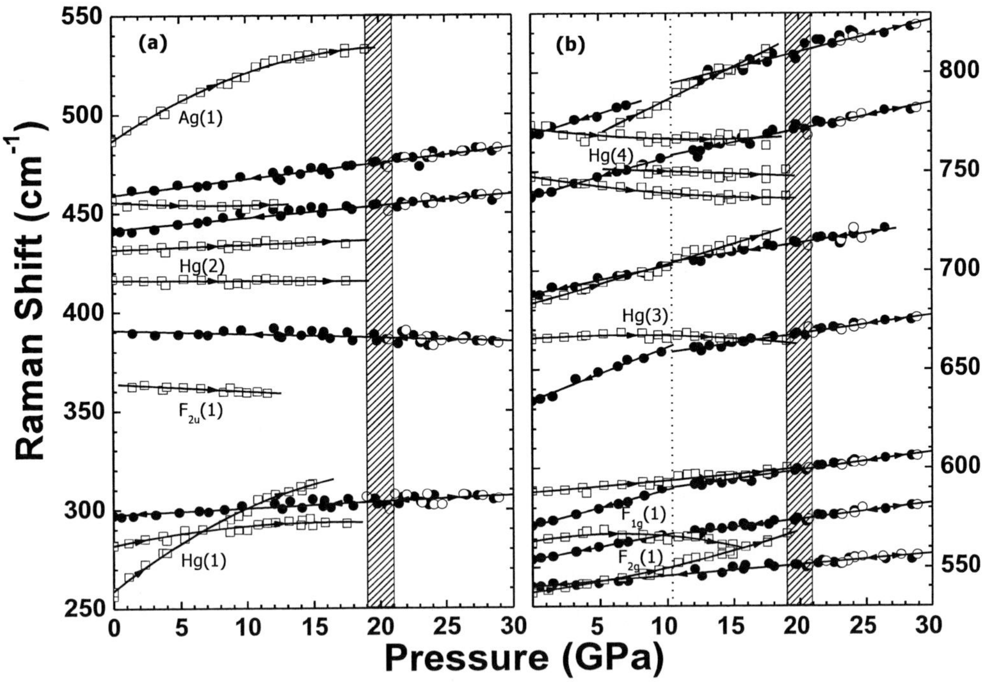

Figure 3. The pressure dependence of the Raman modes of the 2D-T polymer in the frequency regions 250-550 cm^{-1} (a) and 530-830 cm^{-1} (b). Squares and circles represent data taken for the 2D-T polymer in the initial and the high-pressure phase, respectively. The open (solid) symbols denote data taken for increasing (decreasing) pressure runs. Shaded area around 20 GPa denotes the pressure range of the phase transformation.

The pressure dependence of the Raman modes of the 2D-T polymer in the initial (squares) and the high-pressure phases (circles) is shown in Figs. 3 and 4. The open (solid) symbols denote data taken for increasing (decreasing) pressure runs. Solid lines are drawn to guide to the eye while arrows indicate the pressure increase or decrease. The mode assignment in these spectra refers to the irreducible representations of the parent C_{60} molecule following, in general, the same designation as in [17], and is given here only for the initial 2D-T phase of the polymer. The pressure dependence of the phonon frequencies of the 2D-T polymer shows linear behavior for almost all modes and is reversible for pressures up to 12 GPa [29]. However, the modes $H_g(1)$ and $A_g(1)$, demonstrate strong sub-linear pressure dependence. The $A_g(1)$ mode is a "breathing" mode of the fullerene molecular cage and is associated with radial displacements in the atomic motions. The $H_g(1)$ mode is also related, to a large extent, to the radial displacements of the carbon atoms. Thus, these two modes are characterized by out-of-plane displacements of carbon atoms and their sub-linear pressure dependence may be associated, in our opinion, with high anisotropy related to the van der Waals intermolecular bonding of adjacent 2D-polymeric layers and the covalent intermolecular bonding within the layers. Such a behavior is typical for the 2D polymeric layers and was also observed in the 2D-R polymer [30]. In addition, the $A_g(1)$ mode fully disappears for $P \geq 20$ GPa which may be the result of the 3D polymeric bonding in the high-pressure phase, which quenches the "breathing" vibration of the fullerene molecular cage.

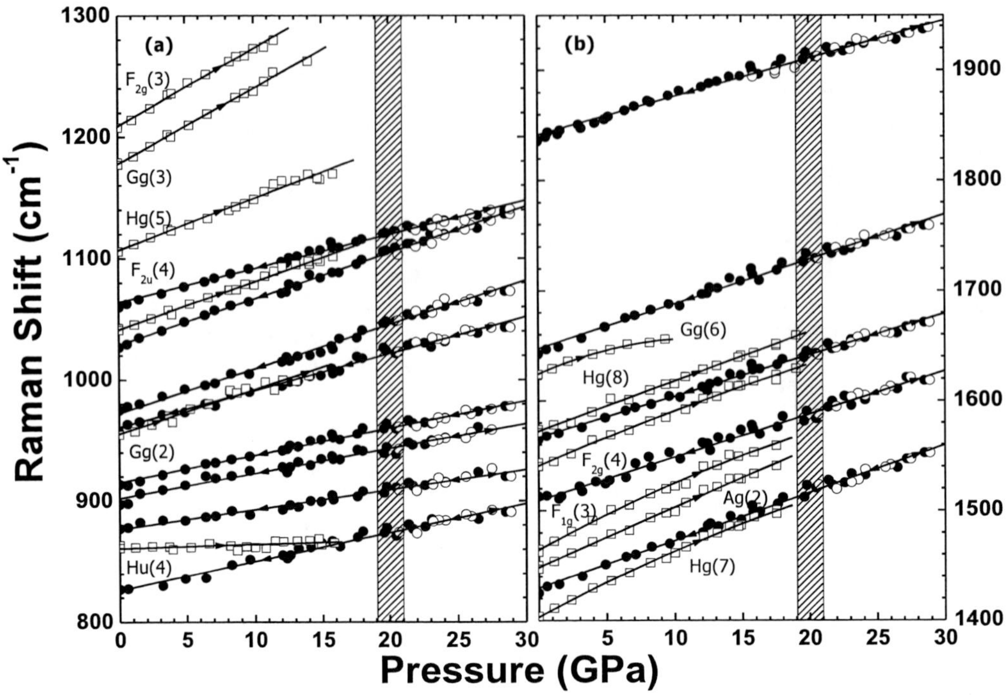

Figure 4. The pressure dependence of the Raman modes of the 2D-T polymer in the frequency regions 800-1300 cm^{-1} (a) and 1400-1950 cm^{-1} (b). Squares and circles represent data taken for the 2D-T polymer in the initial and the high-pressure phase, respectively. Open (solid) symbols denote data taken for increasing (decreasing) pressure runs. Shaded area around 20 GPa denotes the pressure range of the phase transformation.

The frequencies of all the observed modes, in the high-pressure phase, increase with increasing pressure, except for the peak at 391 cm^{-1} which shows a small negative pressure slope. The pressure coefficients $\partial\omega_i / \partial P$ of the Raman modes in the initial phase of the 2D-T polymer vary from –1.2 up to 7.6 cm^{-1}/GPa, while the pressure slopes of the high-pressure phase are ranging from –0.2 up to 4.1 cm^{-1}/GPa. For comparison, the pressure coefficients of the Raman modes in the pristine C_{60} vary in a larger range from –4.1 up to 9.8 cm^{-1}/GPa. These data are in accordance with the fact that the polymerized materials become harder as the degree of polymerization increases [6, 15, 31]. It is interesting to note that the pressure coefficients of the Raman peaks at 1029 and 1064 cm^{-1}, associated with the *sp*^{3}-like coordinated carbon atoms, are comparable to that of the T_{2g} mode of the crystalline diamond (3.8, 2.8 and ~2.7 cm^{-1}/GPa, respectively) [32]. Finally, it is important to note that several Raman modes of the high-pressure phase, located in the frequency region 550-800 cm^{-1}, exhibit changes in their pressure slopes to higher values as the pressure decreases below 10 GPa. These changes may be related with the theoretically predicted [15] relaxation of the tetragonal lattice parameters in the high-pressure phase after pressure release. According to this study, the lattice parameter *a* of the high-pressure structure is expected to enlarge at ambient pressure by 0.3 Å becoming 9.4 Å. We think also that the relaxation of the lattice parameter in the recovered high-pressure phase is responsible for the softening of the 1040 cm^{-1} mode, in the initial 2D-T polymer, down to 1029 cm^{-1} in the new high-pressure phase (the low frequency split component). The Raman modes of the high-pressure phase are related to those of the initial 2D-T polymer, as can be seen from Figs. 3 and 4, showing that they originate from the C_{60} molecular cage vibrations. The nature of some phonon modes in the initial phase of the 2D-T polymer of C_{60}, in particular the Raman peak at ~1040 cm^{-1}, is related to the covalent intermolecular bonding within the 2D polymeric layers [6, 17]. More specifically, this is associated with the vibrations of the sp^3-like coordinated carbon atoms while the much lower frequency of this peak in comparison with that of the T_{2g} mode of diamond [28], is attributed to the different lengths of the sp^3-like bonds in the 2D-T polymer (1.64 Å) and diamond (1.54 Å). In the recovered high-pressure phase, this mode appears to have two components with frequencies 1029 and 1064 cm^{-1}. By assuming, that the high-pressure phase is related with the formation of a 3D polymeric phase of C_{60} proposed in [15], these two Raman peaks might be associated with the existence of two types of sp^3-like coordinated carbon atoms with slightly different bond lengths. Another important feature in the phonon spectrum of the high-pressure phase is the change in the region of the $A_g(2)$ pentagonal-pinch (PP) mode with respect to the pristine C_{60} and the initial 2D-T polymeric phase. The increase of the number of the sp^3-like coordinated carbon atoms in the high-pressure phase results in drastic changes in the PP-mode region. Namely, in the Raman spectrum of the high-pressure phase appear five strong peaks, with the most intense of them located at ~1842 cm^{-1}. Thus, the breakdown of a large number of double C=C bonds in the high-pressure phase leads to the quenching of the PP-mode, and as a result, a number of new Raman peaks appear, which are possibly related to the stretching vibrations of the remaining isolated double C=C bonds.

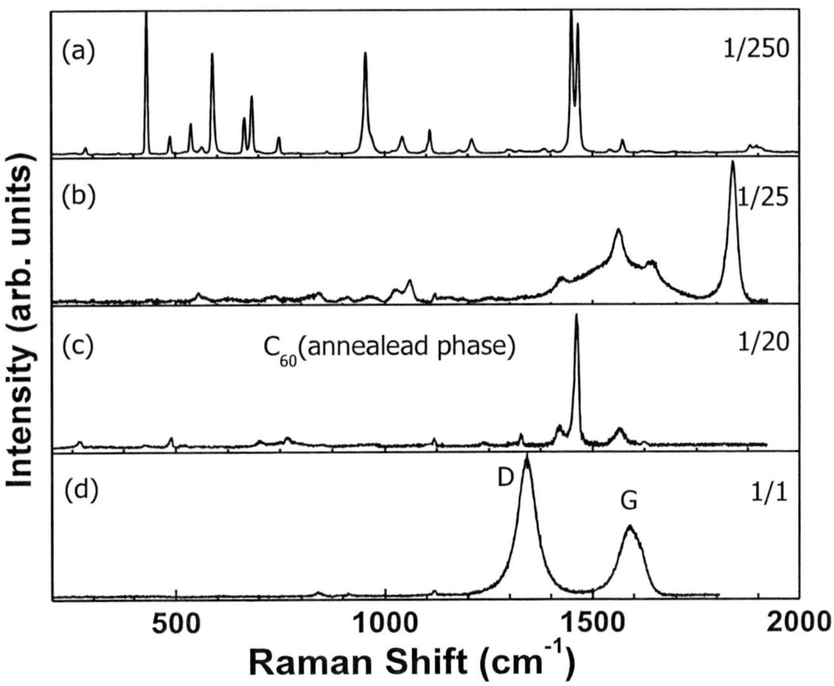

Figure 5. Raman spectra of the initial 2D-T polymer and of the recovered high-pressure phase after pressure release, recorded at ambient conditions. The numbers 1/x indicate the relative scale of the spectra. (a) The initial 2D-T polymer. (b) The high-pressure phase of the 2D-T polymer. (c) The main component among the pieces of the "detonated" sample identified as a mixture of the C_{60} monomer and dimer. (d) The "diamond-like" carbon phase identified among the pieces of the "detonated" sample.

In Figure 5 the Raman spectrum of the 2D-T polymer, recorded at ambient conditions {Figure 5 (a)}, along with that of the high-pressure phase of the recovered material {Figure 5 (b)} are illustrated for comparison. The spectra in Figure 5 were recorded out of the DAC and therefore it was possible to record also the spectrum of each material in the region of the T_{2g} mode of diamond. The recovered high-pressure phase of the 2D-T polymer exhibits a peculiar metastable behavior. More specifically, after a period of a few days from the moment of its exposure to air, the recovered sample was exploded up in pieces, sort of being "detonated", upon laser irradiation. The "explosive" process was probably initiated by the thermal energy deposited by the probing laser beam causing the rapid relaxation of the built-in strain in the sample. Note that the laser power density deposited on the sample, under these conditions, is higher than the density reaching the sample inside the high-pressure cell due to the different objective lenses used. In addition, thermal dissipation conditions are different in the two cases, as the sample in the cell is surrounded by the pressure-transmitting medium. After the "detonation" there have been identified two phases among the pieces of the recovered sample, characterized by their completely different Raman spectra presented in Figure 5 (c) and (d). In Figure 5 (c) the spectrum of the main part of the "detonated" sample is illustrated and as can be seen is similar to that of a mixture of dimers and monomers of C_{60} [1]. The presence of this phase in the "detonated" sample gives a definite proof that the C_{60} molecular cages are preserved in the high-pressure phase of the 2D-T polymer. Finally, in Figure 5 (d) the Raman spectrum of another phase, which seems to be the lesser quantity - of the "detonated" sample, is given. This spectrum is rather weak, consisting of two relatively broad peaks at 1342 and 1591 cm^{-1}. There have been observed phases in C_{60}, treated at a pressure of 12.5 GPa and temperature higher than 700 °C [33], as well as in C_{60} treated at a pressure of 9.5 GPa and temperature higher than 1500 °C [9], having similar Raman spectra. The X-ray and microhardness studies of these phases have shown that they are related to disordered carbon phases having high density and hardness [9, 33, 34] associated with the break-down of C_{60} molecular cages and the formation of a cross-linked structure of graphite-like layers [9]. Indeed, the peak positions in the Raman spectra of these carbon phases, as well as the spectrum presented in Figure 5 (d), are similar to those of the amorphous carbon containing a significant amount of sp^2 bonded carbon atoms [35] and to those of the microcrystalline graphite or the diamond-like carbon films, mostly comprising from sp^3 hybridized carbon atoms [36, 37]. The Raman peaks of this phase at 1591 cm^{-1} (C-band) is related to the stretching vibrations of the sp^2-like coordinated carbon atoms within the graphite layers, while the peak at 1342 cm^{-1} (D-band) appears in disordered carbon materials and is related to the vibrations of the sp^3-like coordinated carbon atoms.

The obtained experimental data provide a strong indication that the 2D-T polymer of C_{60} undergoes an irreversible phase transition above 20 GPa. The transformation takes place via a highly disordered pre-transitional state extending in a pressure range of ~4 GPa and having a rather diffuse Raman spectrum. The prominent Raman peaks of the high-pressure phase associated with the C_{60} molecular cage, as well as the irreversibility of the observed transformation, support the assumption of a further pressure-induced 3D-polymerization, which is rather a solid-state chemical reaction than a structural phase transition. The Raman spectrum of the high-pressure phase is dominated by the very strong Raman peak at ~1842 cm^{-1}, which cannot be related with any internal vibrational mode of the C_{60} molecular cage. The strong Raman peaks in the region of 1600÷1900 cm^{-1}, observed in some chemical compounds of carbon, are related to the stretching vibrations of isolated double C=C bonds

[38]. In analogy to that, the strong peak at 1842 cm^{-1} can be attributed to the destruction of a number of double C=C bonds under further polymerization of 2D-T polymer leaving several isolated C=C bonds in the three-dimensional network of the C_{60} polymeric material. The results obtained by Raman measurements at high pressure have been verified by X-ray diffraction [39, 40, 41], where the crystal structure and stability at the high pressure of the 2D-T polymer were studied. The detailed study of the crystal lattice parameters of the 2D tetragonal and rhombohedral polymers of C_{60} at pressures up to 7 GPa by X-ray diffraction was performed in [40] but the crucial pressure range was not achieved. The first successful examination of the high-pressure Raman data performed in the X-ray study of the 2D-T polymer at pressure up to 37 GPa has revealed a transition at ~24 GPa associated with the formation of interlayer 3+3 cycloaddition along the body diagonal [39]. As a result, the transformation from 2D to 3D polymer was clearly demonstrated, while the structural model obtained in [39] differs from that predicted theoretically [15]. It is interesting to note, that the compressibility of the initial 2D-T polymer obtained by X-ray studies is highly anisotropic; the ambient pressure compressibilities along and perpendicular to the polymeric layers differ more that 20 times [39, 40]. After the transition to the high-pressure phase, the crystal structure remains tetragonal, the lattice parameter ratio *c/a* decreases abruptly from 1.66 to 1.36, while the anisotropy of compressibility disappears. Finally, the bulk modulus of the high-pressure 3D polymeric phase, determined from X-ray measurements, is 407 GPa which is slightly smaller that that of the diamond (443 GPa). Thus, the X-ray study [39] has confirmed that the transformation found for the first time by Raman measurements [42, 43] is indeed of structural nature and is related to further 3D-polymerization of the initial 2D-T polymer. However, such a structural transition was not found in an earlier structural study of the 2D-T polymer at high pressure [41], which has shown a gradual amorphization of the initial polymeric phase at pressures higher than 20 GPa. A possible reason for this disagreement is the strong dependence of the pressure-induced 3D-polymerization of the 2D-T polymer on the structural details of the initial polymeric phase. As it was mentioned above, there are two crystal structures of the 2D-T polymer, characterized by the space groups of $P\ 4_2/mmc$ and *Immm*. Since the initial 2D-T polymer in [39] was *Immm* with ~20% impurity of $P\ 4_2/mmc$, the 3D polymerization, which is characteristic of *Immm* structure, did take place. However, in samples with $P\ 4_2/mmc$ space group as a majority phase, a different transition is expected at a different pressure. Because this is a competing process with amorphization due to the nonhydrostaticity conditions in the diamond anvil cell, this may be the reason for the results observed by Leger *et al.* [41].

4. Raman Spectra of the 2D-R Polymer and Pressure Induced Phase Transitions

The Raman spectra of the 2D-R polymeric phase of C_{60}, in the frequency region 100-2050 cm^{-1} and for various pressures up to ~30 GPa, are illustrated in Figure 6a. The spectra were recorded upon pressure increase, while the spectral region around the strong triply degenerate T_{2g} mode of diamond, appearing at 1332 cm^{-1} at ambient pressure [28], is excluded. The background, which is gradually increasing with pressure, has been subtracted from the experimental spectra. The initial spectrum, taken at ~0.9 GPa, is similar to that recorded at ambient conditions and exhibits all typical Raman features of the 2D-R polymer [17]. The spectrum contains some additional peaks, indicated by arrows, related most likely with the presence of oligomers in the material. The Raman spectrum of the polymer is richer in structure than that of pristine C_{60} due to the splitting of the Raman active five-fold degenerate H_g phonon modes and/or to the activation of silent modes. Table 2 shows the correlation of the symmetry groups I_h (molecular symmetry of pristine C_{60}) and D_{3d} (molecular symmetry of C_{60} in the 2D-R polymer) [17]. As in the case of the 2D-T polymeric phase, defects and stresses in the crystal structure of the polymer, as well as natural isotopic ^{13}C substitution, may facilitate the appearance of some u-modes in the Raman spectrum of the material. The frequencies and the assignment of the Raman modes of the 2D-R polymer of C_{60} are shown in Table 3. The mode assignment refers to the irreducible representations of the C_{60} symmetry group [17]. As the pressure increases, the Raman peaks shift to higher energy and their bandwidth gradually increases. The pressure shift of the Raman peaks is accompanied by a gradual redistribution of the intensities among the various Raman modes: the relative intensities of the $H_g(3)$, $H_g(4)$, $G_g(2)$, $F_{1g}(2)$, $H_g(8)$, and $G_g(6)$ modes increase, while the intensities of the $H_g(1)$ and $A_g(1)$ modes decrease. These features in the pressure behavior of the Raman intensities are typical of the various polymeric phases of C_{60} and of some other fullerene-related materials [19, 29, 30]. The broadening of the Raman peaks is further enhanced above 10 GPa, most probably due to the solidification of the pressure-transmitting medium leading to pressure gradients.

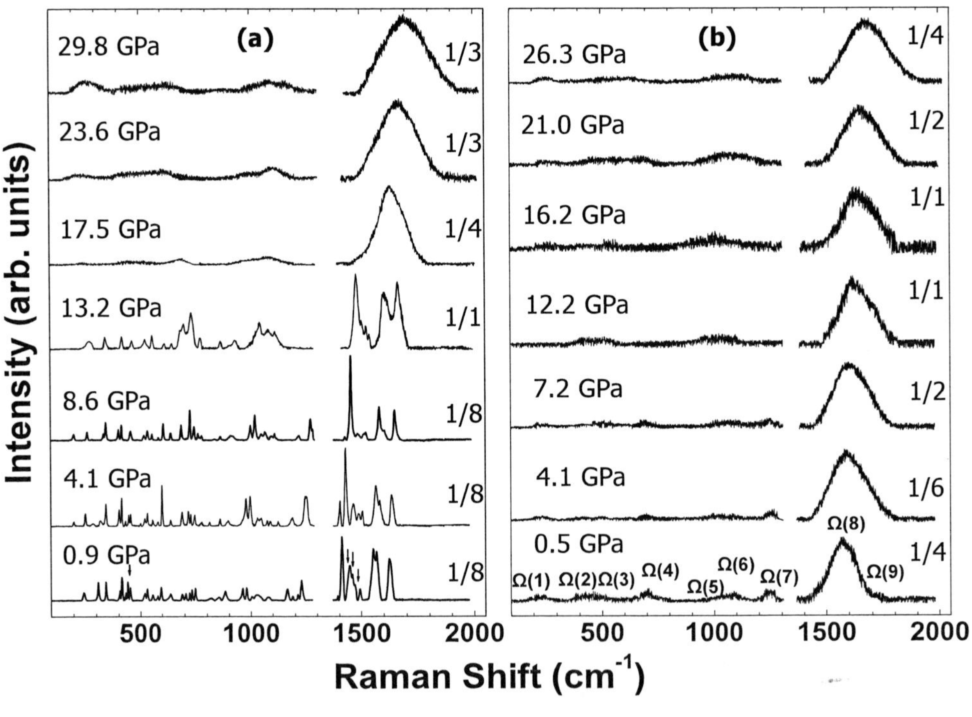

Figure 6. Raman spectra of the 2D-R polymer at room temperature and various pressures, recorded for (a) increasing and (b) decreasing pressure runs. The vertical arrows indicate peaks related to the presence of oligomers in the material. The numbers 1/x indicate the relative scale of the spectra.

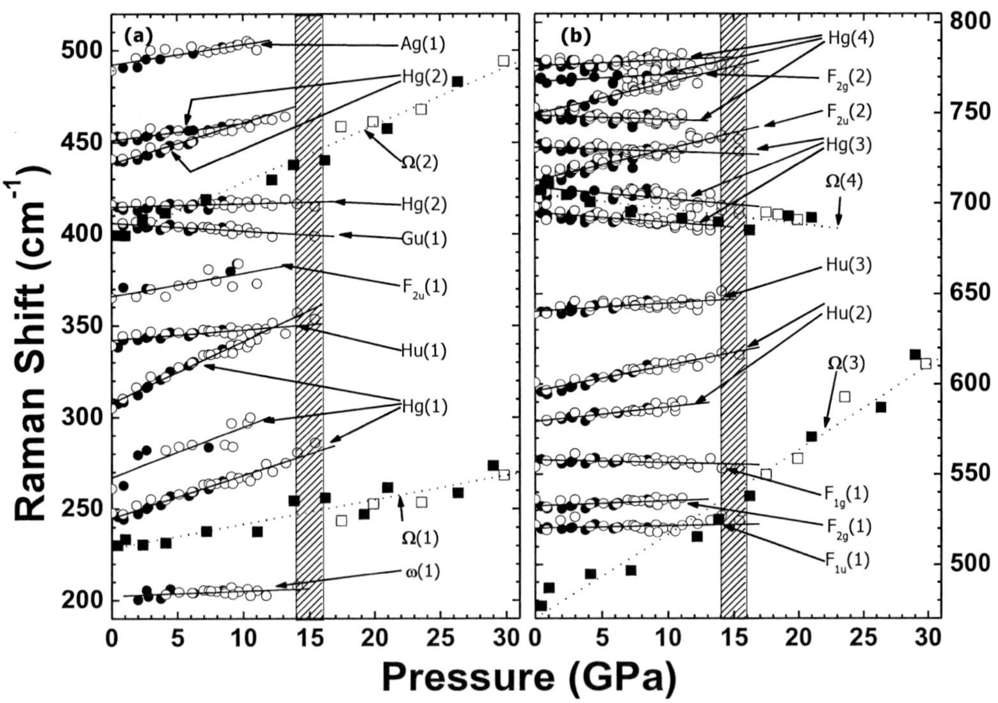

Figure 7. The pressure dependence of the Raman modes of the 2D-R polymer in the energy regions 190-520 cm^{-1} (a) and 470-810 cm^{-1} (b). Circles (squares) represent data taken for the 2D-R polymer in the initial phase (high-pressure phase). The open (solid) symbols denote data taken for increasing (decreasing) pressure runs. Shaded area around 15 GPa denotes the pressure range of the phase transformation.

Drastic changes in the Raman spectrum are first observed near 15 GPa, it becomes very diffuse and loses its fine structure in all frequency regions. This transformation in the Raman characteristics is preceded by a rapid decrease in the intensity of the $A_g(2)$ PP-mode which vanishes at pressures above 15 GPa. The Raman spectra at P $\geq$ 15 GPa differ significantly from the spectrum of the 2D-R phase in the pre-transitional pressure range, exhibiting a considerable broadening of the Raman peaks, while the relative intensity distribution among the various phonon bands is preserved. The broad Raman features above P $\geq$ 15 GPa, designated as $\Omega(1)$-$\Omega(9)$ in the fourth column of Table 3 can be tracked back to the initial 2D-R polymeric phase and seem to incorporate the corresponding group of the well resolved Raman peaks of this material at ambient pressure. The transformed material shows spatial uniformity as it was verified by its Raman response by probing various spots all over the surface of the sample.

Figure 6(b) shows the Raman spectra of the high-pressure phase of the 2D-R polymer upon pressure release. Decrease of pressure, from ~30 GPa down to ambient pressure, results in the gradual shift of the Raman peaks to lower energies without any observable changes in the Raman intensity distribution. The high-pressure phase remains stable down to ambient conditions. We should remark that the cycle of the pressure decrease lasted for a period of about one month. The number of the diffuse Raman bands and their positions in the recovered high-pressure phase are different from those in the initial 2D-R phase (see Table 3). The diffuse Raman spectra of the recovered material are typical of a highly disordered state, while the spectra taken at various sites of the sample are identical showing the spatial uniformity of the recovered material. The Raman spectrum of the high-pressure phase differs significantly from that of the amorphous carbon with respect to both the number of peaks and their frequencies. Our observations from several pressure runs show that the recovered sample is rather stable, not changing at ambient conditions at least for a period exceeding one week.

The pressure dependence of the Raman modes of the 2D-R polymer of C_{60} for the initial (circles) and the high-pressure phase (squares) is shown in Figures 7(a) and 7(b) in the energy region 190–810 cm^{-1}. The open (solid) symbols denote data taken for upstroke (downstroke) pressure runs. Solid lines are guides to the eye. The mode assignment in these figures is related only to the initial 2D-R polymer and refers, as in Table 3, to the irreducible representations of the symmetry group of the parent C_{60} molecule (I_h symmetry) [44], following in general the same designation as in [17] with minor changes based on the present high-pressure Raman data. As it can be seen from Figures 7(a) and 7(b) (see also Table 3), the pressure-induced shift of the majority of the Raman modes of the initial 2D-R polymer is linear and positive, with exception of six modes with symmetries $G_u(1)$, $F_{1g}(1)$, $H_g(3)$, and $H_g(4)$, which display small negative pressure shifts. The pressure dependence of all Raman modes is reversible with pressure at least up to ~10 GPa, the highest pressure reached during the first experimental pressure run (solid circles in Figs. 7(a) and 7(b)). The shaded area near 15 GPa indicates the pressure range where drastic changes in the Raman spectra take place. The transition to a new high-pressure phase results in the appearance of the diffuse Raman features related to a highly disordered state. The frequencies of the bands in the high-pressure phase were defined with somewhat lower accuracy because of their diffuse nature. Their pressure dependence shows large dispersion and is positive for almost all modes, except for the $\Omega(4)$ mode, which shows a small negative pressure slope.

Figures 8(a) and 8(b) show the pressure dependence of the Raman modes of the 2D-R polymer for the initial (circles) and the high-pressure (squares) phases in the energy region 820-1720 cm^{-1}. All Raman modes of the initial 2D-R polymeric phase in this energy region show positive pressure shift with linear dependence, while this behavior is reversible with pressure for at least up to ~10 GPa. The shaded area near 15 GPa also indicates the pressure region where the irreversible transition to the disordered high-pressure phase takes place. The pressure dependence of the diffuse Raman bands in the high-pressure phase is also positive except for the $\Omega(5)$ mode. As it can be seen from Figs 7 and 8, the pressure dependence of the $\Omega(8)$ and $\Omega(9)$ modes, for the downstroke pressure run, is close to the pressure dependence of the corresponding group of modes $H_g(8)$ and $G_g(6)$ of the initial 2D-R polymer located in this frequency region. On the contrary, the pressure dependence of the $\Omega(1)$-$\Omega(7)$ modes differs significantly from that of the group of modes of the initial 2D-R polymer which are located in this frequency region. This may be partially related to the relatively high uncertainty in the determination of the frequency positions of the $\Omega(1)$-$\Omega(7)$ bands due to their low intensity in comparison with the more intense $\Omega(8)$-$\Omega(9)$ modes. The pressure coefficients of the Raman modes in the initial 2D-R phase range between –0.6 and 7.2 cm^{-1}/GPa, while those in the high-pressure phase vary between –1.4 and 4.7 cm^{-1}/GPa. At the same time, the pressure coefficients of the Raman modes in pristine C_{60} vary between –4.1 and 9.8 cm^{-1}/GPa. These results are compatible with the known experimental data and theoretical predictions that the polymerized fullerenes become harder as the degree of polymerization increases [6, 15, 31, 45].

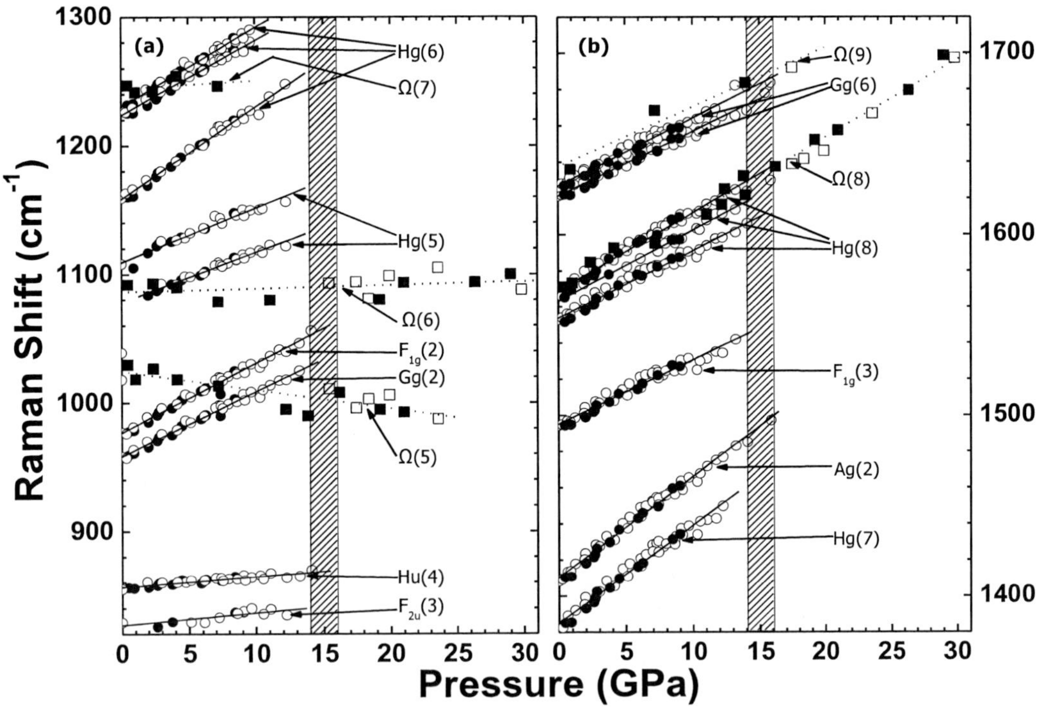

Figure 8. The pressure dependence of the Raman modes of the 2D-R polymer in the energy regions 820-1300 cm^{-1} (a) and 1380-1720 cm^{-1} (b). Circles (squares) represent data taken for the 2D-R polymer in the initial phase (high-pressure phase). The open (solid) symbols denote data taken for increasing (decreasing) pressure runs. Shaded area around 15 GPa denotes the pressure range of the phase transformation.

Pressure-induced imperfections and structural defects may contribute to the broadening of the Raman bands at higher pressures. In order to explore such contributions the recovered samples were annealed under various temperature conditions. The Raman spectra of the annealed samples show that the material undergoes a transformation when subjected to a temperature higher than 550 K. Below this temperature no visible changes in the Raman response of the recovered sample were observed during the annealing procedure. The results of micro-Raman probing of the recovered sample at ambient conditions, before and after annealing are shown in Figure 9. The Raman spectrum of the recovered sample after the pressure cycle, the Raman spectrum of the sample during partial annealing and after complete annealing at 550K are presented in Figs. 9(b), (c), and (d), respectively, along with the initial spectrum of the 2D-R phase in Figure 9(a). All spectra were recorded in the energy region 1350-1800 cm^{-1}, which includes the most intense Raman peaks. The Raman spectrum of the completely annealed material is spatially uniform and differs considerably from that of the high-pressure phase as well as from that of the initial 2D-R polymeric phase. This spectrum contains a relatively narrow and intense Raman band near 1464 cm^{-1}, related to the PP-mode in the case of a mixture of monomers and dimers of C_{60} and is similar to the spectra of various annealed polymeric phases of C_{60} [6, 43]. Note that the Raman response of the partially annealed sample in Figure 9(c) contains mainly the diffuse band related to the high-pressure phase, while the prominent Raman bands (marked by arrows) are related to the inclusion of the C_{60} dimers and monomers.

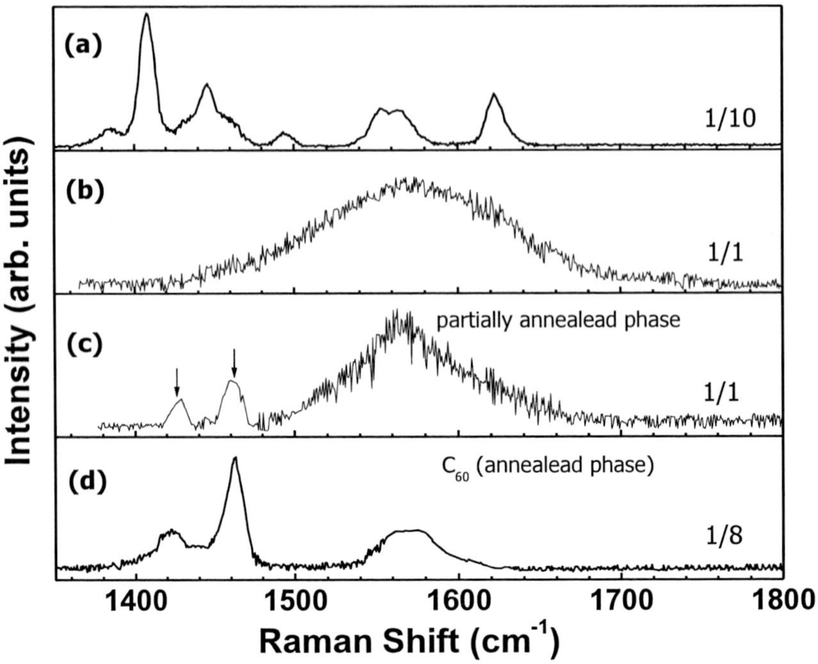

Figure 9. Raman spectra of the 2D-R polymer in the frequency region 1350-1800 cm^{-1} recorded at ambient conditions after release of pressure. (a) The initial 2D-R polymeric phase. (b) The "high-pressure" phase of the polymer. (c) The Raman spectrum of the "high-pressure" phase during partial annealing. Arrows indicate the Raman modes of the C_{60} dimer (d) The Raman spectrum of the "high-pressure" phase after being annealed at ~550 K. The spectrum is similar to that expected for a sample consisting of a mixture of C_{60} monomers and dimers.

The pressure behavior of the Raman modes provides strong indication that the 2D-R polymeric phase of C_{60} undergoes an irreversible transformation at a pressure of ~15 GPa. The initial well resolved Raman spectrum of the 2D-R polymer transforms to a diffuse one that is typical of a disordered state. The pressure dependence of the phonon modes is reversible up to ~10 GPa as it is unambiguously shown by the downstroke pressure data of the experimental run up to 10 GPa. At higher pressure, near 15 GPa, an irreversible transition to a new high-pressure phase takes place. The important feature in the pressure behavior of the phonon modes is the drastic changes in the region of the PP-mode related to the rapid decrease of the intensity of the $A_g(2)$ mode and the enhancement of the neighboring $H_g(8)$ and $G_g(6)$ modes in the pre-transitional pressure range. This behavior is reminiscent of an analogous behavior exhibited by these modes in the 2D-T polymeric phase of C_{60} before its further polymerization under high pressure [18,43, 46]. The PP-mode in pristine C_{60} is related to the in-phase stretching vibration of the five double C=C bonds, which run away from the vertices of each pentagon in the fullerene molecular cage. The frequency of the PP-mode in the polymeric fullerenes decreases as the number of the polymeric covalent bonds per molecular cage increases (see Figure 1). As it was shown in the previous paragraph, the breakdown of a large number of double C=C bonds leads to the quenching of the initial PP-mode in the Raman spectrum of the high-pressure phase of the 2D-T polymer. The attenuation of the intensity of the PP-mode in the pre-transitional pressure range is related to the destruction of a number of double C=C bonds followed by subsequent creation of covalent links between molecules belonging to adjacent polymeric layers in the planar polymers of C_{60}. The application of high pressure decreases preferentially the distance between the polymeric layers due to the anisotropic compressibility of these materials, but the creation of covalent bonds is possible only between C atoms belonging in molecules of adjacent layers and having optimal relative orientations. The X-ray studies of the planar polymers of C_{60} at pressure up to 6 GPa have shown that the center-to-center distances of the nearest C_{60} cages between adjacent polymeric layers in the 2D-R polymer decreases more rapidly than those in the 2D-T polymer. However, the relative molecular orientations and atom-to-atom distances in the 2D-T polymer are more preferable for the formation of regular covalent bonds between molecules in adjacent polymeric layers [40]. Thus, we may assume that the new bonds in the highly compressed 2D-R polymeric phase of C_{60} are formed in a random way because of the presence of non-optimal molecular orientations. As a result, the new high-pressure phase exhibits a high degree of disorder characterized by random out-of-plane polymerization. This behavior differs from that of the 2D-T polymeric phase of C_{60} in which the *Immm* body-centered pseudo-tetragonal crystal structure provides optimal relative orientations of C_{60} molecules in adjacent sheets and leads to a high degree of regularity in the formation of out-of-plane covalent bonds [15]. This regularity in the covalent bonding is manifested in the Raman features of the high-pressure phase of the 2D-T polymer giving well-resolved and rich Raman spectra [42, 45]. It is important to note that inhomogeneity of the 2D-T samples that generally consist of a mixture of the *Immm* and $P\ 4_2/mmc$ tetragonal phases may result also in spatially non-uniform Raman response of the high-pressure phase. The highly ordered high-pressure phase in compressed 2D-T polymer appears in a number of small islands having *Immm* structure and dispersed in the sample [43]. The rest of the sample having $P\ 4_2/mmc$ structure is characterized by a rather diffuse Raman spectrum of the high-pressure phase, which is somewhat reminiscent that of the 2D-R polymer of C_{60}. Thus, in the primitive truly tetragonal structure with $P\ 4_2/mmc$ symmetry, the C_{60} molecules in adjacent polymeric layers do not have optimal relative orientations for the creation of covalent out-of-plane bonds in a regular way similar to that of the 2D-R polymer.

5. Photo- and Pressure-Induced Transformations in the Linear Orthorhombic Polymer of C_{60}

5.1. Photo-Induced Transformation

The Raman spectrum of the pristine 1D-O polymer recorded at ambient conditions is shown in Figure 10(a). This spectrum coincides with the earlier reported spectrum of the 1D-O polymeric phase, especially the number of the Raman active modes, the peak positions and their intensities are practically the same within the accuracy of the measurements [17]. Figure 10(b) shows the Raman spectrum of the 1D-O polymer recorded in the first run of the high-pressure Raman measurements at a starting pressure of ~0.3 GPa obtained upon loading the DAC. The spectrum in Figure 10(b) has quite different structure: the majority of the bands are split and the total number of the peaks is increased. The same behavior, characteristic for pressure-induced transformation, is observed for a number of pressure runs with different specimens of the 1D-O polymer at any starting pressure obtained upon the sample loading into the DAC. The observed pressure-driven transformation is not typical: we have not been able to find the threshold pressure because it occurs even at pressure as low as 0.1 GPa, a different pressure generation technique, with higher pressure resolution in this range, is needed in order to check for this pressure threshold. Figure 10 (c) and 10 (d) show the Raman spectra of the planar 2D-T and 2D-R polymeric phases of C_{60}. The comparison of the Raman spectra of the transformed 1D-O polymer {Figure 10(b)}, to those of the 2D-T {Figure 10(c)} and 2D-R {Figure 10(d)} polymers clearly shows the differences between these polymeric forms in the number, position and relative intensities of the peaks. The increase of the number of peaks testifies that the molecular symmetry of the C_{60} cage in the new phase of the 1D-O polymer is lower than in the pristine orthorhombic polymer (symmetry D_{2h}). In the case of the cross-linking of the pristine 1D-O polymer, the lowering of symmetry could be associated with the formation of new inter-cage covalent bonds between the C_{60} clusters that belong to the neighboring polymeric chains in the structure of the orthorhombic phase.

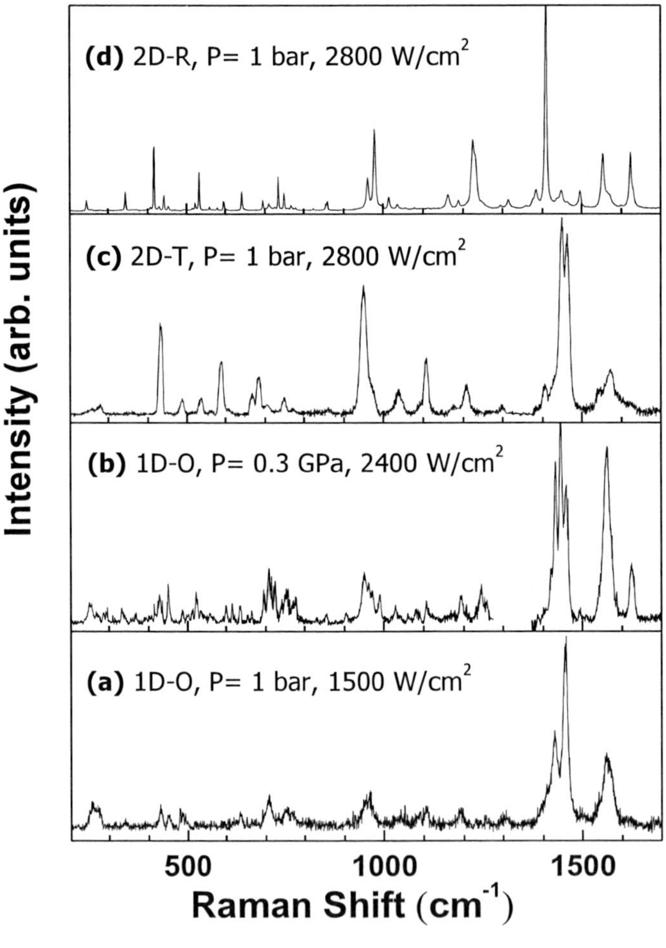

Figure 10. Raman spectra of the initial 1D-O polymer at normal conditions (a) and the transformed 1D-O polymer at 0.3 GPa (b) recorded with different excitation power densities. Raman spectra of the 2D-T (c) the 2D-R and (d) at ambient conditions are shown for comparison.

Table 2. Correlations of the symmetry groups Ih (molecular symmetry in monomeric C60) and D3d (molecular symmetry in the 2D-R polymer of C60) after [17]

I_h	D_{3d}	Splitting ($I_h \rightarrow D_{3d}$)
A_g	A_{1g}	1→ 1
F_{1g}	$A_{2g} + E_g$	0→ 1
F_{2g}	$A_{2g} + E_g$	0→ 1
G_g	$A_{1g} + A_{2g} + E_g$	0→ 2
H_g	$A_{1g} + 2E_g$	1→ 3
A_u	A_{1u}	0→ 0
F_{1u}	$A_{2u} + E_u$	1→ 2
F_{2u}	$A_{2u} + E_u$	0→ 2
G_u	$A_{1u} + A_{2u} + E_u$	0→ 2
H_u	$A_{1u} + 2E_u$	0→ 2

The distinction of the Raman spectrum of the transformed 1D-O polymer with respect to those of the 2D polymerized tetragonal and rhombohedral phases of C_{60} indicates that the new chemical bonds between neighboring polymeric chains of the 1D-O polymer are different from those of the cross-linked polymeric bonds in the planar polymers of C_{60}. The detailed data related to the Raman mode frequencies of these polymeric phases, as well as the Raman frequencies and the mode assignment of the pristine C_{60} are summarized in the Table 4.

To study the structural aspects of the observed transformation we have compared the X-ray diffraction pattern of the initial 1D-O polymer at normal conditions with that of the pressure-treated 1D-O polymer after pressure release. The pressure, up to 3 GPa, was generated in a "toroid"-type high-pressure cell while the duration of treatment was 10 minutes at 300 K. The X-ray diffraction pattern of the pristine material at ambient pressure and the pressure-treated 1D-O polymer after pressure release are shown in Figure 11(a) and Figure 11(b), respectively. As it can be seen from Figure 11, there are no significant differences in the X-ray diffraction patterns of the 1D-O polymer before and after pressure treatment. The positions of all observed peaks are the same, while small differences in peak intensities may be related to the powder material preparation. In addition, the *ex-situ* Raman spectrum of the pressure-treated 1D-O polymer taken after pressure release is the same with the spectrum of the initial polymer. Thus, the *ex-situ* X-ray and Raman data of the preliminary high-pressure treated 1D-O polymer do not show any changes in the crystal structure and the phonon spectrum of the material. On the contrary, the transformation is clearly observed in the *in-situ* high-pressure Raman study, which implies that the material transformation is related to the combined effect of laser irradiation and high pressure application. Therefore, the observed transition is related to pressure-enhanced photo-induced transformation of the material.

Table 3. Phonon frequencies and pressure coefficients of the 2D-R polymer of C60 and its "high-pressure" phase

2D-R polymer of C_{60}			"High-pressure" phase		
Mode[a]	ω_i cm^{-1}	$\partial\omega_i/\partial P$ cm^{-1}/GPa	Mode	ω_i cm^{-1}	$\partial\omega_i/\partial P$ cm^{-1}/GPa
$H_g(1)$	245	2.3	$\Omega(1)$	228	1.3
$H_g(1)$	267	2.8			
$H_g(1)$	308	3.4			
$H_u(1)$	342	0.6			
$G_u(1)$	406	-0.4	$\Omega(2)$	397	3.1
$H_g(2)$	415	0.2			
$H_g(2)$	438	2.2			
$H_g(2)$	451	0.9			
$A_g(1)$	492	1.1	$\Omega(3)$	470	4.7
$F_{1u}(1)$	520	0.1			
$F_{2g}(1)$	532	0.3			
$F_{1g}(1)$	558	-0.2			
$H_u(2)$	579	0.8			
$H_u(2)$	596	1.4			
$H_u(3)$	640	0.4			
$H_g(3)$	695	-0.5	$\Omega(4)$	705	-0.8
$H_g(3)$	709	-0.6			
$F_{2u}(2)$	712	1.8			
$H_g(3)$	731	-0.2			
$H_g(4)$	749	-0.2			
$F_{2g}(2)$	749	1.8			
$H_g(4)$	767	0.4			
$H_g(4)$	776	0.3			
$F_{2u}(3)$	827	1.0			
$H_u(4)$	856	0.8			
$G_g(2)$	958	5.0	$\Omega(5)$	1025	-1.4
$F_{1g}(2)$	977	5.4			
$H_g(5)$	1078	3.9	$\Omega(6)$	1086	0.3
$H_g(5)$	1109	4.2			
$G_g(3)$	1158	7.2			
$H_g(6)$	1224	5.8			
$H_g(6)$	1230	6.2	$\Omega(7)$	1244	0.6
$H_g(7)$	1385	5.4			
$A_g(2)$	1410	5.6			
$F_{1g}(3)$	1495	3.5			
$H_g(8)$	1554	3.7			
$H_g(8)$	1563	3.9	$\Omega(8)$	1568	4.3
$H_g(8)$	1569	4.3			
$G_g(6)$	1621	3.6			
$G_g(6)$	1627	3.8	$\Omega(9)$	1638	3.3

Table 4. Phonon frequencies for the 2D-R, 2D-T, 1D-O and the transformed 1D-O polymeric phase of C_{60}. The corresponding values for monomer C_{60} are also included

2D-R polymer [30]		2D-T polymer [29]		1D-O transformed [47]		1D-O polymer [47]		Monomer C_{60} [48]	
Mode	ω_i (cm^{-1})	Mode	ω_i (cm^{-1})	Mode*	ω_i (cm^{-1})	Mode	ω_i (cm^{-1})	Mode	ω_i (cm^{-1})
$H_g(1)$	245	$H_g(1)$	259		248	$H_g(1)$	251	$H_g(1)$	273
$H_g(1)$	267		280		266		270		
$H_g(1)$	308				288				
$H_u(1)$	342				333		340		
$F_{2u}(1)$	366				366				
					389				
$G_u(1)$	406								
$H_g(2)$	415				411				
$H_g(2)$	438	$H_g(2)$	431		427	$H_g(2)$	425	$H_g(2)$	437
$H_g(2)$	451						450		
$A_g(1)$	492	$A_g(1)$	481		484	$A_g(1)$	486	$A_g(1)$	496
$F_{1u}(1)$	520				521	$\Omega(x)$	523		
$F_{2g}(1)$	532	$F_{2g}(1)$	536		527				
$F_{1g}(1)$	558	$F_{1g}(1)$	563		561				
$H_u(2)$	579	$F_{1g}(1)$	588						
$H_u(2)$	596				598				
		$\Omega(x)$	610		614				
$H_u(3)$	640				634	$H_g(3)$	635		
					654				
		$H_g(3)$	666		662				
$H_g(3)$	695				694				
$F_{2u}(2)$	709				707		707	$H_g(3)$	710
$H_g(3)$	712				722				
$H_g(3)$	731				739				
$H_g(4)$	749	$H_g(4)$	747		752	$H_g(4)$	752		
$F_{2g}(2)$	767								
$H_g(4)$	776		772		774		769	$H_g(4)$	774
$F_{2u}(3)$	827								
$H_u(4)$	856				853	$\Omega(x)$	843		
$H_u(4)$	868	$\Omega(x)$	864						
					903	$\Omega(x)$	897		
$G_g(2)$	958	$G_g(2)$	951		947	$\Omega(x)$	957		
					959				
$F_{1g}(2)$	977	$F_{1g}(2)$	970		969				
					987				
$F_{2u}(4)$	1016								
$F_{2u}(4)$	1037				1027	$\Omega(x)$	1034		
$H_g(5)$	1042	$\Omega(x)$	1041						
$H_g(5)$	1078	$H_g(5)$	1090		1082	$H_g(5)$	1082		
$H_g(5)$	1109		1107		1105		1105	$H_g(5)$	1100
$G_g(3)$	1158	$G_g(3)$	1176						
$G_g(3)$	1195				1190		1190		
$F_{2g}(3)$	1204	$F_{2g}(3)$	1206		1205				
$H_g(6)$	1224								
$H_g(6)$	1230				1241	$H_g(6)$	1240	$H_g(6)$	1243
$H_g(6)$	1260				1257		1258		
$G_g(4)$	1314	$G_g(4)$	1299				1307		

Table 4. (continued)

2D-R polymer [30]		2D-T polymer [29]		1D-O transformed [47]		1D-O polymer [47]		Monomer C_{60} [48]	
Mode	ω_i (cm^{-1})	Mode	ω_i (cm^{-1})	Mode*	ω_i (cm^{-1})	Mode	ω_i (cm^{-1})	Mode	ω_i (cm^{-1})
Hg(7)	1385	Hg(7)	1404		1386	Hg(7)	1398		
Ag(2)	1410				1423		1416		
			1428		1429		1430	$H_g(7)$	1428
		$A_g(2)$	1447		1442		1442		
		$F_{1g}(3)$	1463		1455	$A_g(2)$	1457	$A_g(2)$	1470
$F_{1g}(3)$	1495								
$H_g(8)$	1554	$F_{2g}(4)$	1543		1559				
$H_g(8)$	1563	$H_g(8)$	1567		1559	$H_g(7)$	1560		
$H_g(8)$	1569	$G_g(6)$	1598				1575	$H_g(8)$	1575
$G_g(6)$	1621				1621		1621		
$G_g(6)$	1627								

*No assignment can be made.

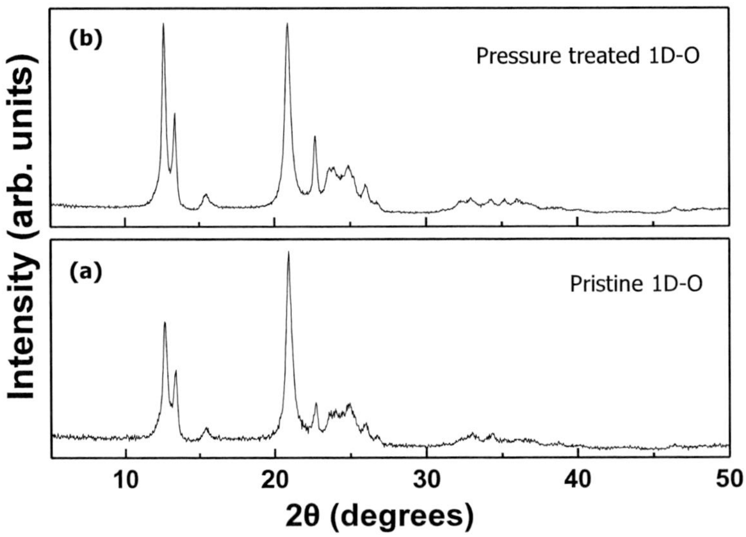

Figure 11. X-ray diffraction pattern at normal conditions of the initial orthorhombic phase of C_{60} (a), and the pressure-treated 1D-O polymer after pressure release (b).

It is known that the HPHT 2D and 3D polymeric phases of fullerene are stable at ambient conditions and withstand the laser irradiation, which does not cause their further polymerization [6]. The main effect of laser irradiation at moderate and high intensities is the overheating of polymeric samples within the laser spot that may destruct the polymeric bonds and recover the initial monomer state. In order to check the stability of the linear orthorhombic polymer of C_{60} under laser irradiation at ambient conditions, we have measured the Raman spectra of the 1D-O polymer at various power densities using the 514.5 nm Ar^+ laser line. The results of these measurements are presented in Figure12. The spectrum in Figure 12(a), measured with a laser power density of 640 W/cm^2, is identical to the Raman spectrum of the pristine 1D-O polymer and does not show any traces of photo-induced transformation. The increase of the laser power density to ~1280 W/cm^2 {Figure 12(b)} does not affect the Raman features within the time scale of the experiment, while at a laser power density of ~3200 W/cm^2 a new Raman band appears near ~1446 cm^{-1}. Further increase of the laser power density to ~6400 W/cm^2 and subsequently to ~12800 W/cm^2 leads to the gradual enhancement of the new band intensity (not shown in the figure). At the same time, the two bands near 1433 cm^{-1} and 1563 cm^{-1} become stronger with respect to the main band at ~1458 cm^{-1}, which is attributed to the PP-mode of C_{60} in the pristine 1D-O polymer [17]. At the highest laser power density of ~25600 W/cm^2 {Figure 12(d)} the Raman bands broaden and shift to lower energies due to sample overheating, which results also in gradual degradation of the polymer under long time exposure. Note, that the latter spectrum is reminiscent of the Raman spectrum of the photo-transformed 1D-O polymer taken at P=0.1 GPa and with a laser power density of 1400 W/cm^2 {Figure 12(e)}.

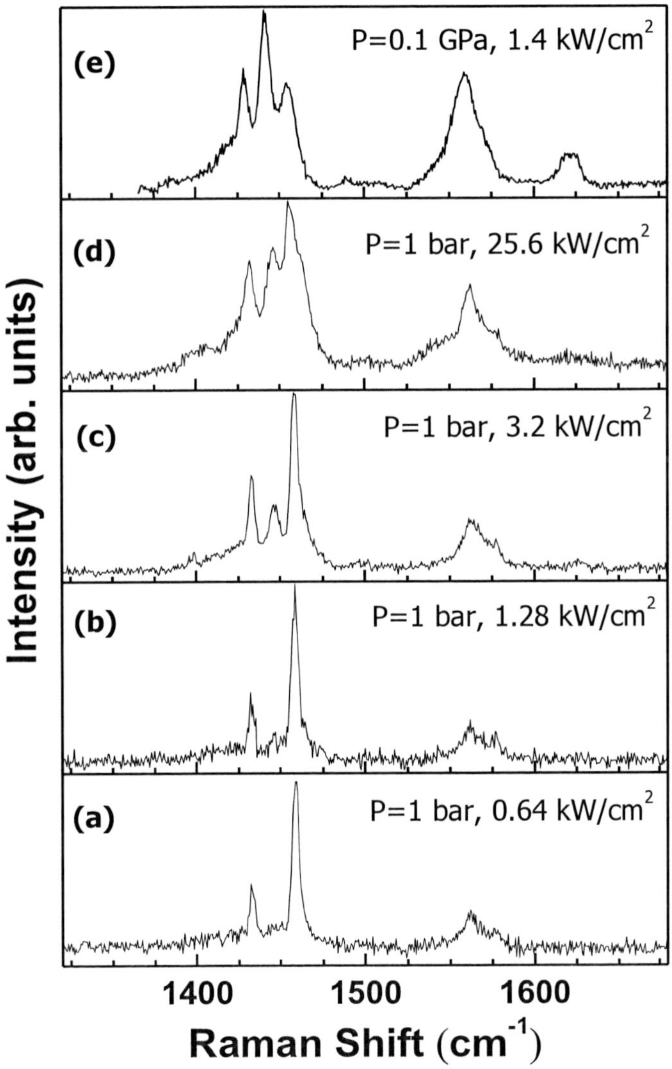

Figure 12. Raman spectra of the pristine 1D-O polymer recorded at ambient conditions and various excitation power densities of the 514.5 nm Ar$^+$ laser line {panels (a)-(d)}. The spectrum of the transformed 1D-O polymer at 0.1 GPa is also included (e).

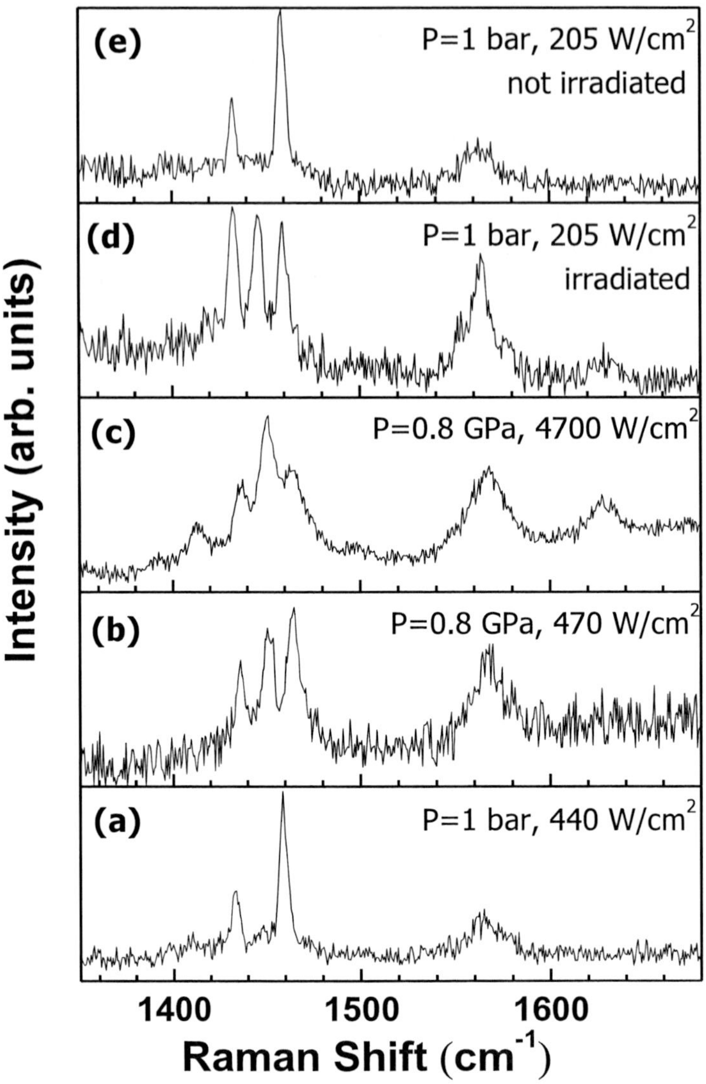

Figure 13. Raman spectra of the 1D-O polymer recorded at various pressures and excitation power densities of the 514.5 nm Ar^+ laser line. The pressure-assisted photo-polymerization starts at laser power density ~470 W/cm^2 (b) and intensifies at ~4700 W/cm^2 (c). After pressure release, the laser treated sites show Raman spectra typical of the photo-polymer (d), while untreated sites do not show photo-polymerization effects (e).

Figure 13 depicts the Raman spectra of the 1D-O polymer measured at various conditions of laser power density and pressure. The Raman spectrum of the 1D-O polymer shown in Figure 13(a) refers to ambient conditions at a laser power density of ~440 W/cm^2. Apparently, these experimental conditions do not cause changes related to photo-polymerization. At the pressure of 0.8 GPa the changes in the Raman spectrum appear at a laser power density of ~470 W/cm^2 {Figure 13(b)}. Further increase of the laser power density to ~4700 W/cm^2 leads to an almost instantaneous transformation of the material to a new phase {Figure 13(c)}. The Raman spectrum recorded in the DAC after pressure release shows the same features as those recorded under pressure of 0.8 GPa {Figure 13(d)}. This indicates that the new phase – appearing at the sample sites treated by laser irradiation at high pressure – remains stable at ambient conditions. On the contrary, the sample sites that were not irradiated by the laser beam at high pressure show typical Raman features of the pristine 1D-O polymer {Figure 13(e)}. In order to avoid any influence of laser irradiation on the Raman spectra acquired after pressure release {Figure 13(d) and Figure 13(e)}, the measurements were performed at a laser power density of ~205 W/cm^2. This value is considerably lower than 3200 W/cm^2 at which the first Raman features related to photo-polymerization at ambient pressure appear. The obtained data imply that the irreversible changes in the Raman spectra of the pristine 1D-O polymer are related unambiguously to the laser irradiation of the samples, which leads to its further photo-induced polymerization. Note that the photo-induced polymerization of the 1D-O polymer, observed for the first time at ambient conditions for an HPHT type of C_{60} polymer, takes place at a laser power density of ~3200 W/cm^2, a value that exceeds more than two orders of magnitude the value of 5 W/cm^2 reported for the photo-polymerization of the monomer C_{60} [1]. The application of high pressure enhances the process, resulting in the drastic increase of the photo-polymerization rate and in the subsequent reduction of the laser power density inducing the transformation. It is important to note that the molecules in the ground state cannot take part in the formation of C_{60} dimers via [2+2] cycloaddition reaction, which is the first step in the polymerization process. According to the Woodward-Hoffmann rule, the straightforward coupling of the C_{60} molecules in their ground state is not favorable due to the symmetry of the highest occupied orbital of C_{60} [49]. However, the molecular orbital of the excited state of C_{60}, being populated by light absorption, has favorable symmetry for dimer formation. On the contrary, the formation of dimers at high pressure takes place even at room temperature without light irradiation [50], which means that the highest occupied molecular orbital of C_{60} at high pressure is out of symmetry limitations on the pair interaction related to the Woodward-Hoffmann rule. In view of this, the simultaneous effect of pressure and light irradiation can stimulate the polymerization process, which results in considerable increase of the polymerization rate. The results of this work confirm that the 1D-O polymer indeed becomes more sensitive to photochemical reactions when high pressure is applied. Concerning the possible mechanism of the high pressure photo-induced polymerization of the pristine 1D-O polymer, one can speculate that it may be similar to the mechanism of high pressure photo-induced transformation of other molecular carbon compounds with unsaturated bonds. In particular, the combined action of pressure and laser irradiation reduces the pressure threshold of the chemical transformation of crystalline benzene from 23 to 16 GPa [51]. According to this report, high pressure induces a distortion of the benzene ring that resembles the molecule in the first excited electronic state S_1. The change of molecular geometry relates to the pressure-induced mixing of the excited S_1 state with the S_0 ground state. That is, the distortion of the molecule at high pressure facilitates the photochemical transformation related to the selective pumping of the system in the S_1 excited state.

The structural features of the pressure-enhanced photo-induced polymerization of the 1D-O polymer may be attributed to the bonding among the linear polymeric chains. The lowering of symmetry, indicated by the appearance of new Raman peaks, can be associated with the formation of new covalent bonds between C_{60} molecules that belong to the neighboring polymeric chains in the structure of the orthorhombic phase. The distinction of the Raman spectrum of the transformed 1D-O polymer from those of the 2D polymerized tetragonal and rhombohedral phases of C_{60} is indicative of new chemical bonds in the transformed orthorhombic phase that are not the typical [2+2] cycloaddition bonds but rather single bonds between polymeric chains. The *in-situ* high pressure X-ray powder diffraction study has clearly shown that the 1D-O polymer transforms to a new polymeric phase, characterized by conjunction of adjacent linear polymeric chains [52]. This study has not revealed the detailed structure of the new polymeric phase since the low-resolution diffraction profiles along with the small number of peaks could not permit further refinement; nevertheless, additional synchrotron radiation X-ray diffraction experiments may clarify this issue.

5.2. Pressure-Induced Transformations

The stability at high pressure and additional pressure-induced transformations of the photo-transformed 1D-O polymer were investigated by means of Raman measurements at pressures up to 29 GPa. The Raman spectra of the photo-transformed 1D-O polymer of C_{60} at various pressures and room temperature are shown in Figure 14. Data recorded upon pressure increase are illustrated in Figure 14(a) whereas those recorded during pressure release in Figure 14(b). The spectral region around the strong triply degenerate T_{2g} mode of diamond, appearing at 1332 cm^{-1} at ambient pressure [28], is again excluded. The background, which increases slightly with pressure, has been subtracted from the spectra for clarity. The initial Raman spectrum of the photo-transformed 1D-O polymer, which consists of a large number of narrow and well-resolved peaks, demonstrates strong pressure dependence. As the pressure increases, the Raman peaks shift to higher energies and their bandwidth gradually increases. The broadening of the Raman bands is further enhanced above 10 GPa due to the solidification of the pressure-transmitting medium.

The most important pressure effects are related to the changes in the number and intensities of the Raman active modes, as well as to their pressure coefficients. Namely, the relative intensities of the Raman modes in the 700 - 1100 cm^{-1} region gradually increase with pressure, while the most important changes were observed near ~15 GPa, where the Raman spectrum loses its fine structure in all frequency regions and becomes very diffuse. This transformation was preceded by a rapid decrease in the intensity of the peaks related to the $A_g(2)$ mode of the pristine C_{60} accompanied by a relative increase in the intensities of the $H_g(8)$ and $G_g(6)$ modes. The broad Raman features in the spectrum of the high-pressure phase above ~15 GPa can be tracked back to the photo-transformed 1D-O polymer of C_{60} and seems that is incorporating the corresponding group of the broad Raman bands of this phase. Despite the similarities in the diffused Raman bands, the spectrum of the high-pressure phase differs significantly from that of the amorphous carbon with respect to the number of peaks as well as to their position. Comparing the Raman spectrum of the high-pressure phase to that of the high-pressure phase of the 2D-T polymeric phase of C_{60} [42, 43], we observe that the former does not contain peaks in the high energy region, like the 1840 cm^{-1} peak observed in the latter phase. In addition, the new phase shows spatially uniform Raman response over all the surface of the sample, as it was documented by probing various places in the sample, a behavior which differs drastically from the 2D-T polymeric phase of C_{60} [42]. Note, that the broad Raman features of the high-pressure phase in the photo-transformed 1D-O polymer of C_{60} resemble the Raman features of the disordered high-pressure phase in the 2D-R polymer, which was also observed above ~15 GPa [30]. The broad Raman bands of the disordered phase shift to lower energies, upon pressure decrease, without any observable changes in their intensity distribution {Figure 14(b)}. The high-pressure phase is recovered and remains stable for several hours at normal conditions.

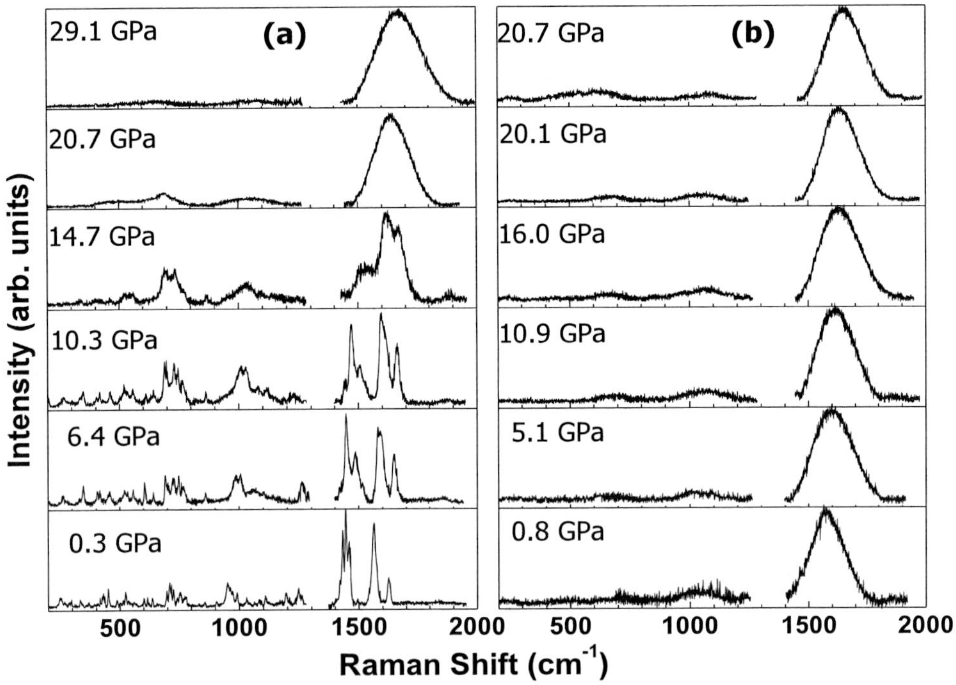

Figure 14. Raman spectra of the photo-transformed 1D-O polymer of C_{60} at room temperature and various pressures, recorded upon pressure increase (a), and upon pressure decrease (b).

The pressure dependence of the Raman mode frequencies of the photo-transformed 1D-O polymer of C_{60} in the energy regions 230 – 790 cm^{-1} and 840- 1720 cm^{-1} is shown in Figure 15 (a) and Figure 15 (b). Circles (squares) represent data recorded in a pressure run up to 12 GPa (29 GPa). Stars denote the Raman frequencies of the pristine 1D-O polymer at normal conditions. Moreover, open (solid) symbols represent data recorded upon pressure increase (decrease). The phonon frequencies obtained upon pressure increase coincide in the two different pressure runs within the accuracy of measurements. The pressure-induced shift of the majority of Raman modes is linear and positive, with exception of a few modes, which display small negative pressure shifts. The pressure coefficients vary between -0.4 cm^{-1}/GPa and 7.0 cm^{-1}/GPa. The pressure dependence of all Raman modes is reversible with pressure at least up to ~12 GPa, the highest pressure reached during the first pressure run. The dashed areas around 4 GPa and 15 GPa indicate the pressure ranges where the transformations of the Raman spectra are taking place. The pressure coefficients of some Raman bands change abruptly at ~4 GPa. The slopes of the low energy Raman modes at 288 cm^{-1} and 333 cm^{-1} change from 8.4 cm^{-1}/GPa to 2.2 cm^{-1}/GPa and from 2.7 cm^{-1}/GPa to 0.04 cm^{-1}/GPa, respectively. For the high frequency modes at 1429 cm^{-1} and 1442 cm^{-1}, the coefficients of the pressure responce change from 3.2 cm^{-1}/GPa to 6.2 cm^{-1}/GPa and from 7.0 cm^{-1}/GPa to 4.0 cm^{-1}/GPa, respectively. In addition, the band at 1559 cm^{-1} splits near ~4 GPa and the pressure coefficients of the split components are slightly different; namely, 4.9 cm^{-1}/GPa and 4.5 cm^{-1}/GPa for the higher and the lower energy component, respectively. The splitting of the mode at 1559 cm^{-1} and the changes in the pressure slopes of a number of modes, along with the reversibility of these effects upon pressure release after reaching 12 GPa, are the indications of a reversible structural phase transition that takes place near ~4 GPa.

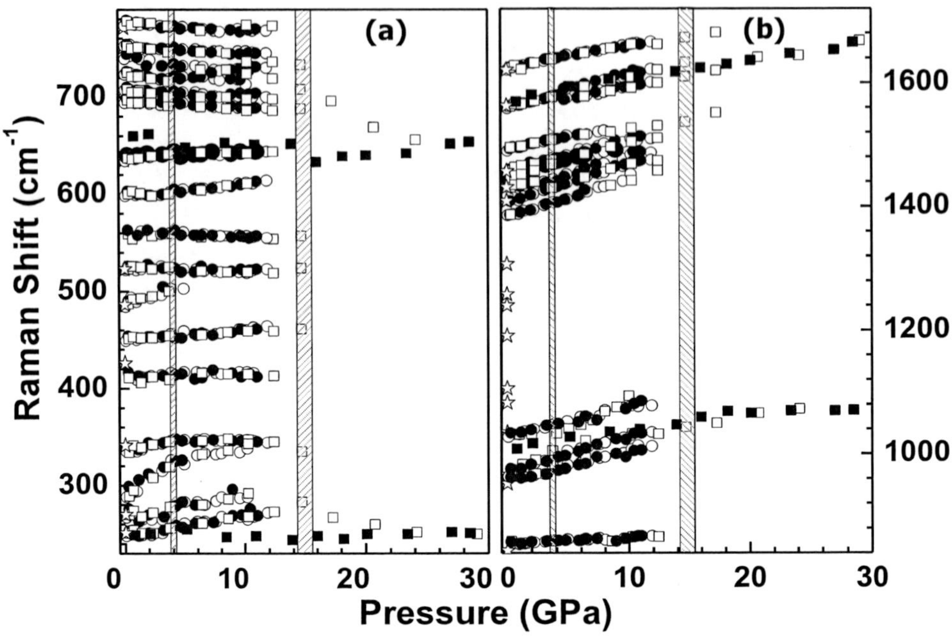

Figure 15. Pressure dependence of the Raman frequencies of the photo-transformed 1D-O polymer of C_{60} in the region 240-790 cm^{-1} (a) and 860 – 1720 cm^{-1} (b). Circles and squares represent data taken upon two different pressure runs up to 12 GPa and 29 GPa, respectively. Stars denote the Raman frequencies of the initial 1D-O polymer at normal conditions. The open (solid) symbols represent data recorded upon pressure increase (decrease). Dashed areas around 4 GPa and 15 GPa mark pressure values where the changes in the pressure-dependence were observed.

Concerning the changes taking place near ~15 GPa, they are apparently related to an irreversible transformation. More specifically, the rapid disappearance of the PP-mode in the pre-transitional pressure range, the drastic broadening of the Raman bands and the irreversibility behavior upon pressure release provide unambiguous evidence of an irreversible transformation to a highly disordered state.

The Raman data support the suggestion for a structural phase transition that may be related to minor changes in the packing of the linked linear polymeric chains. However, to prove that the singularities observed in the high-pressure Raman study are associated to a phase transition, an X-ray study of the structural aspects is necessary. A high-pressure X-ray powder diffraction study of the linear orthorhombic polymer of C_{60} has not revealed clear changes in the diffraction patterns near ~4 GPa [52]. In our opinion, this study is not conclusive since the low-resolution of the data along with the small number of peaks did not permit the complete refinement of the structure. The irreversible changes at ~15 GPa, from the well-resolved Raman spectrum of the photo-transformed 1D-O polymer to a diffuse one, is typical for a transition to a disordered phase. The rapid decrease of the $A_g(2)$ PP-mode intensity and the enhancement of the neighboring $H_g(8)$ and $G_g(6)$ modes, in the pre-transitional pressure regime, are reminiscent to the analogous behavior exhibited by these modes in the 2D-T and 2D-R polymeric phases of C_{60} before their further polymerization under high pressure [30, 42, 43]. Although pressure decreases preferentially the distance between the linear polymeric chains rather than the intercage distance within a chain, the bond formation between the chains is not always preferable. Namely, the C_{60} molecular cages belonging to adjacent polymeric chains may not have optimal relative orientations for the formation of new covalent bonds. Therefore, we expect that the new bonds are being formed in a random way due to some distortion in the molecular orientations after photo-polymerization. As a result, the new high-pressure phase exhibits a high degree of disorder characterized by a random polymerization. Similarly, in the case of 2D-R polymer, the diffuse Raman spectrum of the high-pressure phase is also related to a disordered polymeric phase of C_{60} characterized by random covalent bonding between molecules belonging to adjacent 2D-polymeric planes of the initial rhombohedral phase. It is worth noticing that this behavior differs significantly from that of the 2D-T polymeric phase of C_{60}, in which the pressure-induced shortening of the intermolecular distances, accompanied by the optimal orientation of molecules, leads to a high degree of regularity in the formation of the out-of-plane covalent bonds. Thus, the decrease of the PP-mode intensity in the pre-transitional pressure range in linear and planar crystalline polymers of C_{60} is related to the destruction of a considerable number of double C=C bonds. The subsequent creation of numerous covalent links between molecules belonging to adjacent polymeric planes or chains results in regular or random cross-link polymerization of the initial polymers.

The recovered sample after pressure release was tested by means of micro-Raman in order to check its stability at ambient conditions.

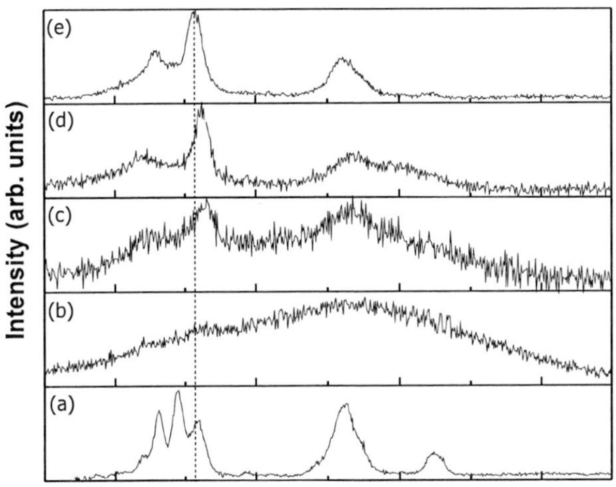

Figure 16. Raman spectra of various phases of the 1D-O polymer of C_{60} in the frequency region 1350-1750 cm^{-1} recorded at normal conditions. (a) The photo-transformed 1D-O polymer. (b) The recovered high-pressure phase immediately after pressure release. (c,d) The Raman spectra of the recovered high-pressure phase at different sites of the specimen recorded ~30 hours after pressure release. (e) The Raman spectrum of the pristine 1D-O polymer. Vertical line shows the position of the PP-mode of the pristine 1D-O polymer.

Figure 16 shows the Raman spectra of various phases of the 1D-O polymer of C_{60} in the frequency region 1350-1750 cm^{-1}, where the changes in the Raman response are more pronounced. The Raman spectrum of the photo-transformed 1D-O polymeric phase shown in Figure 16(a) differs significantly from that of the pristine 1D-O polymer shown in Figure 16(e). The Raman spectrum of the high-pressure phase of the photo-transformed 1D-O polymer immediately after pressure release is shown in Figure 16(b), whereas the in Figs. 16(c) and 16(d) are shown the Raman spectra of the high-pressure phase at different sites of the sample ~30 hours after pressure release. These spectra indicate that the high-pressure phase is metastable and transforms rather quickly to a phase which demonstrates Raman features that resemble those of the initial 1D-O polymer of C_{60} {Figure 16(e)}. Nevertheless, the spectrum in Figure 16(d) is characteristic of a mixture of monomer and dimer forms of C_{60} as it follows from the frequencies of the $A_g(2)$ PP-mode which is shifted to higher energy with respect to that in the initial 1D-O polymer (indicated by vertical dashed line). It is worth noticing that this behavior is similar to that exhibited also by the 2D-T polymeric phase of C_{60} [42, 43]. The transformation of the recovered 1D-O material was observed at normal conditions without any special heat treatment of the sample except that due to the excitation beam during the Raman probing. The behavior of the recovered high-pressure phase of the photo-transformed 1D-O polymer differs from that of the high-pressure phase of the 2D-R polymer, which is more stable and transforms to a mixture of pristine and dimerized C_{60} only after sample annealing [30].

6. PHOTOLUMINESCENCE OF THE C_{60} POLYMERS AT HIGH PRESSURE

The optical transitions to the lowest excited singlet state of the C_{60} molecule (symmetry $^1T_{1g}$) are dipole forbidden, whereas the first allowed transition to $^1T_{1u}$ state has noticeably higher energy [53]. The fluorescence of the C_{60} molecule is related to the vibronically assisted electron-phonon transitions. As a result, the quantum yield of the fluorescence is relatively small: about 10^{-5} in solution and $7x10^{-4}$ in solid, slightly higher due to the presence of impurities and defects [54, 55]. As it follows from the calculations of the electronic structure and absorption spectra measurements the solid C_{60} is a direct-gap semiconductor with gap value in the range 1.5 - 1.8 eV [56, 57]. The PL spectrum of the single crystal of pristine C_{60} at low temperature is well-resolved, exhibiting fine structure associated with optical transitions from shallow defect levels, the so-called X-traps [58]. The polymerization of C_{60} leads to drastic changes in the crystal structure and the phonon spectrum of the C_{60} monomer [4, 5]. The electron energy spectra of the polymeric phases differ considerably from that of the pristine material, this has been predicted in a number of numerical calculations of electronic structure of the planar polymers of C_{60} [56, 59-61]. The decrease of the intermolecular distances, the deformation of the fullerene molecule cage and the lowering of molecular symmetry in polymeric phases of C_{60} affect significantly the electron energy spectrum of the pristine C_{60}. Numerical calculations, performed using the local-density approximation, predict that the tetragonal and rhombohedral polymeric phases of C_{60} are indirect low gap semiconductors and their electronic structure differs significantly from that of pristine C_{60} [53, 59-61]. The optical study of the C_{60} polymers has revealed some important changes in the PL spectrum of pristine C_{60} related to its polymerization, while the specimens identified as orthorhombic and tetragonal polymers were mainly a mixture of various planar polymeric phases of C_{60} [62]. In this paragraph, we present an overview of the results of the PL spectra measurements of perfect crystalline samples of the 1D-O, 2D-R and 2D-T polymeric phases of C_{60}, at normal conditions and as a function of pressure up to 4.0 GPa. The comparison of the PL spectra of the linear and planar polymers with those of the pristine C_{60}, as well as with the results of numerical calculations of the electronic structure, provide an insight in the changes associated with the polymerization of C_{60}.

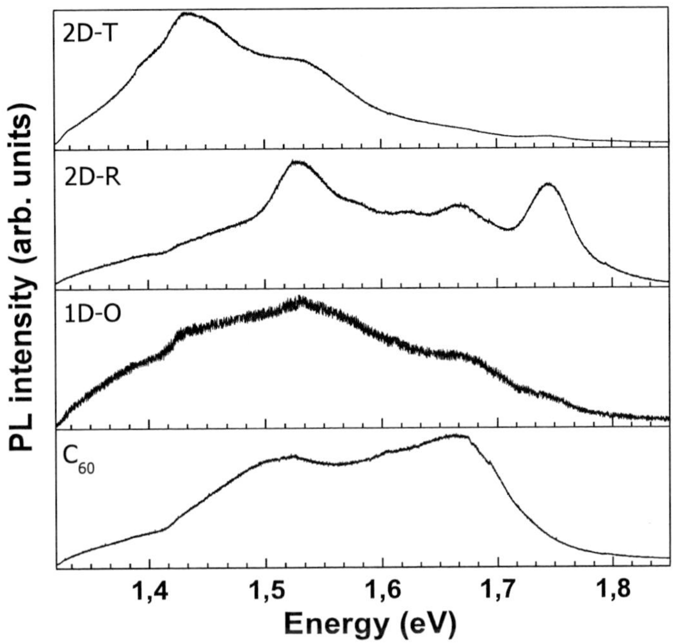

Figure 17. Photoluminescence (PL) spectra of the pristine C_{60}, 1D-O, 2D-R and 2D-T polymeric phases of C_{60} at ambient conditions.

The PL spectra of the C60 single crystals, 1D-O, 2D-R and 2D-T polymers of C60 are shown in Figure17. The PL spectrum of C_{60} single crystals, at normal conditions, consists of two broad bands. The rather diffuse spectrum becomes well resolved when recorded at liquid helium temperature. The optical transitions from shallow defect levels located near the bottom of the conduction band show narrow bands at low temperatures [58]. The measurements at low temperature and high pressure have also revealed a band related with the transitions of the free excitons. The intensity of this band increases with pressure and its pressure coefficient is noticeably higher than those related to the localized excitons [63]. The PL spectrum of the linear 1D-O polymer is also diffuse but its structure differs from that of the pristine C_{60} in both the number of bands and the intensity distribution among them. The planar polymers of C_{60} show additional changes in their PL spectra in comparison to the corresponding pristine C_{60} and 1D-O polymer recorded at normal conditions [64]. The differences refer to the onsets of the spectra, their structure, and the number of bands as well as to their intensity distribution. The PL spectrum of the 2D-T polymer consists of two main bands similar to that of the pristine C_{60}, but its onset is shifted to lower energy by ~0.14 eV and the intensity distribution between these bands is reversed. In addition, the PL spectrum of the 2D-T polymer contains two very weak bands at the higher energy, which coincide with bands observed in the 2D-R polymer, therefore may be attributed to minute impurity of this phase in the 2D-T phase as we shall discuss below. On the contrary, the onset of the PL spectrum of the 2D-R polymer is slightly higher than that of the pristine C_{60}, while the spectrum is better resolved even at room temperature and contains four relatively narrow intense bands. Note, that the quality of the PL spectra is better than those reported earlier [62] that is related, in our opinion, to higher purity and better crystallinity of the polymeric specimens used in the present study.

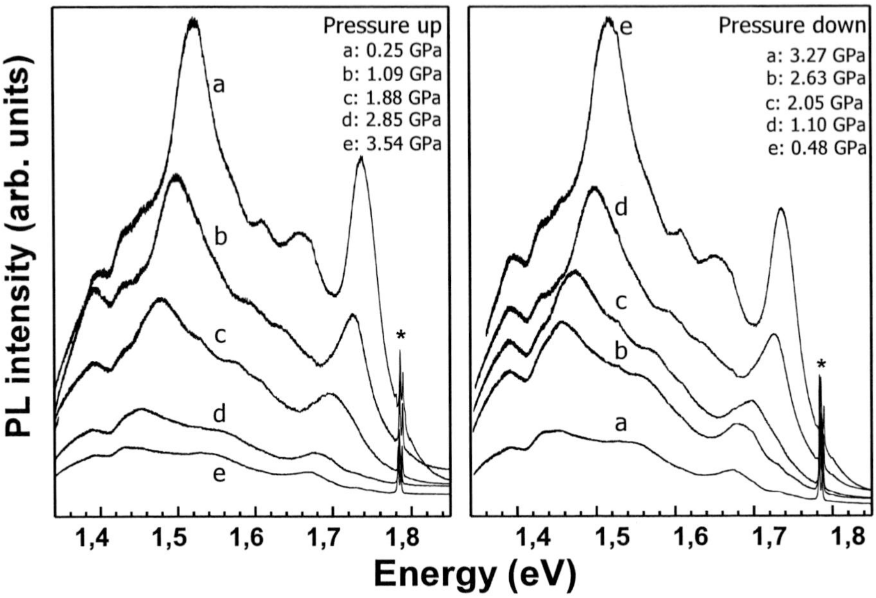

Figure 18. PL spectra of the 2D-R polymer at various pressures and room temperature for upstroke (left panel) and downstroke (right panel) pressure cycles. The ruby peaks are marked by asterisk.

The important difference between the PL spectra of the two planar polymers to that of the pristine C_{60} refers to their pressure behavior. Figure18 shows the PL spectra of the 2D-R polymer at various pressures up to ~3.5 GPa for increasing (left panel) and decreasing (right panel) pressure cycles. As the pressure increases, the bands gradually shift to lower energies and the intensity of the PL spectrum decreases. The red pressure shift in the optical spectra of the fullerene-based materials is typical for solids with van der Waals intermolecular interaction [57, 63]. The pressure coefficients for the various bands of the 2D-R polymer vary in the region -0.022 eV/GPa to -0.028 eV/GPa. For comparison, the pressure coefficient of the free exciton band in the PL spectrum of the pristine C_{60} is about -0.086 eV/GPa [63]. The intensity distribution in the PL spectra is shown as measured: it is not corrected for the spectral response of the CCD. Thus, the rapid decrease of the PL intensity near the low energy part of the spectrum, below 1.4 eV, is related to the spectral cut-off of the CCD. On the other hand, the pressure-induced decrease of the PL intensity at the high-energy part of the spectrum, common characteristic for the pristine C_{60} and its polymeric phases, has another origin. It is associated, in our opinion, with the decrease of the fluorescence quantum yield, which is related to the pressure-induced enhancement of the singlet-triplet conversion. The proximity of the singlet and triplet electronic states in the C_{60} molecule [53] results in the relatively high phosphorescence of the fullerene that reduces the quantum yield of fluorescence. The energy gap between the singlet and triplet electronic states in the molecular solids decreases with the increase of pressure. This results in the enhancement of the singlet-triplet conversion and the increase of the phosphorescence intensity.

The PL spectra of the 2D-T polymer at pressure up to 3 GPa for the increasing (left panel) and decreasing (right panel) pressure cycles are shown in Figure 19. The pressure behavior of the PL spectra is similar, in general, to that of the 2D-R polymer, but the pressure coefficients for the main bands of the 2D-T polymer are smaller and vary between -0.013 eV/GPa and -0.015 eV/GPa. Note that the pressure behavior of the PL spectra for both planar polymers is reversible and all details of the spectra are restored after pressure release. The PL spectra of the 1D-O polymer for increasing (left panel) and decreasing (right panel) pressure cycles are shown in Figure 20. The intriguing feature, in the pressure behavior of the 1D-O polymer, is the irreversibility of the transformation in the PL spectra during the increasing-decreasing pressure cycle. The structure of the PL spectra after pressure release differs from that of the initial material. The difference is most pronounced at the high-energy part of the PL spectrum, which resembles, after pressure release, somehow the spectrum of the 2D-R polymer. This behavior is in agreement with the data, reported above in the high-pressure Raman study of the 1D-O polymer that demonstrates the irreversible photo-induced pressure-enhanced transformation associated with conjugation of the adjacent polymeric chains [47].

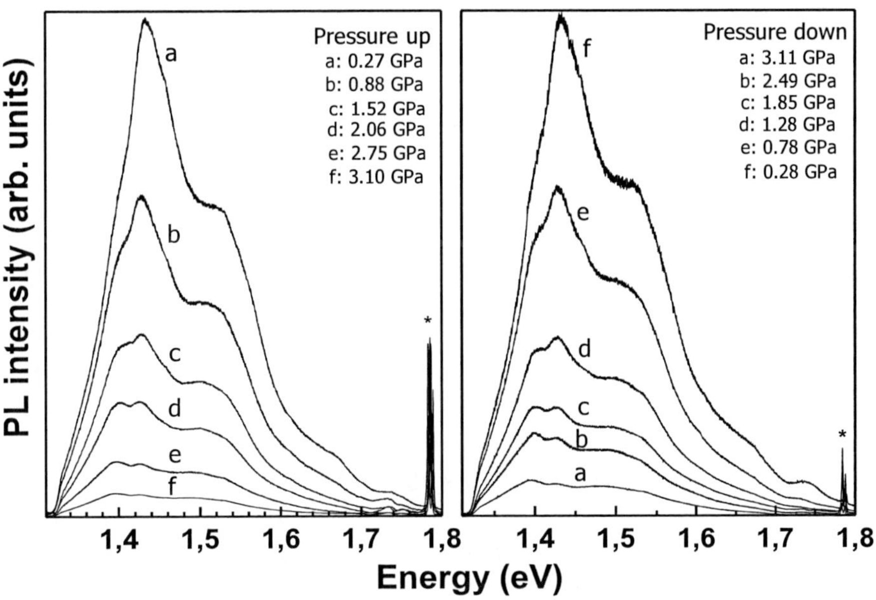

Figure 19. PL spectra of the 2D-T polymer at various pressures and room temperature for upstroke (left panel) and downstroke (right panel) pressure cycles. The ruby peaks are marked by asterisk.

The pressure dependence of the band positions in the PL spectra of the 2D-T (left panel) and 2D-R (right panel) polymers is shown in Figure 21. The open (closed) circles refer to the increasing (decreasing) pressure cycles, respectively, while the stars show the pressure dependence of the free exciton band of the pristine C_{60}. The solid and dashed lines are linear fittings to the experimental data related to the bands of the polymers and pristine C_{60}, respectively. Note that the slopes of the pressure dependencies differ for almost all bands of the two planar polymers except for the high-energy band. The initial position and the pressure dependence of this band coincide in the spectra of the 2D-R and 2D-T polymers, while its pressure coefficient ∂E/∂P= –0.022 eV/GPa, being the same for the two polymers, differs from those of the other two bands of the 2D-T polymer. This means that the specimens of the 2D-T polymer contain small inclusions of the rhombohedral polymer, which result in the appearance of the weak first band at E=1.747 eV in the PL spectra. The fact, that the preliminary Raman characterization of the samples does not show the presence of the other polymeric phase, means that the PL measurements are more sensitive to the purity of the specimens.

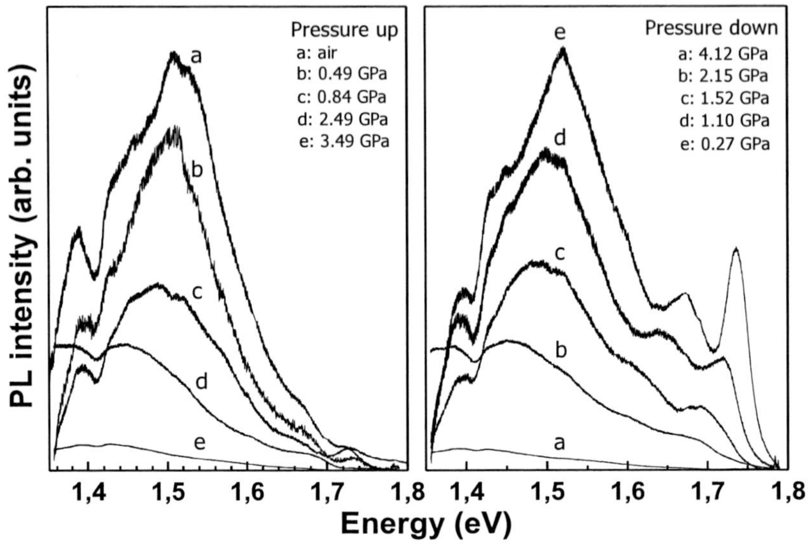

Figure 20. PL spectra of the 1D-O polymer at various pressures and room temperature for upstroke (left panel) and downstroke (right panel) pressure cycles. The ruby peaks are marked by asterisk.

The characteristics of the PL spectra of the planar polymeric phases of C_{60} and the data related to the pressure dependence of the PL bands are strong indications that the electronic structure in these polymeric phases is different with respect to that of the pristine C_{60}. Numerical calculations of the electronic structure predict that the planar 2D-R and 2D-T polymers are indirect gap semiconductors with gap values of 0.35 eV and 0.72 eV for the rhombohedral and tetragonal polymeric phases, respectively [60, 61]. On the contrary, pristine C_{60} is a direct-gap semiconductor with gap energy ~1.5 eV as it follows from the local density approximation (LDA) numerical calculations [56]. The calculated gap values are noticeably smaller than the experimental values determined from the PL spectra of the polymeric phases. This discrepancy tends to be smaller for the pristine C_{60} in which the difference between the calculated and the experimental determination are smaller than those corresponding to the polymers. Furthermore, the most intriguing aspect in the PL behavior is that the onset of the PL spectrum of the 2D-R polymer is shifted to higher energy with respect to the position of the free exciton band of the pristine C_{60}. Normally the electronic spectra of the polymers should be shifted to lower energies: this is related to the decrease of the intermolecular distances in polymeric phases in comparison to those of the pristine C_{60} under high pressure. In fact, the situation is more complicated due to the deformation of the fullerene molecular cage, which can affect significantly the electronic structure of the polymer. The numerical LDA calculations take into account the deformed fullerene molecule cage and the reduced intermolecular distances in the polymeric phases, nevertheless the experimental results of the present work are rather far apart from the calculated data. A possible reason for this discrepancy may be related, in particular, with the structure of the experimental PL spectra of polymers, which represent mainly the direct phonon-assisted electronic transitions due to their relatively large intensity with respect to the indirect ones.

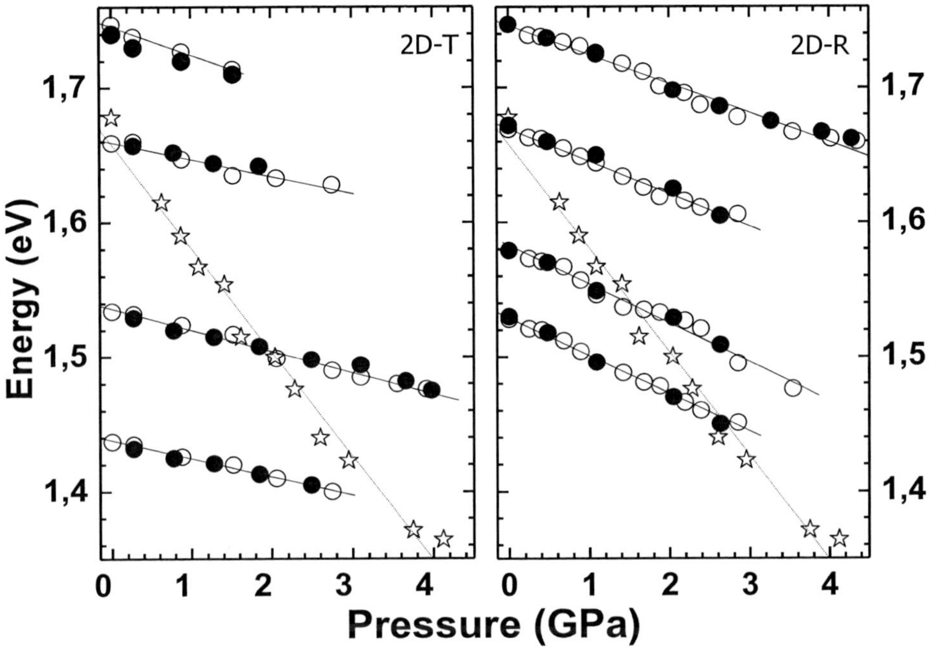

Figure 21. The pressure dependence of the band energies in the PL spectra of pristine C60 (stars), the 2D-T (left panel) and the 2D-R (right panel) polymers (circles). Open (closed) symbols represent data recorded for increasing (decreasing) pressure runs.

The important feature in the pressure behavior of the PL spectra is the noticeable difference in the pressure coefficients of the 2D-R and 2D-T polymers despite the fact that their bulk moduli B_0 are very close (the values of B_0 are 14.4, 28.1 and 29.9 GPa for the pristine C_{60}, 2D-R and 2D-T phases, respectively) [6, 40]. According to the X-ray diffraction measurements the compressibility of the 2D-R polymer within the polymeric sheets is almost three times larger than that of the 2D-T polymer, whereas the out-of-plane compressibility is almost the same [40]. The large difference in the pressure coefficients of the two polymers may be related to the difference in the in-plane compressibility of the polymers. Summarizing this paragraph, significant differences between the PL spectra of the linear 1D-O polymer, 2D-R, 2D-T planar polymers and solid C_{60} were found. These differences along with the difference in their pressure behavior provide the experimental proof that the polymerization of C_{60} results in considerable changes of the electronic structure of the pristine C_{60}. In addition, the experimental data regarding the energy gap values are qualitatively compatible with the results of numerical calculations, however they are rather far from quantitative agreement.

7. Thermal Stability and Decomposition Kinetics of the 2D-R Polymer of C_{60}

The stability of the fullerene polymers at ambient and elevated temperatures was studied by differential scanning calorimetry (DSC) and IR spectroscopy. These studies have shown that fullerene polymers are not stable at elevated temperature; heating to ~560 K results in the destruction of intermolecular C-C bonds and reversion to the initial C_{60} monomer phase [65, 66]. The DSC studies performed for various polymers under a heating rate of 10-20 K/min show a strong endothermic peak between 525 and 565 K that was not observed during the cooling scan. The temperature of the irreversible transition depends on the polymeric phase and somewhat on the scanning rate, indicating that the polymer decomposition process is controlled by kinetics. The change of enthalpy related to the complete decomposition of polymers is the highest for the C_{60} dimers, its value decreases for the linear polymeric chains and becomes smallest for the planar polymeric phases [65]. The differences in the enthalpy change between the various polymeric phases and the kinetics of the polymer decomposition suggest the possible formation of intermediate polymeric/oligomeric states during the process of the temperature-induced decomposition of the polymeric fullerene networks. To study the polymer decomposition process and intermediate states formed, we have measured the spatially-resolved Raman spectra of the single crystalline 2D-R polymer of C_{60} after its treatment at various temperatures up to 600 K. The distinct difference between the Raman band frequencies of the various polymeric/oligomeric phases, namely the PP-mode position, allows their identification in the intermediate state that may appear upon the polymer decomposition. Raman spectra of the 2D-R polymer measured in the region of the $A_g(2)$ mode, after sample treatment for 0.5 hour at various temperatures, are presented in Figure 22. The spectra were measured at room temperature to avoid the sample damage that was observed within the laser spot for $T \geq 430$ K even at laser intensity as low as 0.005 mW. The Raman spectra show that the 2D-R polymer is stable up to ~510 K, while at higher temperature a material transformation takes place as it can be inferred from the decrease of the PP-mode intensity, the increase of the background and the appearance of new Raman peaks. The transformation takes place through an intermediate state; the Raman spectrum of the material treated at 523 K has a relatively large background and new peaks typical for the PP-mode of the 2D-T and the 1D-O polymer as well as of the C_{60} dimers and monomers, which coexist with the PP-mode of the initial 2D-R polymer. Inclusion of 2D-T-like oligomers in the intermediate state can be deduced from the Raman line at ~1447 cm^{-1}, characteristic of the PP-mode in the 2D-T polymer. Their presence is related to the dissociation of 4 intermolecular C-C bonds of the initial 2D-R polymer, resulting in the creation of 2D-T-like oligomers having 8 intermolecular C-C bonds per C_{60} molecule. The Raman peak at ~1459 cm^{-1}, typical for linear polymeric chains, suggests the presence of inclusions related to the dissociation of 8 intermolecular C-C bonds of the initial 2D-R polymer and the formation of 1D-O-like oligomers having 4 intermolecular C-C bonds per C_{60} molecule.

The intermediate state of the partially decomposed 2D-R polymer was observed up to 560 K where the material changes drastically its composition resulting in the domination of C_{60} monomers with some inclusion of C_{60} dimers. Figure 23 shows the relative intensities of the $A_g(2)$ PP-mode as a function of the treatment temperature of the initial state of the 2D-R polymer (open circles), the final state of the material after high-temperature treatment (HTT) comprising of C_{60} monomers and dimers (closed circles) as well as the intermediate state comprising of 2D-T- and 1D-O-like oligomers (diamonds). The Raman spectra were measured after a number of 0.5 hour heat treatments at various temperatures on pieces of fresh sample. For most of the treatments, two or more spectra were averaged from different sample sites and/or sample pieces. The relative intensities of the various components, defined as the intensity of the $A_g(2)$ peak of each component normalized to their sum, reflect their relative concentration in the intermediate state. Thus, the concentration of the 2D-T- and 1D-O-like oligomers in the intermediate state increases with the increase of the treatment temperature up to its maximum at ~525 K, while at higher treatment temperature it gradually decreases to zero at ~560 K.

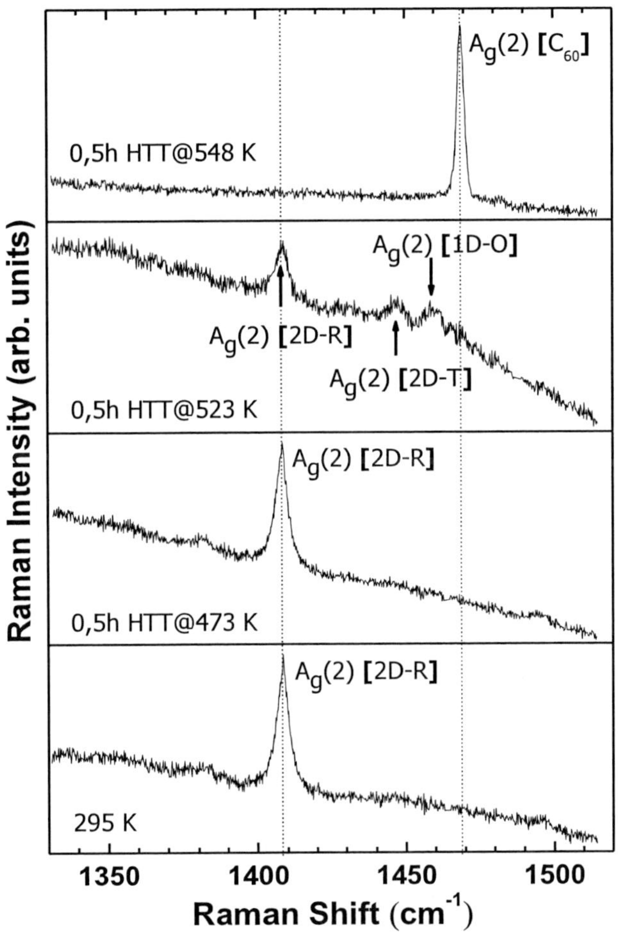

Figure 22. Raman spectra of the 2D-R polymer measured at room temperature in the frequency region of the $A_g(2)$ mode (left vertical line) after HTT for 0.5 hour at various temperatures. The arrows mark the PP-modes related to oligomers in the partially decomposed intermediate state. After complete decomposition at 548 K only the PP-mode of monomer C_{60} remains in the spectrum (right vertical line).

The existence of 2D-T- and 1D-O-like oligomers in the intermediate state underlines the fact that the 12 intermolecular C-C bonds of the 2D-R polymer do not break simultaneously, in spite of their equivalence related to the hexagonal symmetry of the planar rhombohedral polymeric sheets. The broad lineshape of the Raman peaks related to the $A_g(2)$ mode of the 2D-T- and the 1D-O-like oligomers indicate either the structural disorder or the small size of the oligomer inclusions. It should be also noted, that the formation of oligomers was observed earlier in the IR spectra of the partially decomposed 2D-R polymer, while the two-step decomposition of the 2D-R polymer through the formation of an intermediate triangular cyclic trimer state was proposed to account for the observations [67]. According to our Raman data, the situation is more complicated; the polymer decomposition takes place through the formation of an intermediate state that is in fact a mixture of the initial 2D-R polymer, 2D-T- and 1D-O-like oligomers of C_{60} as well as monomer and dimers of C_{60}. Nevertheless, the long time HTT results in the complete decomposition of the material and the final state of the decomposed 2D-R polymer is pure monomer C_{60}, possibly with small inclusions of C_{60} dimers. This final state of the decomposed 2D-R polymer is structurally ordered as its Raman spectrum is characterized by relatively low background and sharp peaks. Moreover, no sign of amorphous carbon formation was evident in the Raman spectra for the temperature and treatment time used in our study.

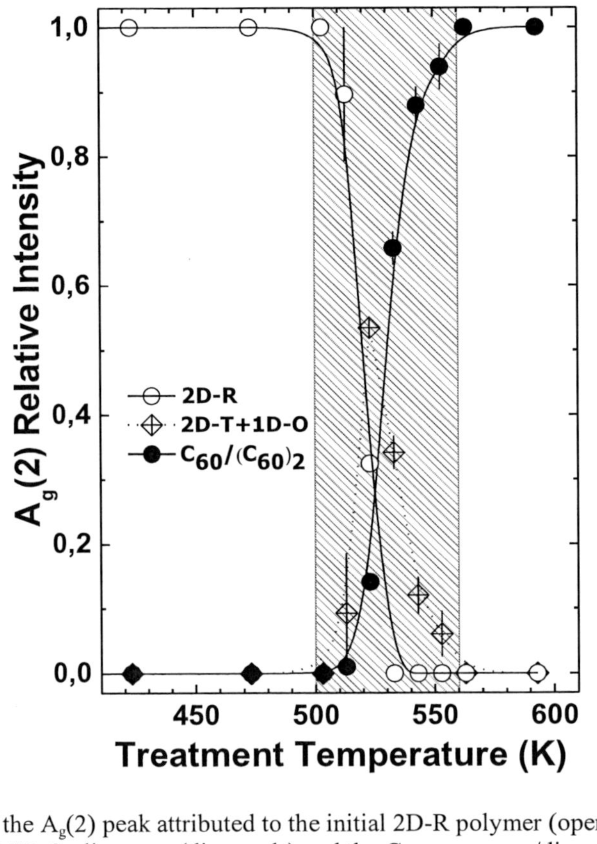

Figure 23. Intensity of the $A_g(2)$ peak attributed to the initial 2D-R polymer (open circles), the intermediate 2D-T and 1D-O oligomers (diamonds) and the C_{60} monomers/dimers (solid circles) normalized to their sum as a function of the treatment temperature for HTT lasted for 0.5 hour on pieces of fresh sample. The spectra were averaged from different sample sites and/or sample pieces, while the error bars refer to the standard error of the mean. The shaded area denotes the temperature region of the polymer decomposition; the lines through the data are guides to the eye.

To study the kinetics of the 2D-R polymer decomposition, the room temperature Raman spectra were measured for samples treated at various temperatures and heat treatment times. The Raman spectra of the 2D-R polymer in the frequency region of the $A_g(2)$ PP-mode after treatment at 513 K for 0.5-2.0 hours are illustrated in Figure 24. The treatment for 0.5 hour results in the decrease of the $A_g(2)$ mode intensity of the 2D-R polymer. The HTT for 1 hour results in further decrease of the $A_g(2)$ peak intensity and the appearance of $A_g(2)$ mode peaks associated with the presence of the 2D-T- and the 1D-O-like oligomers as well as with the monomeric C_{60}. Finally, the heat treatment for 2 hours leads to the practically complete decomposition of the 2D-R polymer as the $A_g(2)$ mode peak, attributed to the C_{60} monomer, dominates the Raman spectrum. Note that, the effect of heating has an additive character as there is no significant difference between the spectra recorded from a sample after continuous HTT at 513 K for a certain treatment time from those of a sample treated at the same temperature for the same total time but with intermediate cooling(s) to room temperature.

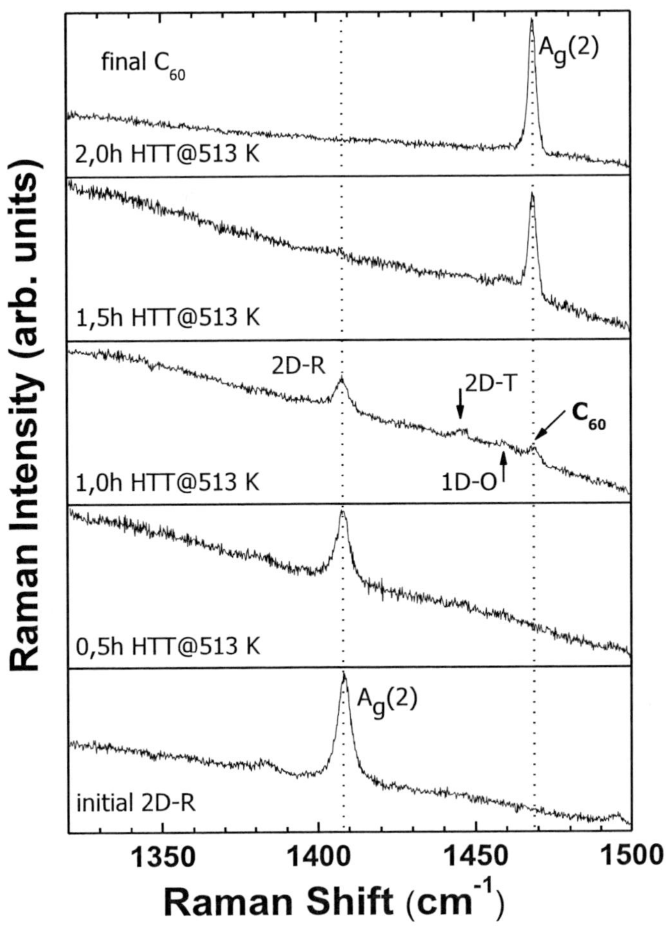

Figure 24. Raman spectra of the 2D-R polymer measured at room temperature in the high frequency region after HTT at 513 K for 0.5-2.0 hours. Arrows indicate the PP-modes of the intermediate 2D-T-like and 1D-O-like oligomers as well as of the monomer C_{60}. Vertical lines indicate the PP-mode of the initial 2D-R polymer and the final monomer C_{60}.

Figure 25 shows the intensity of the $A_g(2)$ mode of the 2D-R polymer normalized to the sum of the corresponding peaks associated with all polymeric/oligomeric phases of C_{60} as a function of the HTT time for different temperatures.

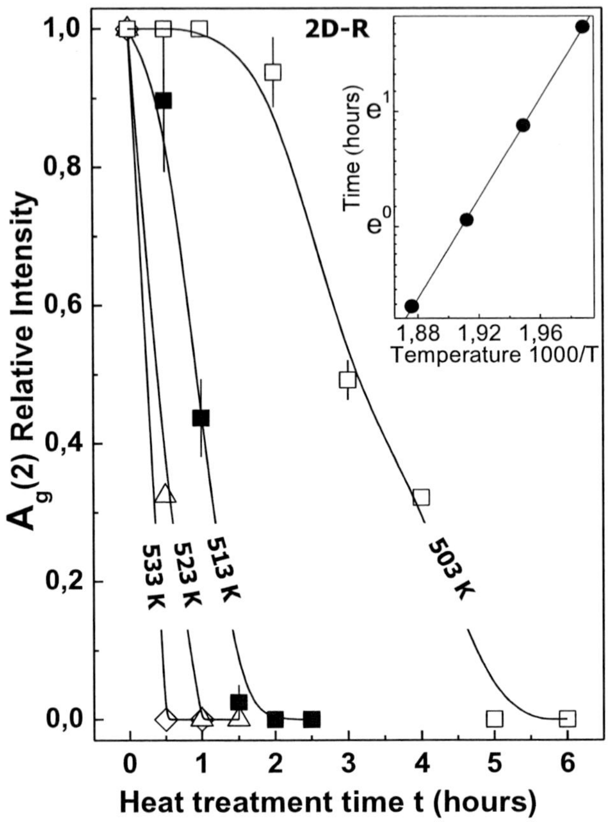

Figure 25. Intensity of the $A_g(2)$ peak of the initial 2D-R polymer normalized to the sum of the $A_g(2)$ peaks associated with the 2D-R polymer, 2D-T-like and 1D-O-like oligomers as well as with the monomer/dimer mixture of C_{60}, as a function of the heat treatment time at various temperatures. Diamonds, triangles, circles and squares correspond to treatment temperatures of 533, 523, 513 and 503 K, respectively. The data were averaged from different samples and the error bars refer to the standard error of the mean, while the lines are guides to the eye. Inset: Arrhenius plot of the polymer's complete decomposition time versus the treatment temperature.

The time required for the complete decomposition of the 2D-R polymer at 533 K is ~0.5 hour, while it increases at lower temperatures. The occurrence of polymer decomposition at temperatures as low as 503 K and the increase of the time needed for complete decomposition with decreasing treatment temperature, indicates the activation-type behavior of the polymer decomposition transition that is typical for chemical reactions. According to DSC measurements, the minimum of the total energy related to the *fcc* monomeric state of C_{60} is higher by ~0.13 eV/molecule than the total energy minimum related to the 2D-R polymer of C_{60} [65]. The stability of the C_{60} monomer at ambient conditions is due to the energy barrier that separates the polymeric state from the monomer state. The theoretically predicted energy barrier for the 2D-R polymer is 1.6 eV/molecule [59]. Another theoretical calculation of the energy barrier yields a value of 1.7 eV/molecule [68]. We can estimate the barrier value using our Raman data taking into account that the polymer decomposition time versus heat treatment temperature can be described by an Arrhenius equation:

$$\tau(T) = A \times \exp(E_A/k_B T)$$

where E_A is the activation energy (in fact the energy barrier between the polymer and monomer states), k_B is the Boltzmann constant, T is the treatment temperature and $\tau(T)$ is the temperature-dependent polymer decomposition time constant. The constant A, measured in time units, is related to the characteristic phonon frequency [69]. The Arrhenius plot of the polymer decomposition time versus the treatment temperature is shown in the inset of Figure 25. The experimental points exhibit a remarkably good linear dependence in logarithmic scale, yielding an activation energy of E_A=1.76±0.07 eV/molecule. This value is smaller than the value of 1.9±0.2 eV/molecule obtained in [69] from thermal expansion measurements on 2D-R polymeric powders. The activation energy obtained from the Raman experiments presented here is in good agreement with the calculated barrier energy; it is ~10% larger than the data reported in [59] and coincides within experimental accuracy with the data reported in [68]. It is important to note, that the enthalpy of the polymer decomposition extracted from DSC measurements, is not related to the energy barrier between the polymeric and the monomeric C_{60} state, but to the difference between the total energy minima. For example, the transition enthalpy for the 2D-R polymer obtained in [65] and [66] is equivalent to an energy difference between the two minima of ~0.13 eV/molecule and ~0.11 eV/molecule, respectively. We can estimate the complete decomposition time of the 2D-R polymer from the obtained data; at 300 K it is equal to ~4.49×10^8 years, but it collapses to ~62 hours at T=473 K. Thus, the fullerene polymers demonstrate rather large fragility at elevated temperatures that may be a reason preventing their practical use.

Conclusion

The crystalline polymers of C_{60} are not stable with respect to the high-pressure and high temperature treatment. The planar polymers show further polymerization at high pressure and room temperature. The Raman data provide a strong indication that the 2D-T polymer of C_{60} undergoes an irreversible phase transition above 20 GPa. The transformation takes place via a highly disordered pre-transitional state extending in a pressure range of ~4 GPa and having a rather diffuse Raman spectrum. The prominent Raman peaks of the high-pressure phase associated with the C_{60} molecular cage, as well as the irreversibility of the observed transformation, support the assumption of a further pressure-induced polymerization resulting in ordered 3D-cross-linked structure. This observation was confirmed in the X-ray study of the 2D-T polymer at pressure up to 37 GPa that revealed a transition at ~24 GPa associated with the formation of interlayer 3+3 cycloaddition along the body diagonal. The high-pressure 3D-polymeric phase demonstrates very high hardness; the bulk modulus determined from X-ray data is 407 GPa which is slightly smaller that that of the diamond (443 GPa).

The pressure behavior of the Raman modes the 2D-R polymer of C_{60} provides strong indication that the polymer undergoes an irreversible transformation at P~15 GPa. The initial well resolved Raman spectrum of the 2D-R polymer transforms to a diffuse one that is typical of a disordered state. Unlike to the 2D-T polymer, the high-pressure phase in 2D-R polymer is formed by the random creation of covalent bonds between adjacent polymeric layers due to various distortions in the molecular orientations.

In contradiction to the planar polymers of C_{60}, the linear orthorhombic polymer of C_{60} is not stable with respect to laser irradiation and transforms to a new polymeric phase. The laser intensity necessary for the photo-induced polymerization at ambient conditions is two orders of magnitude higher than that needed for the polymerization of the C_{60} monomer. The application of high pressure drastically increases the photo-polymerization rate and the transformation becomes almost instantaneous. The structural aspect of the photo-polymerization of the 1D-O polymer is related to conjugation of the adjacent polymeric chains. The high-pressure Raman study of the photo-transformed 1D-O polymer of C_{60} shows a possible structural phase transition at ~4 GPa, whereas further increase of pressure to ~15 GPa causes a transformation to a high-pressure phase with a diffuse spectrum typical of a disordered phase. As in the case of the 2D-R polymer, the high-pressure phase in photo-transformed 1D-O polymer is formed by the random creation of covalent bonds between polymeric layers due to some distortion in the molecular orientation. The high-pressure phases of linear and planar crystalline polymers of C_{60} recovered to ambient conditions are metastable and transform to a mixture of pristine and dimerized C_{60} under heating. This fact confirms the retention of the fullerene molecular cage in the high-pressure phases, while the similarity of the pre-transitional behavior of Raman spectra in linear and planar polymeric phases of C_{60} shows the common nature of the high-pressure phases related to their 3D cross-linked polymerization.

The crystalline linear and planar polymers of C_{60} show a gradual decomposition at high temperature. The 2D-R polymer composition changes drastically after treatment at ~560 K, resulting in the domination of C_{60} monomers with some inclusion of C_{60} dimers. The decomposition takes place through an intermediate state that, in fact, is a mixture of the initial 2D-R polymer, 2D-T- and 1D-O-like oligomers of C_{60} as well as monomer and dimer C_{60}. The final state of the decomposed 2D-R polymer is pure monomer C_{60} with, possibly, small inclusions of C_{60} dimers. The decomposition of the 2D-R polymer may occur at relatively low temperatures. The polymer decomposition time dependence on the treatment temperature is of the Arrhenius type while the energy barrier between 2D-R polymer and C_{60} monomer is 1.76±0.07 eV/molecule. The estimation of the complete decomposition time of the 2D-R polymer at 300 K gives ~4.49×10^{8} years while at T=473 K it collapses to ~62 hours. Thus, the fullerene polymers demonstrate rather large fragility at elevated temperatures that may be a reason preventing their practical use.

ACKNOWLEDGMENTS

The support of the Russian foundation for basic research, (Russia) and the General Secretariat for Research and Technology, (Greece) is gratefully acknowledged. The authors would like to thank Drs. J. Arvanitidis, D. Christofilos, K. Papagelis, S. Assimopoulos, A. Soldatov, T. Wägberg, S. Yamanaka, V. Davydov and V. Agafonov for their help with the preparation and characterization of various polymers of C_{60}, and Profs. Y. Iwasa, B. Sundqvist, K. Prassides and S. Ves for helpful collaboration.

References

[1] Rao, A. M.; Zhou, P.; Wang, K.-A.; Hager, G. T.; Holden, J. M.; Wang, Y.; Lee, W.-T.; Bi, X.-X.; Eklund, P. C.; Cornett, D. S.; Duncan, M. A.; Amster, I. J. *Science* 1993, 259, 955-957.

[2] Winter, J.; Kuzmany, H. *Solid State Commun.* 1992, 84, 935-938.

[3] Stephens, P. W.; Bortel, G.; Faigel, G.; Tegze, M.; Janossy, A.; Pekker, S.; Oszlanyi, G.; Forro, L. *Nature* 1994, 370, 636-639.

[4] Iwasa, Y.; Arima, T.; Fleming, R. M.; Siegrist, T.; Zhou, O.; Haddon, R. C.; Rothberg, L. J.; Lyons, K. B.; Carter, H. L.; Hebard, A. F.; Tycko, R.; Dabbagh, G.; Krajewski, J. J.; Thomas, G. A.; Yagi, T. *Science* 1994, 264, 1570-1572.

[5] Nunez-Regueiro, M. ; Marques, L.; Hodeau, J.-L. ; Bethoux, O.; Perroux, M. *Phys. Rev. Lett.* 1995,74, 278-281.

[6] Sundqvist, B. *Advances in Physics* 1999, 48, 1-134.

[7] Brazhkin, V. V.; Lyapin, A. G.; Popova, S. V.; *JETP Lett.* 1996, 64, 802-807.

[8] Blank, V.; Popov, M.; Buga, S.; Davydov, V.; Denisov, V. N.; Ivlev, A. N.; Marvin, B. N.; Agafonov, V.; Ceolin, R.; Szwarc, H.; Rassat, A. *Phys. Lett.* A 1994, 188, 281-286.

[9] Blank, V. D.; Buga, S. G.; Dubitsky, G. A.; Serebryanaya, N. R.; Popov M. Yu.; Sundqvist, B. Carbon 1998, 36, 319-343.

[10] Marques, L.; Mezouar, M.; Hodeau, J.-L.; Nunez-Regueiro, M.; Serebryanaya, N. R.; Ivdenko, V. A; Blank, V. D.; Dubitsky, G. A. *Science* 1999, 283, 1720-1723.

[11] Chernozatonskii, L. A.; Serebryanaya, N. R.; Mavrin, B. N.; *Chem. Phys. Lett.* 2000, 316, 199 -204.

[12] Moret, R. *Acta Cryst.* A 2005, 61, 62-76.

[13] Dzyabchenko, A. V.; Agafonov, V.; Davydov, V. A. *Crystallogr. Rep.* 1999, 44, 18-24.

[14] Ruoff, R. S.; Ruoff, A. L. *Nature* 1991, 350, 663-664.

[15] Okada, S.; Saito, S.; Oshiyama, A. *Phys. Rev. Lett.* 1999, 83, 1986-1989.

[16] Burgos, E.; Halac, E.; Weht, R.; Bonadeo, H.; Artacho, E.; Ordejon, P. *Phys. Rev. Lett.* 2000, 85, 2328-2331.

[17] Davydov, V. A.; Kashevarova, L. S.; Rakhmanina, A. V.; Senyavin, V. M.; Ceolin, R.; Szwarz, H.; Allouchi, H.; Agafonov, V. *Phys. Rev.* B, 2000, 61, 11936-11945.

[18] Arvanitidis, J.; Papagelis, K.; Meletov, K. P.; Kourouklis, G. A.; Ves, S.; Kordatos, K.; Wudl, F.; Prassides, K. *Phys. Rev.* B 1999, 59, 3180-3183.

[19] Kourouklis, G. A.; Ves, S.; Meletov, K. P. *Physica* B 1999, 265, 214-222.

[20] Wägberg, T.; Soldatov, A.; Sundqvist, B. *Eur. Phys. J.* B 2006, 49, 59-65.

[21] Senyavin, V. M.; Davydov, V. A.; Kashevarova, L. S.; Rakhmanina, A. V.; Agafonov, V.; Allouchi, H.; Ceolin, R.; Sagon, G.; Szwarc, H. *Chem. Phys. Lett.* 1999, 313, 421-425.

[22] Jayaraman, A. *Rev. Sci. Instrum.* 1986, 57, 1013-1031.

[23] Barnett, D.; Block, S.; Piermarini, G. J. *Rev. Sci. Instrum.* 1973, 44, 1-9.

[24] Davydov, V. A.; Kashevarova, L. S.; Rakhmanina, A. V.; Agafonov, V.; Allouchi, H.; Ceolin, R.; Dzyabchenko, A. V.; Senyavin, V. M.; Szwarc, H. *Phys. Rev.* B 1998, 58, 14786-14790.

[25] Winter, J.; Kuzmany, H.; Soldatov, A.; Persson, P.-A.; Jacobsson, P.; Sundqvist, B. *Phys. Rev.* B 1996, 54, 17486-17492.

[26] Persson, P.-A.; Edlund, U.; Jacobsson, P.; Johnels, D.; Soldatov, A.; Sundqvist, B. *Chem. Phys. Lett.* 1996, 258, 540-546.

[27] Porezag, D.; Pedersson, M. R.; Frauenheim, T.; Kohler, T. *Phys. Rev.* B 1995, 52, 14963-14970.

[28] Solin, S. A. ; Ramdas, A. K. *Phys. Rev.* B 1970, 1, 1687-1698.

[29] Arvanitidis, J.; Meletov, K. P.; Papagelis, K.; Ves, S.; Kourouklis, G. A.; Soldatov, A.; Prassides, K. *J. Chem. Phys.* 2001, 114, 9099-9104.

[30] Meletov, K. P.; Arvanitidis, J.; Kourouklis, G. A.; Iwasa, Y.; Prassides, K. *Chem. Phys. Lett.* 2002, 357, 303-315.

[31] Haines, J.; Leger, J. M. *Solid State Commun.* 1994, 90, 361-363.

[32] Goncharov, A. F.; Makarenko, I. N.; Stishov, S. M. *JETP Letters* 1985, 41, 184-187.

[33] Brazhkin, V. V.; Lyapin, A. G.; Popova, S. V.; Voloshin, R. N.; Antonov, Yu. V.; Lyapin, S. G.; Klyuev, Yu. A.; Naletov, A. M.; Mel'nik, N. N. *Phys. Rev.* B 1997, 56, 11465-11471.

[34] Brazhkin, V. V.; Lyapin, A. G.; Popova, S. V.; Klyuev, Yu. A.; Naletov, A. M. *J. Appl. Phys.* 1998, 84, 219-226.

[35] Weiler, M.; Sattel, S.; Giessen,T.; Jung, K.; Ehrhardt, H.; Veerasamy, V. S.; Robertson, *J. Phys. Rev.* B 1996, 53, 1594-1608.

[36] Nemanich, R. J.; Solin, S. A. *Phys. Rev.* B 1979, 20, 392-401.

[37] Shroder, R. E.; Nemanich, R. J.; Glass, J. T. *Phys. Rev.* B 1990, 41, 3738-3745. 38.

[38] Long, D. A. Raman Spectroscopy, McGraw-Hill: London, UK, 1976, p. 158-159.

[39] Dam Hieu Chi; Iwasa, Y.; Takano, T.; Watanuki, T.; Ohishi, Y.; Yamanaka, S. *Phys. Rev.* B 2003, 68, 153402/1-153402/4.

[40] Kawasaki, S.; Yao, A.; Matsuoka, Y.; Komiyama, S.; Okino, F.; Touhara, H.; Suito, K. *Solid State Commun.* 2003 125, 637-640, (2003).

[41] Leger, J. M.; Haines, J.; Davydov, V. A.; Agafonov, V. *Solid State Commun.* 2002, 121, 241-244.

[42] Meletov, K. P.; Arvanitidis, J.; Tsilika, I.; Assimopoulos, S.; Kourouklis, G. A.; Ves, S.; Soldatov, A.; Prassides, K. *Phys. Rev.* B 2001, 63, 054106/1-054106/4.

[43] Meletov, K. P.; Tsilika, I.; Assimopoulos, S.; Arvanitidis, J.; Kourouklis, G. A.; Ves, S.; Sundqvist, B.; Wägberg, T. *Chem. Phys. Lett.* 2001, 341, 435-441.

[44] Martin, M. C.; Du, X.; Kwon, J.; Mihaly, L. *Phys. Rev.* B 1994, **50**, 173-183.

[45] Meletov, K. P.; Arvanitidis, J.; Assimopoulos, S.; Kourouklis, G. A.; Sundqvist, B. *JETP* 2002, 95, 736-747.

[46] Kourouklis, G. A.; Meletov, K. P. New Diamond and Frontier Carbon Technology 2002, 12, 303-314.

[47] Meletov, K. P.; Davydov, V.A.; Rakhmanina, A. V.; Agafonov, V.; Kourouklis, G. A. *Chem. Phys. Lett.* 2005, 416, 220-224.

[48] Bethune, D. S.; Meijer, G.; Tang, W. C.; Rosen, H. J.; Golden, W. G.; Seki, H.; Brown, C. A.; de Vries, M. S. *Chem. Phys. Lett.* 1991, 179, 181-186.
[49] Fagerstrom, J.; Stafstrom, S. *Phys. Rev.* B 1996, 53, 13150-13158.
[50] Davydov, V.A.; Kashevarova, L. S.; Rakhmanina, A. V.; Agafonov, V.; Ceolin, R.; Szwarc, H. *JETP Lett.* 1998, 68, 928-933.
[51] Ciabini, L.; Santoro, M.; Bini, R.; Schettino, V. *Phys. Rev. Lett.* 2002, 88, 085505/1-085505/4.
[52] Le Parc, R.; Levelut, C.; Haines, J.; Davydov, V.A.; Rakhmanina, A. V.; Papoular R. J.; Belova, E. E.; Chernozatonskii, L. A.; Allouchi, H.; Agafonov, V. *Chem. Phys. Lett.* 2007, 438, 63-66.
[53] Negri, F.; Orlandi, G.; Zerbetto, F. *Chem. Phys. Lett.* 1998, 144, 31-37.
[54] Wang, Y. J Phys Chem 1992, 96, 764-767.
[55] Lane, P. A.; Swanson, L. S.; Ni, Q.-X.; Shinar, J.; Engel, J. P.; Barton, T. J.; Jones, L. *Phys. Rev. Lett.* 1992, 68, 887- 890.
[56] Saito, S.; Oshiyama, A. *Phys. Rev. Lett.* 1991, 66, 2637-2640.
[57] Meletov, K. P.; Dolganov, V. K.; Zharikov, O. V.; Kremenskaya, I. N.; Ossip'yan, Yu. A. *J. Phys. France* 1992, 2, 2097-2105.
[58] Guss, W.; Feldman, J.; Göbel, E. O.; Taliani, C.; Mohn, H.; Müller, W.; Häussler, P.; ter Meer, H.–U. *Phys. Rev. Lett.* 1994, 72, 2644-2647.
[59] Hui, C.; G. Scuseria, E. *Phys. Rev. Lett.* 1995, 74, 274-277.
[60] Okada, S.; Saito, S. *Phys. Rev.* B 1997, 55, 4039-4041.
[61] Okada, S.; Saito, S. *Phys. Rev.* B 1999, 59, 1930-1936.
[62] Venkateswaran, U. D.; Sanzi, D. S.; Krishnappa, J.; Marques, L.; Hodeau, J.-L.; Nunes-Regueiro, M.; Rao, A. M.; Eklund, P. C.; *Phys. Stat. Sol.* (b) 1996, 198, 545-552.
[63] Meletov, K. P.; Negrii, V. D. *JETP Lett.* 1998, 68, 248-252.
[64] Meletov, K. P.; Kourouklis, G. A. *Chem. Phys. Lett.* 2005, 403, 338-342.
[65] Iwasa, Y.; Tanoue, K.; Mitani, T.; Yagi, T. *Phys. Rev.* B 1998, 58, 16374-16377.
[66] Korobov, M. V.; Senyavin, V. M.; Bogachev, A. G.; Stukalin, E. B,; Davydov, V. A.; Kashevarova, L. S.; Rakhmanina, A. V.; Agafonov, V.; Szwarc, H. *Chem. Phys. Lett.* 2003, 381, 410-415.
[67] Korobov, M. V.; Bogachev, A. G.; Popov, A. A.; Senyavin, V. M.; Stukalin, E. B.; Dzyabchenko, A. V.; Davydov, V. A.; Kashevarova, L. S.; Rakhmanina, A. V.; Agafonov, V. Carbon 2005, 43, 954-961.
[68] Saito, S.; Okada, S. *AIP Conf. Proc.* 1998, 442, 198-202.
[69] Nagel, P.; Pasler, V.; Lebedkin, S.; Soldatov, A.; Meingast, C.; Sundqvist, B.; Persson, P.-A.; Tanaka, T.; Komatsu, K.; Buga, S.; Inaba, A. *Phys. Rev.* B 1999, 60, 16920-16927.

In: Polymer Phase Behavior
Editors: T. P. Ehlers and J. K. Wilhelm, pp. 123-169 ISBN: 978-1-61324-336-7
© 2011 Nova Science Publishers, Inc.

Chapter 3

POLYMER PHASE BEHAVIOR IN NANOCOMPOSITES

G. V. Kozlov[*]
Kabardino-Balkarian State University,
Russian Federation

ABSTRACT

Polymer phase (polymer matrix) behavior in nanocomposites in many respects defines the behavior of nanocomposite as a whole independently from the used nanofiller type (disperse particles, organoclay, nanotubes and so on). In connection with this it is necessary to account for polymer matrix structure changes at nanofiller introduction in initial matrix polymer. These changes can be realized with the aid of different processes, namely, crystallization, amorphous polymer phase structure change, interfacial regions formation. In its turn, such factors as polymer matrix chain flexibility, interfacial adhesion level, nanofiller particle shape and so on influence on characteristics and realization possibility of the mentioned processes. Hence, at polymer nanocomposites structure formation complex dynamics of polymer phase behavior in them is observed, that defines in the long run a nanocomposite properties. In this aspect particularly important is the role of interfacial regions, which are the same reinforcing element of structure in polymer nanocomposites as actually nanofiller. For semicrystalline polymer matrix nanofiller can play a nucleator role changing in reality the indicated matrices crystallinity degree. In the present review the quantitative relationships of the initial polymer characteristics and their modification at nanofiller introduction and their influence on nanocomposite final structure are considered.

Keywords: Nanocomposite, nanofiller, crystallization, polymer phase changes, interfacial adhesion, polymer chain flexibility.

[*] Chernyshevskiy st., 173, Nal'chik – 360004, KBR, Russian Federation. e-mail: i_dolbin@mail.ru.

INTRODUCTION

Polymer phase (polymer matrix) behavior has critical significance for the behavior of polymer nanocomposite as a whole. Widely applied earlier microcomposite models [1] showed their scanfiness, that resulted to the development of percolation [2] and fractal [3] models for polymer composites reinforcement description in the last years. The most essential distinction of the mentioned models is that circumstance, that the latter do not take into consideration filler (nanofiller) elasticity modulus at polymer composites most important characteristic - reinforcement degree determination, i.e. composites (nanocomposites) elasticity modulus enhancement in comparison with corresponding parameter for matrix polymer. It is obvious, that in such treatment polymer phase role in nanocomposites properties complex increases repeatedly.

The other important aspect of polymer nanocomposites theoretical description is complex dynamical nature of polymer phase in these materials. As a rule, this circumstance at present is not taken into consideration (in any case, properly). A nanofiller introduction in the initial matrix polymer always results to the essential changes of its structure on all its levels: molecular, topological and supramolecular (or, more precisely, suprasegmental). This process can be expressed by a number of effects: interfacial regions formation, crystallinity degree change, variation of local order level in polymer matrix and so on. The additional aspect of the problem is interconnection of the changes on all indicated structural levels. It is natural, that it complicates the theoretical description of polymer phase in nanocomposites behaviour. Let us also note, that interfacial regions in polymer nanocomposites represent, actually, a separate phase, whose characteristics are differed essentially from the bulk polymer matrix properties. So, the authors [4] showed experimentally, that for nanocomposites butadiene-styrene rubber/shungite interfacial regions elasticity modulus exceeded in 8 times the corresponding parameter for the bulk polymer matrix.

The adduced above a brief list of problems, arising at the description of polymer phase in nanocomposites structure and properties, makes obvious their solution complexity. Nevertheless, the development lately of such new for this field theoretical approaches as solid body synergetics, fractal analysis, cluster model of polymers amorphous state structure, percolation theory and so on makes possible quantitative description of all structural changes of matrix polymer at nanofiller introduction. The mentioned approaches would be used in the present review at the description of three main effects, realized in polymer matrix (crystallization, amorphous polymer phase structure change and interfacial regions formation) on the example of three main classes of polymer nanocomposites, filled with disperse particles, organoclay and carbon nanotubes.

FEATURES OF NANOCOMPOSITES SEMICRYSTALLINE MATRIX CRYSTALLIZATION

At present there are well known strong enough and diverse changes of polymers crystalline structure, occurring at the introduction in to them fillers of various sorts [5]. As a rule, these changes are described within the frameworks of polymers crystalline morphology. However, lately the fractal model was developed, which takes into consideration the whole complexity of polymers structure. It is supposed, that changes occur not only on supramolecular level of structure, but on molecular and topological levels [6]. Simultaneously it is necessary to take into account, that crystalline phase morphology variation causes noncrystalline regions structure changes [7]. As a rule, the introduction in semicrystalline polymer of particulate nanofillers results to polymer matrix crystallinity degree increase, since nanofiller particles serve as nucleators. Such effect was observed in nanocomposites high density polyethylene/calcium carbonate (HDPE/$CaCO_3$) [8, 9]. Let us note an important feature of semicrystalline polymers filling: filler introduction can result to both reduction and enhancement of polymer matrix crystallinity degree K. So, the authors [10] found out K decrease from 0.72 up to 0.38 at the introduction of carbon fibers (CF) in HDPE at its volume content φ_f=0.303. Therefore the authors [11] performed quantitative description of HDPE crystalline and amorphous phases structural changes at the introduction in to it disperse nanofiller $CaCO_3$.

Polymers crystallinity degree K change is closely connected with their crystallization kinetics, which can be described quantitatively by well-known Kolmogorov-Avrami equation [12]:

$$K = 1 - e^{-zt^n}, \tag{1}$$

where z is crystallization rate constant, t is crystallization process duration, n is Kolmogorov-Avrami exponent, characterizing nucleation type and growing crystalline structures form for the given polymer.

As it has been shown in paper [13], the exponent n is connected with fractal dimension D_{ch} of a chain part between local order domains (clusters) as follows:

$$n = 3(D_{ch} - 1) + 1. \tag{2}$$

D_{ch} value, which characterizes polymer molecular mobility level, can be calculated with the aid of the following equation [14]:

$$\frac{2}{\varphi_{cl}} = C_\infty^{D_{ch}}, \tag{3}$$

where φ_{cl} is local order domains (clusters) relative fraction, C_∞ is characteristic ratio, which is equal to 5.7 for HDPE [15].

In its turn, the parameter φ_{cl} was determined according to the following percolation equation [14]:

$$\varphi_{cl} = 0.03(1 - K)(T_m - T)^{0.55}, \tag{4}$$

where T_m and T are melting and testing temperatures, which are equal to 396 and 293 K, respectively.

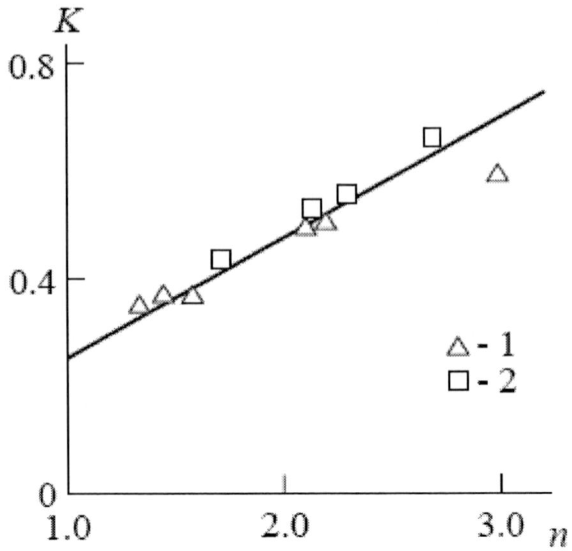

Figure 1. The dependence of crystallinity degree K on Kolmogorov-Avrami exponent n for carbon plastics HDPE/CF (1) and nanocomposites HDPE/$CaCO_3$ (2) [11].

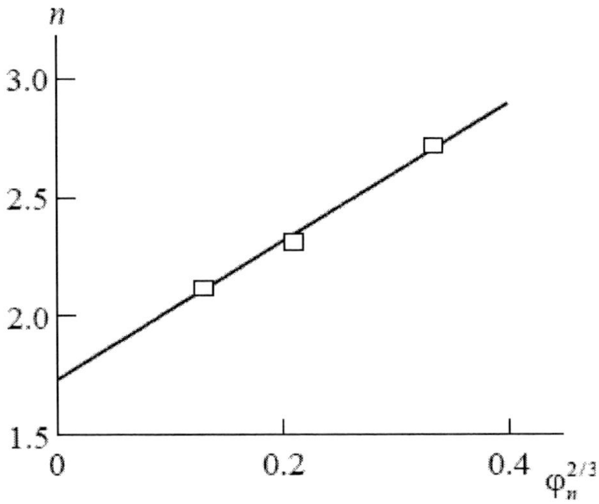

Figure 2. The dependence of Kolmogorov-Avrami exponent n on $CaCO_3$ contents φ_n for nanocomposites HDPE/$CaCO_3$ [11].

In Figure 1 the dependence $K(n)$ is adduced for microcomposites HDPE/CF according to the data of paper [10] and nanocomposites HDPE/$CaCO_3$. As one can see, despite the opposite K changes at filler contents increase (CF contents increase results to K reduction [10] and $CaCO_3$ contents increase – to its enhancement [9]), the indicated dependence is described by a common linear correlation, which allows to determine the greatest possible K variation for the considered materials: K=0.25 at n=1 (or D_{ch}=1.0) and K=0.90 at n=4 (or D_{ch}=2.0).

Since the nanofiller $CaCO_3$ is simultaneously a nucleator in nanocomposite matrix crystallization process then one should expect the exponent n increase at its contents φ_n growth. In Figure 2 the dependence $n(\varphi_n^{2/3})$ is adduced, where the value $\varphi_n^{2/3}$ characterizes particles $CaCO_3$ surface total area, which has shown linear growth n at $\varphi_n^{2/3}$ increase (or φ_n enhancement). At φ_n=0 the dependence $n(\varphi_n^{2/3})$ extrapolates to $n\approx1.72$, that is Kolmogorov-Avrami exponent for the initial HDPE and at $\varphi_n\approx0.7$ the greatest value n=4 is reached [12]. It is significant that the indicated value $\varphi_n\approx0.7$ corresponds well to the magnitude φ_n=0.6-0.7, which gives top nucleation effect [8]. Let us note, that for CF in paper [10] the opposite effect was obtained, namely, n reduction at φ_f growth. Such discrepancy is explained by different surface structure of using fillers – CF and $CaCO_3$. If in the first case fibers have smooth surface with dimension $d_{surf}\approx2.0$, then in the second case $CaCO_3$ disperse particles possess rough enough surface with dimension d_{surf}, close to 2.5 [16]. As Pfeifer [17] has shown, such difference of filler surface structure defines conformations difference of adjoining to surface macromolecular coils – they are stretched (are straightened) on smooth surface and maintain the initial conformation of statistical coil on a rough one. In its turn, this defines D_{ch} reduction in the first case and constant value or this dimension increase – in the second one. For $CaCO_3$ the dependence of n on its contents φ_n can be described by the following empirical equation [11]:

$$n = n_{\text{HDPE}} + 2.90\varphi_n, \tag{5}$$

where n_{HDPE} is n value for the initial HDPE, which is equal to 1.72.

In paper [10] for K calculation according to Kolmogorov-Avrami equation the constants following values were used: z=0.145 and t=2. In Figure 3 the comparison of experimental K and calculated according to the equation (1) K^T crystallinity degree is adduced for microcomposites HDPE/CF and nanocomposites HDPE/$CaCO_3$, moreover for both indicated materials the same constants z and t, adduced above, were used. As one can see, the theory and experiment good correspondence was obtained (the average discrepancy of K^T and K makes up ~ 10 %). From the equations (1) and (2) it follows, that K^T value is this case is defined by only fractal dimension D_{ch}, which, in its turn, depends on polymer matrix macromolecular coils conformation, which is formed at contact with filler surface [17]. Let us note that in this case filler type is of no importance. So, n reduction effect at φ_n increase (and, accordingly, K reduction) was observed for the system polyamide-6/quartz powder (particulate filler) [18].

As it is known [12], the exponent n defines the forming crystalline phase morphology of polymers. In case of athermic nucleation at $n\leq2$ ribbons are formed by a two-dimensional growth mechanism, at $n\leq3$ – circles and at $n>3$ – spheres. Fractional values n mean a combined mechanism of thermal/athermic nucleation, the fractional part decreasing indicates on athermic mechanism role enhancement, at that i.e. intensification of all crystallites growth simultaneous beginning [12]. As it follows from the data of Figure 2, precisely such effect is observed at φ_n growth, as was expected, since $CaCO_3$ is a nucleator [8, 9]. As electron microscopy data showed in case of nanocomposites HDPE/$CaCO_3$, the formation of large spherolites from flat fan-like lamellas was observed for the initial HDPE, whereas for nanocomposite HDPE/$CaCO_3$ with $CaCO_3$ content of 20 mass % a small pronounced bulk spherolites growth was realized [8].

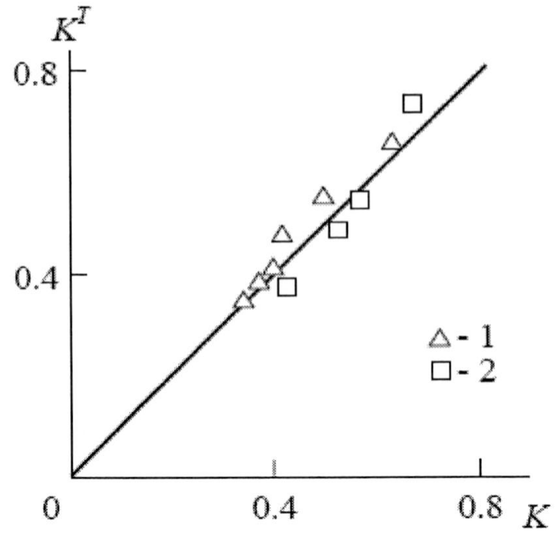

Figure 3. The comparison of experimental K and calculated according to the equation (1) K^T crystallinity degree for carbon plastics HDPE/CF (1) and nanocomposites HDPE/$CaCO_3$ (2) [11].

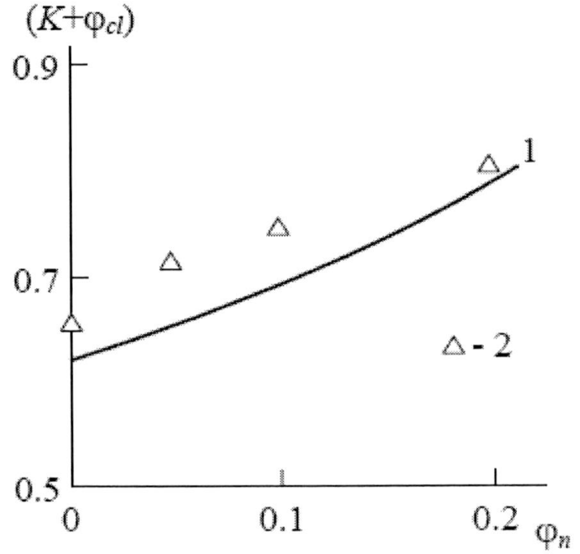

Figure 4. The dependences of densely-packed regions relative fraction $(K+\varphi_{cl})$ (1) and reduced sum $(K+\varphi_{cl})^{red}$ on $CaCO_3$ contents φ_n for nanocomposites HDPE/$CaCO_3$ [11].

The value φ_{cl} calculation according to the equation (4) proved found out its reduction from 0.23 up to 0.14 at $CaCO_3$ contents growth from zero up to 20 mass %, that is accompanied by crystallinity degree enhancement from 0.43 up to 0.67 [9]. In Figure 4 the dependence of the sum $(K+\varphi_{cl})$ on φ_n is adduced, which shows weak increase at φ_n growth. As it is known [19], the greatest relative fraction of densely-packed domains, i.e. in our case $(K+\varphi_{cl})$, can be calculated within the frameworks of thermal cluster concept:

$$(K+\varphi_{cl})=\left(\frac{T_m-T}{T_m}\right)^{0.37},\tag{6}$$

where the exponent 0.37 is equal to minimum value of percolation cluster order exponent [20].

The calculation of $(K+\varphi_{cl})$ value according to the equation (6) should take into consideration nanofiller availability in polymer nanocomposite and then this sum reduced value $(K+\varphi_{cl})^{red}$ was used:

$$(K+\varphi_{cl})^{red}=\frac{K+\varphi_{cl}}{1-\varphi_n}.\tag{7}$$

In Figure 4 the dependence of $(K+\varphi_{cl})^{red}$ on φ_n is shown by points and its comparison with the experimental dependence gives a good correspondence (the average discrepancy is within the limits of 7 %).

If for nanocomposites HDPE/$CaCO_3$ K increase at φ_n growth is observed (Figure 2), then for nanocomposites HDPE/Na^+-montmorillonite (HDPE/MMT) the opposite tendency was found, namely, crystallinity degree reduction from 0.50 up to 0.37 at organoclay mass contents enhancement up to 10 mass % [21]. Therefore the authors [22] studied the causes of nanocomposites HDPE/MMT crystallinity degree reduction at nanofiller contents growth within the frameworks of the considered above model [6, 11].

For analysis of value K variation the equations (1)-(3) were used, in the last of them the value φ_{cl} was replaced on densely-packed regions relative fraction φ_{dens}. As it is known [16], in nanocomposites polymer/organoclay local order domains (clusters) and interfacial regions, which are formed on organoclay platelets surface, with relative fractions φ_{cl} and φ_{if}, accordingly. The value φ_{cl} was calculated according to the percolation relationship (4) and interfacial regions relative fraction φ_{if} can be determined with the aid of the equation [16]:

$$\frac{E_n}{E_m}=1+11(\varphi_n+\varphi_{if})^{1.7},\tag{8}$$

where E_n and E_m are elasticity moduli of nanocomposite and matrix polymer, accordingly, and the value φ_n can be estimated as follows [23]:

$$\varphi_n=\frac{W_n}{\rho_n},\tag{9}$$

where W_n is organoclay mass contents, ρ_n is nanofiller density, which is equal to about 1800 kg/m^3 for Na^+-montmorillonite [16].

In Figure 5 the dependence of crystallinity degree K on the obtained experimentally Kolmogorov-Avrami exponent values n, accepted according to the data of paper [21], is adduced. As one can see, the linear increase K at n growth is observed, which allows to determine the limiting values K for the considered nanocomposites: K=0.33 at n=1 (or D_{ch}=1.0) and K=0.58 at n=4 (or D_{ch}=2.0).

In Figure 6 the comparison of the obtained experimentally n and calculated according to the equation (2) n^T Kolmogorov-Avrami exponent values for nanocomposites HDPE/MMT is adduced. As one can see, the good enough correspondence between theory and experiment was obtained (the average discrepancy of n and n^T makes up 12 %). The comparison of Figures 5 and 6 demonstrates clearly the cause of K reduction at W_n growth in case of the nanocomposites, filled with organoclay.

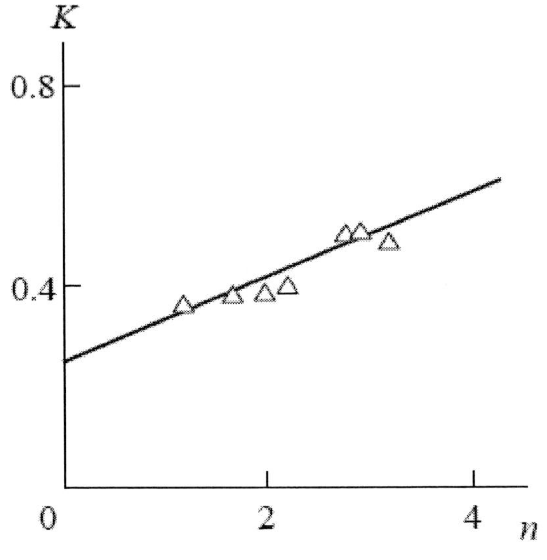

Figure 5. The dependence of crystallinity degree K on Kolmogorov-Avrami exponent n for nanocomposites HDPE/MMT and HDPE-M/MMT (HDPE-M is maleated HDPE) [22].

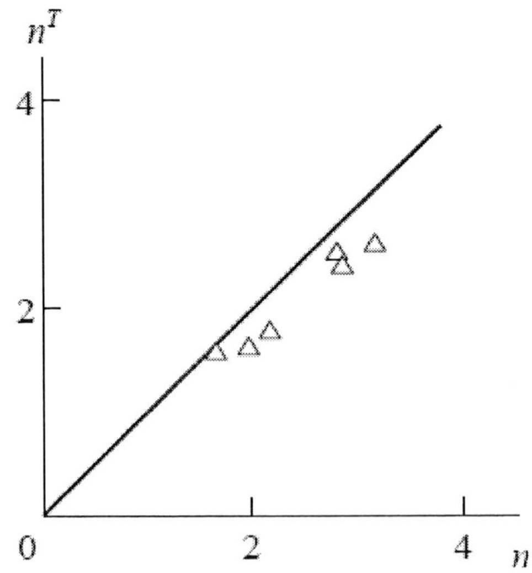

Figure 6. The comparison of experimental n and calculated according to the equation (2) n^T Kolmogorov-Avrami exponent values for nanocomposites HDPE/MMT and HDPE-M/MMT [22].

Organoclay introduction in matrix polyethylene results to densely-packed interfacial regions formation, relative fraction of which increases essentially at φ_n growth. So, the authors [16] obtained the following relationships between φ_{if} and φ_n:

$$\varphi_{if} = 1.910\varphi_n \tag{10}$$

for exfoliated organoclay and

$$\varphi_{if} = 0.955\varphi_n \tag{11}$$

for intercalated one.

The increase φ_{if} results to densely-packed regions relative fraction φ_{dens} (φ_{dens}=φ_{cl}+φ_{if}) growth, reduction of polymer phase molecular mobility level, characterized by dimension D_{ch}, according to the equation (3) and the exponent n decrease according to the equation (2), that in its turn reduces K value (see the equation (1) and Figure 5). Simple empirical correlation can be obtained between parameter n and φ_n (Figure 3) [22]:

$$n = n_{\text{HDPE}} - 40.5\varphi_n, \tag{12}$$

where n_{HDPE} is Kolmogorov-Avrami exponent for matrix polyethylene, which is equal to 3.25 in the considered case.

The equations (5) and (12) comparison shows, that unlike disperse nanofiller ($CaCO_3$), which is a nucleator [8, 9, 11], organoclay reduces the value n and, hence, K (Figure 5). Besides, organoclay action at n reduction is stronger by more than an order of the same $CaCO_3$ effect in the Kolmogorov-Avrami exponent enhancement.

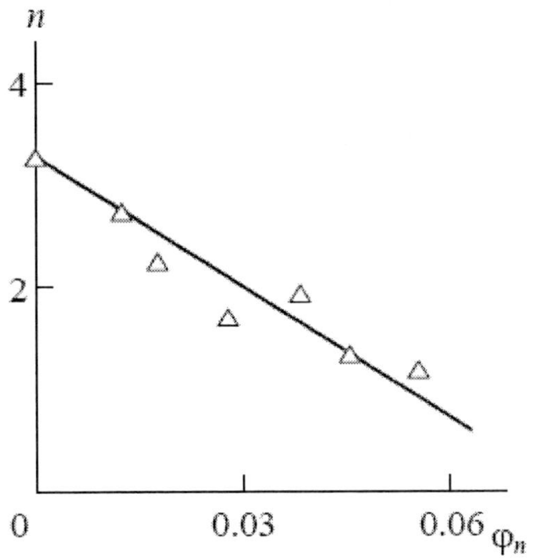

Figure 7. The dependence of Kolmogorov-Avrami exponent n on MMT volume contents φ_n for nanocomposites HDPE-M/MMT [22].

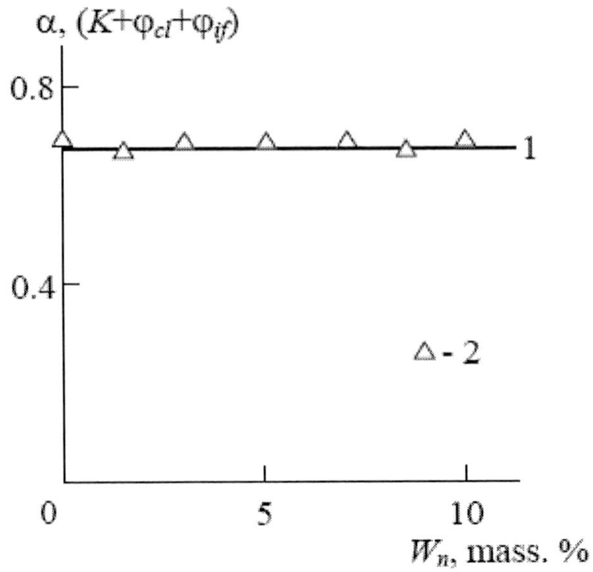

Figure 8. The dependence of parameters α (1) and $(K+\varphi_{cl}+\varphi_{if})$ (2) on MMT mass contents W_n for nanocomposites HDPE-M/MMT [22].

As it was noted above, the exponent n defined polymers forming crystalline phase morphology [12]. As it follows from the data of Figure 7, at Na^+-montmorillonite contents increase crystalline phase morphology transition from three-dimensional spherolites to one-dimensional ribbon structures is observed. This is explained by a flat structure of silicate platelets, on the surface of which crystallites are formed by one-dimensional growth mechanism, i.e. by epitaxial crystallization [24].

There is one more structural treatment, explaining nanocomposites crystalllinity degree reduction at organoclay contents growth – a thermal cluster model [25]. According to this model densely-packed regions contents, in the capacity of which the sum $(K+\varphi_{cl}+\varphi_{if})$ should be considered, cannot exceed definite value α, given by the modified equation (6) [25]:

$$\left(K+\varphi_{cl}+\varphi_{if}\right)\leq\alpha=\left(\frac{T_m-T}{T_m}\right)^{0.37}. \tag{13}$$

In Figure 8 the comparison of parameters $(K+\varphi_{cl}+\varphi_{if})$ and α as a function W_n for the considered nanocomposites is adduced, from which their approximate equality follows. Hence, with the appreciation of the condition $(K+\varphi_{cl}+\varphi_{if})\approx\alpha=$const φ_n increase, resulting to φ_{if} growth (see the equations (10) and (11)) should give K reduction, that is observed experimentally. Let us note, that at W_n=10 mass % calculation according to the equation (10) shows φ_{if} increase on the value 0.11, that corresponds to K decrease on the value 0.13.

Hence, the stated above results demonstrate decisive influence of molecular mobility level on crystallinity degree value for nanocomposites with semicrystalline polymer matrix. In its turn, the fractal dimension of a chain part between clusters, characterizing the mentioned mobility, is defined by nanofiller particles surface structure and interfacial regions formation mechanism. The offered fractal model of crystallization process is universal for polymer composites (nanocomposites) with semicrystalline matrix irrespective of nanofiller type [11, 22].

As it was noted above, in the filling process changes occur and in the molecular level, that cannot influence on polymer nanocomposites crystallization process. The authors [26] studied polymer phase chains statistical flexibility on $CaCO_3$ particles efficiency as nucleator on the example of nanocomposites HDPE/$CaCO_3$. For nucleation efficiency by $CaCO_3$ particles N determination the method, based on the application of usual cycles of DSC cooling was used. The value N was determined according to the equation [8]:

$$N = 100 \times \frac{T_{\mathrm{CNA}} - T_c}{T_{c\max} - T_c}, \qquad (14)$$

where T_{CNA} is the peak crystallization temperature of the polymer with the nucleating agent, T_c is the crystallization temperature, and $T_{c\max}$ is the optimum self-nucleation temperature.

As the results of papers [8, 9] showed, the introduction in HDPE of both micro- and nanoparticles $CaCO_3$ results to simultaneous increase of both crystallinity degree K (K=0.43-0.67) and nucleation efficiency N (N=35-52 %). Let us note, that $CaCO_3$ microparticles with diameter of 1200 nm give higher values N (N=44-52 %) and nanoparticles $CaCO_3$ with diameter of 50-60 nm give higher crystallinity degree. The common physical principles of these changes of parameters K and N can be determined within the frameworks of the model [6], according to which the value K depends on polymer chain flexibility and is determined as follows:

$$K = 0.32 C_\infty^{1/3}, \qquad (15)$$

where C_∞ is characteristic ratio, which serves as polymer chain statistical flexibility indicator [27].

The change of the molecular characteristic C_∞ at $CaCO_3$ particles introduction in HDPE can be estimated as follows. For this purpose it is necessary to calculate fractal (Hausdorff) dimension d_f of nanocomposites structure, that can be made according to the equation [28]:

$$d_f = (d-1)(1+\nu), \qquad (16)$$

where d is dimension of Euclidean space, in which a fractal is considered (it is obvious, that in our case d=3), ν is Poisson's ratio, which is estimated by mechanical tests results with the aid of the relationship [29]:

$$\frac{\sigma_Y}{E_n} = \frac{1-2\nu}{6(1+\nu)}, \qquad (17)$$

where σ_Y is yield stress, E_n is nanocomposite elasticity modulus.

The value C_∞ can be calculated according to the following equation [14]:

$$C_\infty = \frac{2d_f}{d(d-1)(d-d_f)} + \frac{4}{3}. \tag{18}$$

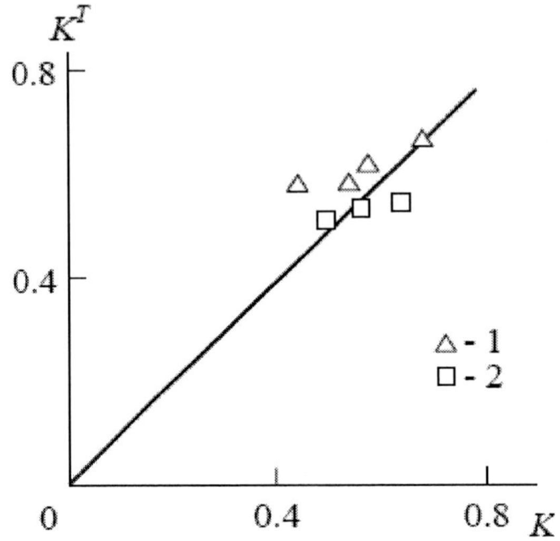

Figure 9. The comparison of experimental K and calculated according to the equation (15) K^T crystallinity degree values for nanocomposites HDPE/$CaCO_3$ with particles size of 1200 nm (1) and 50-60 nm (2) [26].

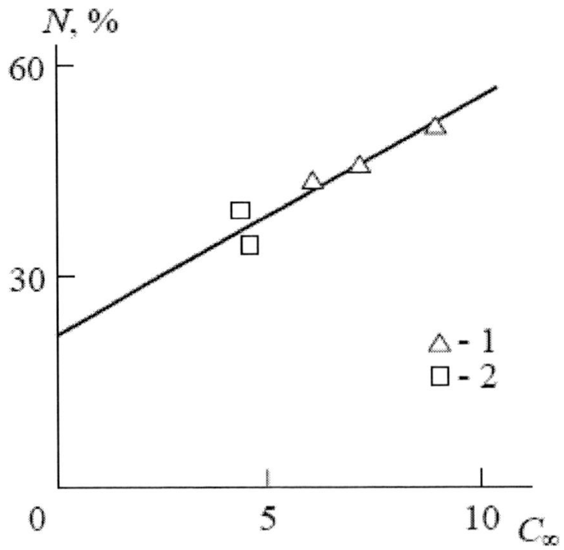

Figure 10. The dependence of nucleation efficiency N on characteristic ratio C_∞ for nanocomposites HDPE/$CaCO_3$ with particles size of 1200 nm (1) and 50-60 nm (2) [26].

The comparison of experimental K and calculated according to the equation (15) crystallinity degree values is adduced in Figure 9. As one can see, the good enough correspondence of theory and experiment is obtained (the average discrepancy of K and K^T makes up 10 %). Thus, the adduced above results demonstrate, that crystallinity degree change is due to molecular characteristic (polymer phase chain flexibility) variation, which is due to the disperse filler (nanofiller) introduction.

In Figure 10 the dependence of nucleation efficiency N on characteristic ratio C_∞ value for nanocomposites HDPE/$CaCO_3$ is adduced. As one can see, the linear correlation $N(C_\infty)$ is obtained, which can be approximated by the following empirical equation [26]:

$$N = 21 + 3.5C_\infty, \%. \tag{19}$$

Let us consider two limiting cases of the dependence $N(C_\infty)$, which can be obtained by the indicated dependence extrapolation to C_∞=0 and N=100 %. The value C_∞ according to the equation (18) cannot be equal to zero (for this the dimension d_f should be a negative one, that has no physical significance) and at d_f=0 its minimum value makes up 4/3. Thus, from the plot of Figure 10 it follows that minimum value of nucleation efficiency for $CaCO_3$, modified by stearic acid, in HDPE matrix makes up ~ 26 %. According to the definition [30] the value C_∞ can be written as follows:

$$C_\infty = \frac{\langle R_g^2 \rangle}{n\langle l_0^2 \rangle}, \tag{20}$$

where $\langle R_g \rangle$ is molecule gyration radius, n and $\langle l_0 \rangle$ are a number and mean length of molecule skeletal bond, accordingly.

The condition d_f=0 at d=3 is true for point object [31]. This, means, that in the equation (20) n=1 and then $C_\infty = \langle R_g^2 \rangle / \langle l_0^2 \rangle$, where l_0=1.54 Å for polyethylenes [32]. At minimum value C_∞=4/3 and l_0=1.54 Å the value R_g=1.778 Å or such low-molecular particle diameter will be equal to ~ 3.56 Å, that is a typical size for particles of such kind.

According to the equation (19), at N=100 % the value C_∞=22.6 and then the calculation according to the equation (18) gives corresponding to the indicated magnitude C_∞ value d_f=2.954. This value d_f practically precisely corresponds to maximum for real solids, i.e. for ideal rubbers, fractal dimension d_f=2.95 [28]. Hence, the equation (19) includes the entire structural range for polymer materials, beginning from low-molecular particles (for example, monomers) up to the ideal rubbers (d_f=0-2.95).

The authors [33] carried out the study of the efficiency of nucleation by nanoparticles $CaCO_3$ on the example of nanocomposites LDPE/$CaCO_3$ [34]. Since the values N and K are defined by polymer phase chain flexibility, characterized by parameter C_∞ (the equations (19) and (15), accordingly), then correlation between them should be expected, the indicated parameters characterize one and the same crystallization process at that. The estimations showed that $CaCO_3$ contents enhacement in nanocomposites LDPE/$CaCO_3$ up to 50 mass % resulted to C_∞ growth from 2.18 (for the initial LDPE) up to 6.0. According to the equations (19) and (15) this means N increase from 28.6 up to 42 % and K – from 0.389 up to 0.545. This assumes, that nanoparticles $CaCO_3$ are nucleation centers for crystalline regions in nanocomposites LDPE/$CaCO_3$ and their efficiency in this quality grows at $CaCO_3$ cotents increase, resulting in the end to crystallinity degree K enhancement.

In Figure 11 the correlation $K(N)$ for the considered nanocomposites is adduced, which has proved to be linear and pass through coordinates origin. The indicated correlation can be expressed analytically as follows [33]:

$$K = 0.013N, \quad (21)$$

where N is given in per cent.

The equation (21) allows to estimate the limiting characteristics of crystallization process for LDPE. At K=1.0 let us obtain the greatest value N=76.9 %, that according to the equation (19) corresponds to the greatest magnitude C_∞=16. According to the equation (15) the greatest value K can be determined as equal to 0.77. The minimum value C_∞ is accepted equal to 2 [27]. This allows to estimate the minimum magnitudes N and K as equal to 28 % and 0.40, respectively. The limiting values K=0.40-0.77 correspond well to the literary data, where K=0.50-0.65 for LDPE [35]. As it was expected, theoretical range of K variation is somewhat wider than a realized one in practice.

Let us estimate further the conditions, at which the greatest crystallinity degree, i.e. K=0.77, can be received for LDPE. Polymers glass transition temperature T_g can be estimated according to the empirical equation [14]:

$$T_g = 129\left(\frac{S}{C_\infty}\right)^{1/2}, \quad \text{K}, \quad (22)$$

where S is macromolecule cross-sectional area, which is equal to 14.35 Å^2 for polyethylenes [36].

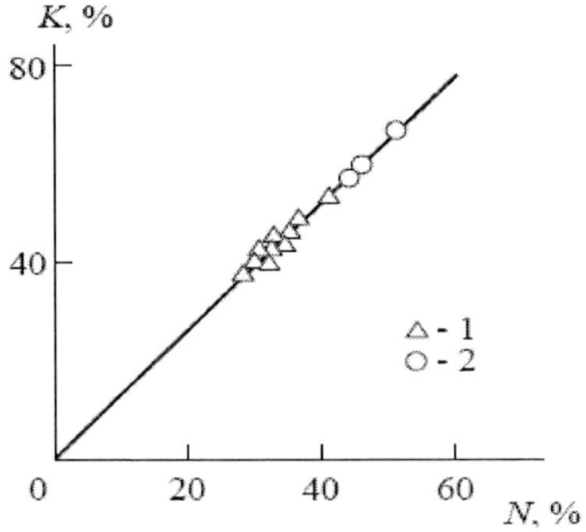

Figure 11. The dependence of crystallinity degree K on nucleation efficiency N for nanocomposites HDPE/$CaCO_3$ (1) and composites HDPE/$CaCO_3$ (2) [33].

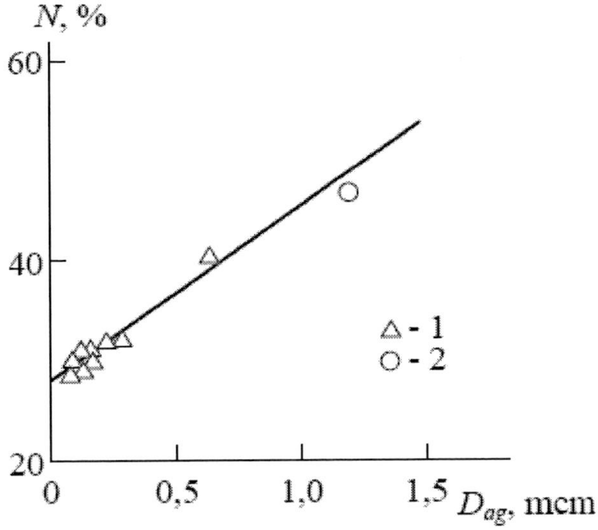

Figure 12. The dependence of nucleation efficiency N on particles (aggregates of particles) size D_{ag} for nanocomposites HDPE/$CaCO_3$ (1) and composites HDPE/$CaCO_3$ (2) [33].

At present T_g estimations for polyethylenes are contradictory enough: T_g=170-250 K. The calculation C_∞ according to the equation (22) in this case gives limiting values 8.25-3.82 and K limiting magnitudes according to the equation (15) 0.50-0.65, that corresponds excellently to the literary data [35]. For the greatest value K=0.77 reaching the condition C_∞=16 (see above) and value T_c=122 K realization is required [33].

Let us note that N enhancement is not due to $CaCO_3$ contents increase, but to its particles aggregates diameter D_{ag} growth, which can be calculated according to the technique [16]. In Figure 11 the experimental data for composites HDPE/$CaCO_3$ [8] are adduced, which correspond well to the present paper results. Although for the latter the values N are higher, than for nanocomposites LDPE/$CaCO_3$, but $CaCO_3$ contents in them is smaller (5-20 mass %) [8]. At the same time $CaCO_3$ particles size in composites HDPE/$CaCO_3$ makes up 1200 nm. In Figure 12 the dependence $N(D_{ag})$ for both types of the indicated composites is adduced, which is approximated well by one linear correlation, described by the equation [33]:

$$N = 28 + 18D_{ag}, \%, \qquad (23)$$

where D_{ag} is given in mcm.

The reasons of the shown in Figure 12 correlation are obvious enough. Firstly, the larger D_{ag}, the is smoother nanofiller particles (aggregates of particles) surface. Secondly, the higher D_{ag} is, the smaller nanofiller particles surface fractal dimension d_{surf} is (for D_{ag}=80 nm d_{surf}=2.6 and for D_{ag}=1.20 mcm d_{surf}=2.2), i.e. the surface is less rough. Both indicated factors create conditions for polymer phase chains “stretching” on filler particles surface, that increases C_∞ and, hence, N.

Thus, the adduced above data have shown, that particles $CaCO_3$ are crystallization effective nucleator for polyethylenes. The contents $CaCO_3$ increase results to nucleation efficiency N growth and crystallinity degree enhancement for nanocomposites polyethylene/$CaCO_3$. The polymer chain flexibility is a decisive factor in this process. The value N grows at nanofiller particles aggregation intensification, which accompanies $CaCO_3$ contents enhancement, and corresponds to the increase of $CaCO_3$ particles aggregates diameter [33].

The Influence of Polymer Phase Molecular Characteristics Change on Nanocomposites Mechanical Properties

At present there are many enough experimental observations, supposing influence of polymer phase molecular characteristics on reinforcement degree E_n/E_m of nanocomposites. It was noticed, that the best results in nanocomposites with layered nanofiller were obtained for matrices, consisting of rigid-chain polymers. So, in Figure 13 the dependences of reinforcement degree E_n/E_m on montmorillonite volume contents φ_n for nanocomposites on the basis of poly(butylene terephtalate) [37] and thermotropic liquid crystalline polyester [38] are adduced. As one can see, the value E_n/E_m increases at φ_n growth much faster for the second of the indicated nanocomposites. The authors [39] received for nanocomposites on the basis of polyimide record reinforcement degree - $E_n/E_m \approx 3.7$ at montmorillonite content 3 mass % only. These results did not obtain proper explanation and even did not attract any attention, since the authors confined themselves to the fact statement and general assumptions.

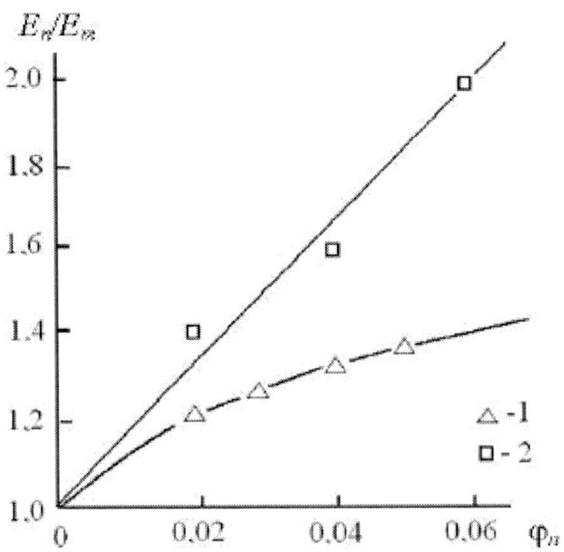

Figure 13. The dependences of reinforcement degree E_n/E_m on montmorillonite contents φ_n for nanocomposites on the basis of poly(butylene terephthalate) (1) [37] and thermotropic liquid crystalline polyester (2) [38].

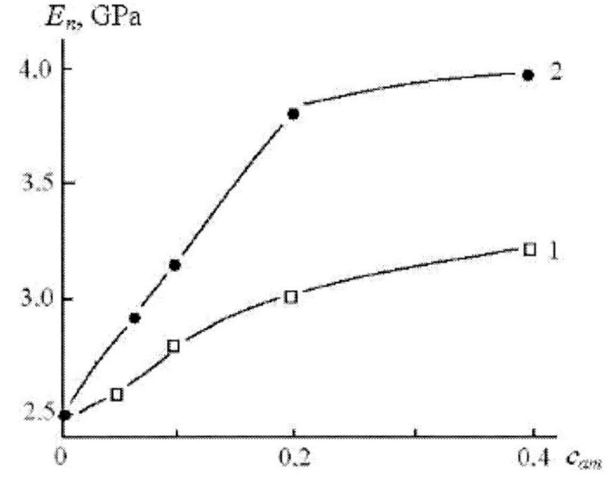

Figure 14. The dependence of elasticity modulus E_n on amine groups ratio c_{am} for nanocomposites on the basis of polyimide at testing temperature 293 (1) and 373 K (2) [41].

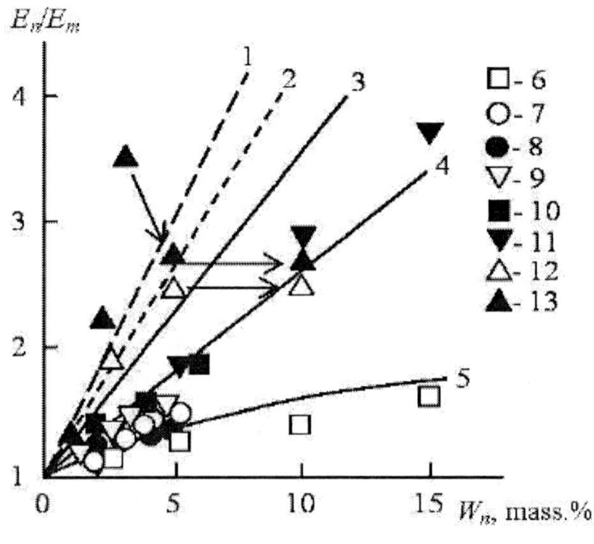

Figure 15. The dependences of reinforcement degree E_n/E_m on filling degree W_n for nanocomposites, filled with Na^+-montmorillonite. 1-5 – the theoretical dependences, corresponding to the equations (24) (1, 2) and (25) (3, 4) at L/d_{pl}=100 (1, 3) and 50 (2, 4) and to the equation (26) (5). 6-13 – the experimental data for nanocomposites on the basis of epoxy polymer at $T<T_g$ (6), polyamide-6 (7), poly(butylene terephthalate) (8), polycarbonate (9), thermotropic liquid crystalline polyester (10), epoxy polymer at $T>T_g$ (11), polypropylene (12) and polyimide (13) [16].

Another interesting experimental observation also remains without proper attention. As it is known, the reinforcement degree of the same polymer is essentially higher in a rubbery state than in a glassy one. For example, for nanocomposites on the basis of epoxy polymer the value E_n/E_m in glassy state does not exceed ~ 1.65, whereas at the temperature, exceeding on 40 K the glass transition temperature, the value E_n/E_m for the same nanocomposite achieves ~ 3.8 [40]. The indicated effect is also observed without nanocomposites polymer matrix phase state change, but simply at testing temperature raising. As the example in Figure 14 the dependences of elasticity modulus E_n on amine groups ratio for nanocomposites on the basis of polyimide for two testing temperatures (293 and 373 K) are adduced. As one can see, the value E_n/E_m is essentially higher at T=373 [16]. The stated results assumed, that polymer phase molecular characteristics is a key parameter at nanocomposites reinforcement description.

In virtue of strong anisotropy of layered silicates particles shape Halpin-Tsai and Mori-Tanaka models are often used for theoretical estimation of filled with them nanocomposites reinforcement degree [23]. Halpin-Tsai equation and the closed-form solution based on Mori-Tanaka model have the following form [23]:

$$\frac{E_n}{E_m} = \frac{1+2(L/d_{pl})\varphi_n\eta}{1-\varphi_n\eta}, \tag{24}$$

$$\eta = \frac{(E_n/E_m)-1}{(E_n/E_m)+2(L/d_{pl})},$$

and

$$\frac{E_n}{E_m} = \frac{1}{1-\varphi_n\xi}, \tag{25}$$

$$\xi = \frac{-2\nu_m A_3 + (1-\nu_m)A_4 + (1+\nu_m)A_5 A}{-2A},$$

where L/d_{pl} is an aspect ratio of anisotropic particles (L is length, d_{pl} is thickness of silicate platelet), ξ is a positive coefficient depending on the matrix Poisson's ratio ν_m and constants A and A_i, which can be calculated from the matrix/particle properties and components of Eshelby tensor, which depends on the particle aspect ratio (L/d_{pl}) and dimensionless elastic constants of the polymer matrix [16].

For the case of isotropic (spherical) filler particles E_n/E_m estimation can be carried out according to the equation [42]:

$$\frac{E_n}{E_m} = 1 + 11.6\varphi_n - 44.4\varphi_n^2 + 96.3\varphi_n^3. \tag{26}$$

The theoretical dependences (lines), corresponding to the equations (24) and (25) at L/d_{pl}=50 and 100, and also to the equation (26) are adduced in Figure 15. As abscissa axis Na^+-montmorillonite contents W_n in mass % is chosen as most often used in practice. Besides, in the same Figure the experimental values E_n/E_m (points) for seven nanocomposites are plotted. Let us note first of all, that the transition from theoretical dependences with larger values L/d_{pl} to smaller ones at W_n increase (in Figure it is indicated by arrows) is observed. Such transition was expected in virtue of the layered silicate platelets aggregation, which is

Na^+-montmorillonite, at W_n increase. The aggregation means platelets number rise in "stack" (tactoid) d_{pl} growth at L=const and, as consequence, the ratio L/d_{pl} decrease. But the most interesting aspect of the dependences of Figure 15 is the fact, that different polymer matrices correspond to different theoretical curves, showing in addition widely variating E_n/E_m values [16].

It is significant that reinforcement degree corresponds to a class of polymer, forming nanocomposites polymer phase. The greatest ratios E_n/E_m are obtained for polymers, whose chains are able to stretch on silicate platelet surface (rigid-chain polyimide, crystallizing polypropylene and thermotropic liquid crystalline polyester), intermediate – for polymers, whose chains are able to stretch only partly (polycarbonate, poly(butylene terephtalate) and amorphous polyamide-6), and the smallest ones – for nanocomposites on the basis of epoxy polymer, the capability of which to chains stretching decreases sharply by the availability of transverse covalent bonds network.

Hence, the data of Figure 15 indicate that the nanocomposite ability to reinforcement is defined not actually by filler particles shape anisotropy, but by polymer phase ability to reflect (reproduce) this anisotropy. In other words, the filler role comes to polymer phase structure modification in comparison with the initial matrix polymer. The similar concept was used for polymer microcomposites reinforcement description [43]. However, this general treatment distinction consists in the fact that in the microcomposites case bulk polymer matrix structure changes (its fractal dimension d_f is increased) [43] and in the nanocomposites case – only interfacial regions structure at common condition d_f=const [44].

The indicated ability of polymer matrix or, more precisely, interfacial regions to reflect filler particles form anisotropy one can express quantitatively with the polymer chain statistical segment length l_{st}: the larger l_{st}, the higher E_n/E_m. Let us remind, that the value l_{st} is connected with characteristic ratio C_∞ by the following equation [30]:

$$l_{st} = C_\infty l_0, \tag{27}$$

where l_0 is the main chain skeletal bond length and since C_∞ is a polymer chain statistical flexibility indicator [27], then l_{st} plays the same role. In Figure 16 the dependences of E_n/E_m on l_{st} for the seven adduced in Figure 15 nanocomposites at two contents of Na^+-montmorillonite (W_n=2 and 5 mass %) are shown. As one can see, these dependences are linear and at l_{st}=0 are extrapolated to E_n/E_m=1.0, that means the reinforcement absence for low-molecular matrices (at any rate, by interfacial regions formation mechanism). These dependences can be described analytically as follows [16]:

$$\frac{E_n}{E_m} = 1 + A l_{st}, \tag{28}$$

where A is constant, which is equal to 1.08 and 1.36 for nanocomposites with W_n=2 and 5 mass %, if l_{st} is given in nm.

The common dependence of E_n/E_m on W_n and l_{st} (with appreciation of A increase with W_n growth) for nanocomposites with exfoliated (nonaggregated) organoclay has the following form [16]:

$$\frac{E_n}{E_m} = 1 + 0.32 W_n^{1/2} l_{st}, \tag{29}$$

where W_n is given in mass %, l_{st} – in nm.

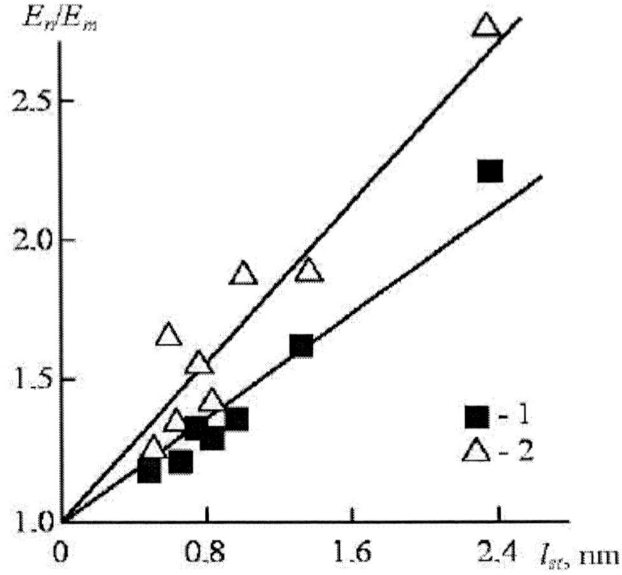

Figure 16. The dependences of reinforcement degree E_n/E_m on statistical segment length l_{st} for nanocomposites with Na^+-montmorillonite contents 2 (1) and 5 (2) mass % [16].

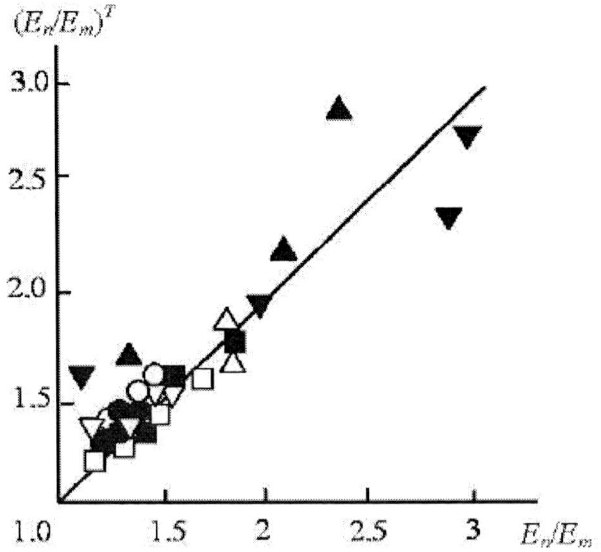

Figure 17. The comparison of experimental E_n/E_m and calculated according to the equation (29) $(E_n/E_m)^T$ reinforcement degree for nanocomposites with $W_n \leq 5$ mass %. Designations are the same as in Figure 15 [16].

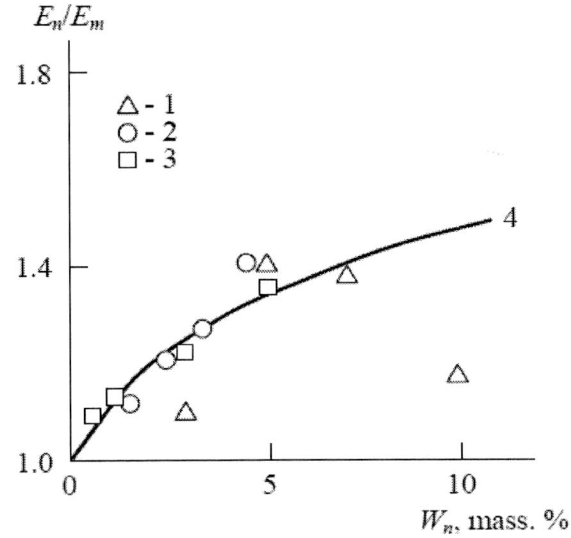

Figure 18. The dependence of reinforcement degree E_n/E_m on nanofiller contents W_n for nanocomposites phenylone/CNT (1), PC/MMT (2) and PA-6/MMT (3). 4 – calculation according to the equation (29) [45].

In Figure 17 calculated according to the equation (29) and obtained experimentally reinforcement degree values for the considered nanocomposites at $W_n \le 5$ mass % are adduced. As one can see, the estimation according to the equation (29) gives a good correspondence to experiment.

In Figure 18 the dependence $E_n/E_m(W_n)$, calculated according to the equation (29) (the solid line) and experimental data (points) for nanocomposites on the basis of phenylone, filled with carbon nanotubes (CNT) is adduced. A good enough correspondence of theory and experiment is observed [45]. Accounting for the fact, that the equation (29) was obtained for nanocomposites, filled with organoclay, the data of Figure 18 allow to confirm, that CNT possess the same ability to rise polymer elasticity modulus as well as exfoliated layered silicates.

From the data of Figure 18 it follows, that at W_n=10 mass % the experimental value E_n/E_m is essentially lower than its theoretical magnitude. The cause of this effect can be elucidated by the equation (8) usage. For nanocomposites phenylone/CNT the values φ_{if}=0.029; 0.096; 0.112 and 0 for CNT contents W_n=3, 5, 7 and 10 mass %, accordingly, were obtained. The condition $\varphi_{if}=0$ is due to nanocomposites structure synergetic behaviour: for W_n=10 mass % its behaviour becomes chaotic one, CNT orientation factor η, which is a governing parameter of the mentioned structure, turns into zero and according to the equation [47]:

$$\varphi_{if} = 1.09\eta \tag{30}$$

the value φ_{if} is also equal to zero.

In Figure 18 the experimental dependences $E_n/E_m(W_n)$ are also adduced for two nanocomposites, filled with organoclay: polycarbonate/Na^+-montmorillonite (PC/MMT) [48] and polyamide-6/Na^+-montmorillonite (PA-6/MMT) [23]. The choice of these nanocomposites is due to closeness of PC, PA-6 and phenylone molecular characteristics [32]. As it follows from the data of Figure 18, the dependences $(E_n/E_m)(W_n)$ for nanocomposites, filled with Na^+-montmorillonite, correspond to both results for nanocomposites phenylone/CNT and theoretical calculation according to the equation (29).

The close results, obtained for the three mentioned above nanocomposites, are due to the identical values of the received values φ_{if} owing to comparable contact area of nanofiller-polymer matrix. The equation (10) gives the relation between φ_{if} and φ_n for exfoliated layered silicates. For CNT the same relation can be obtained from the following considerations. As it is known [49], nanocomposites polymer matrix and nanofiller particles surface, interacting at interfacial regions formation, are fractal objects. At their interaction there is a sole linear scale l, defining such objects interpenetration distance [50]. Since in polymer composites filler elasticity modulus, as a rule, is higher than the corresponding parameter for polymer matrix, it is assumed [49], that in this case filler indentation in polymer matrix occurs and then l is equal to interfacial layer thickness l_{if}. It can be written [50]:

$$l_{if} \sim a\left(\frac{r_n}{a}\right)^{2(d-d_{surf})/d}, \tag{31}$$

where a is a lower linear scale of polymer matrix fractal behaviour, r_n is a nanofiller particles radius, d_{surf} is fractal dimension of nanofiller particles surface.

For CNT the mean value r_n=17.5 nm, the value a is accepted equal to l_{st} and d_{surf} magnitude can be accepted in the first approximation equal to 2.0 [16]. As it is known, d_{surf}=2.0 means nanofiller smooth surface, that results to stretching of macromolecular coils on it [17] and, as consequence, to C_∞ growth in comparison with the similar parameter for bulk polymer matrix. Therefore for the estimations according to the equation (31) C_∞=9 was accepted [51] and then a=l_{st}=1.26 nm. At the indicated parameters the equation (31) gives l_{if}=7.34 nm. The equation can be used further [16]:

$$\varphi_{if} = \varphi_n\left[\left(\frac{r_n + l_{if}}{r_n}\right)^3 - 1\right], \tag{32}$$

that at r_n=17.5 nm and l_{if}=7.34 nm gives [45]:

$$\varphi_{if} = 1.86\varphi_n. \tag{33}$$

The comparison of the equations (10) and (33) shows approximate equality of the coefficients in them, that defines close values of reinforcement degree for nanocomposites, filled with organoclay and carbon nanotubes.

It is necessary to indicate, that high values E_n/E_m for nanocomposites phenylone/CNT are due to nanotubes size nanometer scale. So, in case of short fibers (r_n=4000 nm [47]) the calculation according to the equations (31) and (32) at other equal conditions gives [45]:

$$\varphi_{if} = 0.18\varphi_n, \quad (34)$$

that defines essentially lower microcomposites reinforcement degree in comparison with nanocomposites at other equal conditions.

And let us note in conclusion matrix polymer molecular characteristics strong influence on nanocomposites reinforcement degree, which follows from the equation (29). The indicated factor acts stronger, than nanofiller content, since the value l_{st} is included in the equation (29) to the first power and W_n – to the half power.

Thus, the stated above results allow to make three main conclusions. Firstly, polymers reinforcement degree by CNT (nonaggregated) is comparable with corresponding characteristic for exfoliated layered silicates (organoclays). Secondly, polymer molecular characteristics, namely, chain statistical flexibility, influence on reinforcement degree is much stronger, than it assumes earlier. And, thirdly, interfacial phenomena, i.e. interfacial regions formation, is the main factor, defining nanocomposites reinforcement [45].

The polyurethanes, widely used in industry, represent two-phase polymer materials, consisting of rigid and flexible blocks. At room temperature the first are in a glassy state, the second – in rubber-like. An organoclay small amounts introduction (within the limits of 1-5 mass %) improves essentially polyurethanes properties [52, 53]. So, these rubber-like polymers modulus at strain of 100 % and organoclay content 5 mass % increases approximately in 3.6 times, and strength – in about 1.5 times [53]. The questions about structure of the received from them nanocomposites arise in connection with polyurethanes two-phase state. The first from them is linked to reinforcement high degree causes. The second question is related with nanofiller (organoclay) concentration in either phase of polyurethane. And at last, the third question is related to reinforcement degree sharp increase at testing temperatures higher than rigid blocks glass transition temperature [53]. The authors [54] gave the answer to the stated above questions within the frameworks of the offered above reinforcement model (see the equation (29)).

The polyurethanes (PU) and nanocomposites on its basis glass transition temperature was determined by the dynamic mechanical spectroscopy method. It has been found out [53], that PU have two glass transition temperatures, which are equal to 233 and 324 K, i.e. the studied polyurethane is a two-phase material, where T_g=233 K is related to flexible blocks and T_g=324 K – to rigid ones. It has been reported, that for PU on the basis of poly(tetramethylene glycol) (PTMG) T_g value is equal to 185 K. Higher values T_g for the considered PU and wide damping peaks, received in dynamical mechanical tests, assume partial mixing of flexible and rigid segments phases. The nanocomposites on the basis of PU glass transition temperatures are approximately equal to these temperatures for the initial matrix polymer [53].

Using the stated above data, the value C_∞ and, accordingly, l_{st} can be calculated according to the equations (22) and (27) for polyurethane each phase separately. For rigid blocks C_∞=9.64, l_{st}=1.47 nm and for flexible ones - C_∞=18.6, l_{st}=2.83 nm. Further the reinforcement degree $(E_n/E_m)^T$ calculation according to the equation (29) gives this parameter values at using l_{st}=1.47 nm, if organoclay is concentrated in rigid blocks phase, and at using l_{st}=2.83 nm, if organoclay concentration is realized in flexible blocks phase. In Figure 19 the comparison of the experimental E_n/E_m and calculated by the indicated mode $(E_n/E_m)^T$ reinforcement degree values for nanocomposites PU/MMT. As it follows from the data of this Figure , the good correspondence of theory and experiment was obtained in that case, if in the equation (29) the value l_{st}=2.83 nm for flexible blocks was used. In case of value l_{st}=1.47 nm for rigid blocks application the equation (29) gives understated values of reinforcement degree. This comparison demonstrates, that for the considered polyurethane organoclay is concentrated in devitrificated at testing temperature (293 K) flexible blocks phase (T_g=233 K) and high reinforcement degree E_n/E_m at organoclay small contents (W_n=1-5 mass %) is defined by high values l_{st} or C_∞ (by polymer chain high flexibility), typical for rubber-like polymer (see above).

Let us consider the effect of reinforcement degree sharp increase at testing temperatures higher than rigid blocks glass transition temperature (T_g=324 K). Such effect was observed earlier as well, for the example, for nanocomposites epoxy polymer/organoclay. The authors [16] explained the observed effect by sharp increasing of chain flexibility in rubber-like state, that is reflected by C_∞ and, respectively, l_{st} enhancement. The flexible blocks phase T_g increase in comparison with the value obtained earlier (from 185 up to 233 K) the authors [53] explained by flexible and rigid blocks mixing definite degree and, accordingly, by the influence of the latter on the first. The rigid blocks phase devitrification removes the indicated limitations and therefore for the value C_∞ calculation in this case (at temperatures higher than 324 K) according to the equation (22) T_g=185 K is accepted. Then the estimations with the indicated equation usage give C_∞=29.7, l_{st}=4.51 nm. Further value $(E_n/E_m)^T$ can be calculated according to the equation (29). The comparison of experimental E_n/E_m and calculated by the indicated mode $(E_n/E_m)^T$ reinforcement degree values for nanocomposites PU/MMT at temperatures higher than rigid block glass transition temperature (T_g=324 K) is also adduced in Figure 19, from which a good correspondence of theory and experiment follows.

Hence, the adduced above data allow to give an answer to the three stated questions. Firstly, organoclay particles are concentrated in devitrificated flexible blocks phase. Secondly, polymer chains high flexibility in the indicated phase defines a large enough reinforcement degree of nanocomposites polyurethane/organoclay. And, thirdly, still larger polymer chain flexibility, realized after rigid blocks devitrification, results to corresponding enhancement of reinforcement degree for this state of the considered nanocomposites [54].

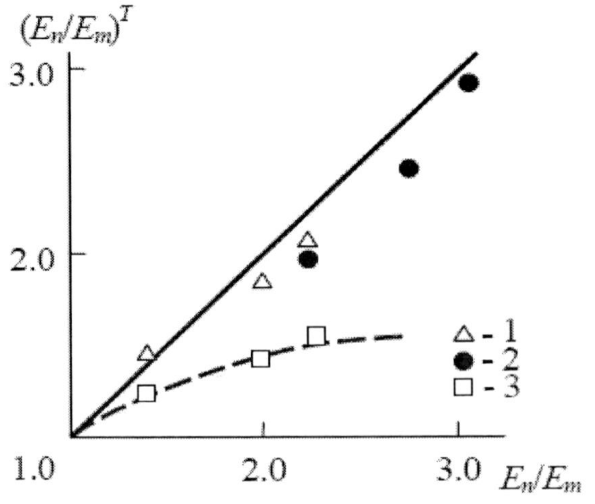

Figure 19. The comparison of experimental E_n/E_m and calculated according to the equation (29) $(E_n/E_m)^T$ at l_{st}=2.83 (1), 4.51 (2) and 1.47 (3) nm reinforcement degree values for nanocomposites PU/MMT [54].

It is obvious, that for particulate nanofiller the equation (29) should be modified by including into it the nanoparticles diameter D_p. Such modification allows to obtain the following equation for particulate-filled polymer nanocomposites reinforcement degree estimation [55]:

$$\frac{E_n}{E_m} = 1 + \frac{0.19 W_n l_{st}}{D_p^{1/2}}, \tag{35}$$

where W_n is given in mass %, l_{st} and D_p – in nm.

The comparison of experimental E_n/E_m and calculated according to the equation (35) $(E_n/E_m)^T$ reinforcement degree for particulate-filled nanocomposites (Figure 20) shows the good correspondence of theory and experiment.

As it is known [56-58], a particulate-filled polymer nanocomposites reinforcement degree is a function of nanofiller particles diameter. In Figure 21 the calculated according to the equation (35) dependence of E_n/E_m on D_p at the following values of the included into this equation parameters: W_n=30 mass %, l_{st}=1 nm is adduced. As one can see, the sharp growth of reinforcement degree at D_p reduction is observed, that was expected [56-58]. This correspondence has not only qualitative, but also quantitative character: in Figure 21 the boundaries of “reinforcing” and “super-reinforcing” nanofillers according to the classification, offered in paper [56], are indicated by vertical shaded lines. As it follows from the plot of Figure 21, these values correspond well to gradient change of reinforcement degree, calculated according to the equation (35).

The dependences of E_n/E_m on W_n, plotted according to the equations (29) and (35) (l_{st}=1 nm, D_p=7 nm) for polymer nanocomposites, filled with layered (or CNT) and particulate nanofiller, accordingly, are presented in Figure 22.

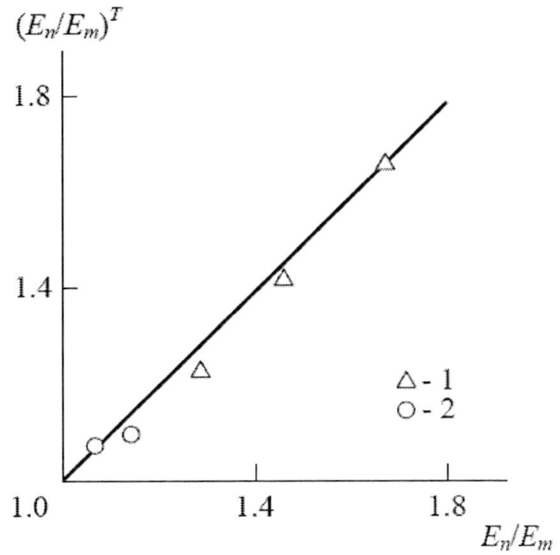

Figure 20. The comparison of experimental E_n/E_m and calculated according to the equation (29) $(E_n/E_m)^T$ reinforcement degree for nanocomposites on the basis of PP (1) and phenylone (2) [55].

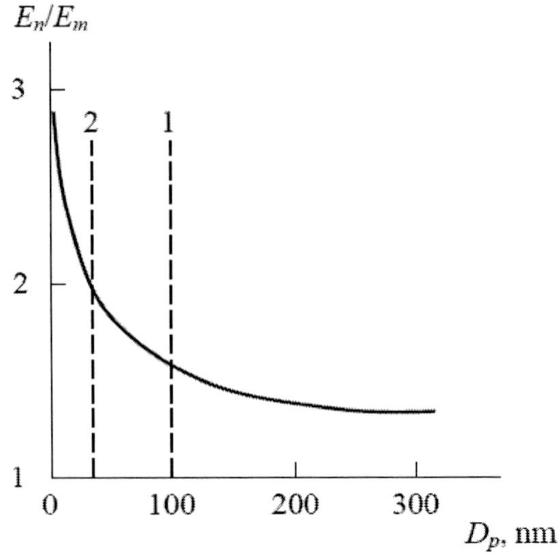

Figure 21. The dependence of reinforcement degree $(E_n/E_m)^T$ on nanofiller particles diameter D_p calculated according to the equation (35) for nanocomposites, filled with disperse particles. The vertical shaded lines indicate the boundary for reinforcing (1) and superreinforcing (2) nanofillers [55].

As one can see, at relatively small W_n (< 20 mass %) the reinforcement degree at the same W_n values is higher for nanocomposites, filled with layered silicate or CNT, but at W_n>20 mass % particulate nanofiller gives a higher reinforcement degree. This transition is explained as follows. It is known [16], that the layered silicates and CNT at contents W_n>10 mass % display strong tendency to aggregation. Since the equations (29) and (35) are empirical ones, which were received on the basis of the experimental data analysis, then they take into consideration this effect by the dependence of E_n/E_m on W_n to the half power for layered nanofillers and CNT (see the equation (29)) and to first power for particulate particles (the equation (35)).

In the general case filled with organoclay polymer nanocomposites plasticity, characterized by strain up to failure, decreases sharply at small (of order of 5-10 mass %) nanofiller contents [24, 48]. For this nanocomposites brittleness effect explanation the authors [59] used the fractal analysis methods, within the frameworks of which the value of strain up to failure ε_f was described according to the following equation [60]:

$$\varepsilon_f = C_\infty^{D_{ch}} - 1, \qquad (36)$$

where the characteristic ratio C_∞ was determined according to the equation (18) and dimension D_{ch} was calculated according to the equation (3) with φ_{cl} replacement by densely-packed regions φ_{dens} of nanocomposite structure relative fraction.

As it is known [16], in nanocomposite structure local order domains (clusters), interfacial regions and actually nanofiller with relative fractions φ_{cl}, φ_{if} and φ_n, accordingly, are densely-packed regions. The value φ_{cl} was calculated with the aid of the equation [14]:

$$d_f = 3 - 6\left(\frac{\varphi_{cl}}{C_\infty S}\right)^{1/2}, \tag{37}$$

where S is macromolecule cross-sectional area, which is equal to 25 Å^2 for PC [36].

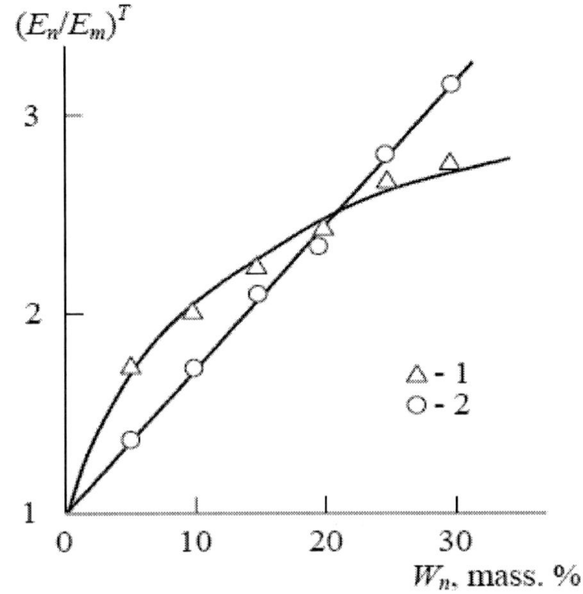

Figure 22. The dependences of reinforcement degree $(E_n/E_m)^T$ on nanofiller contents W_n for nanocomposites with layered (1) and disperse (2) nanofiller, calculated according to the equations (29) and (35), accordingly [55].

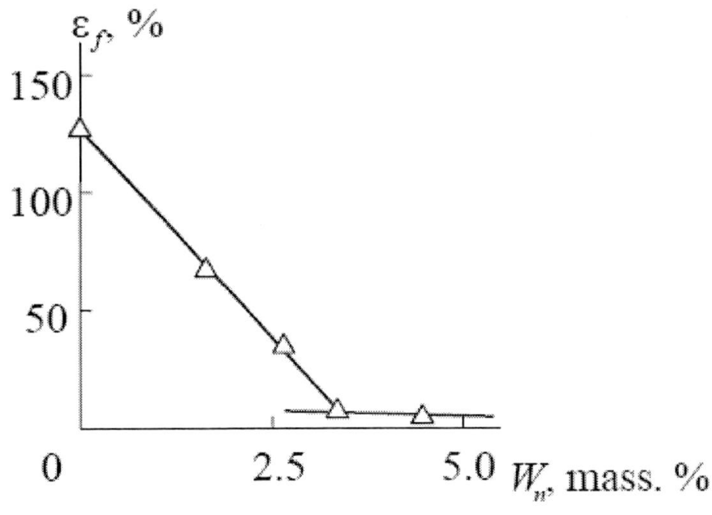

Figure 23. The dependence of strain up to failure ε_f on MMT contents W_n for nanocomposites PC/MMT [59].

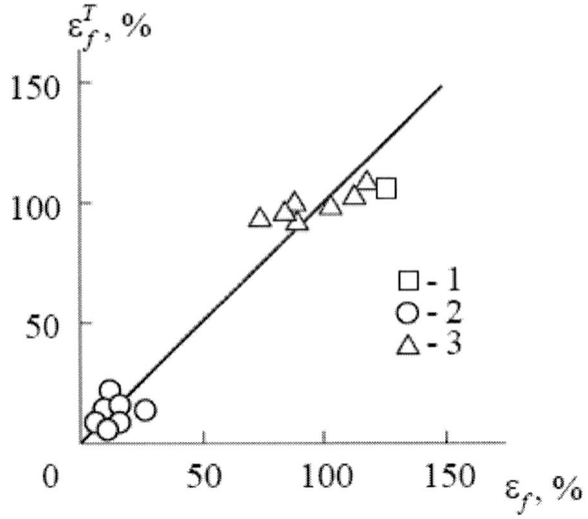

Figure 24. The comparison of experimental ε_f and calculated according to the equation (36) ε_f^T strain up to fracture values for PC (1) and nanocomposites PC/MMT with MMT contents 2.4 (2) and 4.5 (3) mass % [59].

As the data of paper [48] showed, interlayer spacing between MMT platelets d_{001} for the considered nanocomposites PC/MMT makes up 17.1-24.3 Å, that allows to attribute them to the exfoliated ones class, for which the value φ_{if} is determined according to the equation (10).

In Figure 23 the dependence of ε_f on MMT contents W_n, received experimentally is adduced. As one can see, for nanocomposites PC/MMT within the range of W_n=2.5-3.4 mass % sharp change of the dependence $\varepsilon_f(W_n)$ character and absolute values ε_f essential reduction is observed. Therefore the assumption was made [59], that at MMT small contents (W_n<3.2 mass %) only the clusters should be accepted as densely-packed regions and at W_n>3.2 mass % as the indicated regions the sum is accepted [59]:

$$\varphi_{dens} = \varphi_{cl} + \varphi_{if} + \varphi_n = \varphi_{cl} + 2.91\varphi_n. \tag{38}$$

In Figure 24 the comparison of ε_f obtained experimentally and calculated according to the considered above fractal model ε_f^T values of limiting strain up to failure for PC and seven nanocomposites PC/MMT with MMT contents 2.4 and 4.5 mass %. As one can see, the well enough correspondence of theory and experiment is obtained (the average discrepancy between ε_f and ε_f^T makes up ~ 20 %). This correspondence supposes correctness of the made above assumption: at small W_n (≤3.2 mass %) availability of MMT and interfacial regions does not impose restrictions on molecular mobility (D_{ch}≈1.39) and, hence, does not decrease essentially the value D_{ch} in comparison with PC. This results to high enough ε_f values comparable with the corresponding parameter for PC. And on the contrary, MMT contents increase larger than 3.2 mass % reduces sharply molecular mobility level (D_{ch}≈1.02-1.05) in virtue of φ_{dens} growth (see the equation (38)), that defines corresponding reduction of nanocomposites plasticity, characterized by strain ε_f. Let us note, that similar sharp ε_f reduction at $W_n \geq 2.5$ mass % was observed for other classes of nanocomposites, as well for example, polypropylene/MMT [24]. The described above technique was applied successfully for plasticity treatment of particulate-filled nanocomposites poly(vinyl chloride)/$CaCO_3$ [61] and nanocomposites polypropylene/MMT [62].

For polymer nanocomposites, filled with CNT, strain up to failure ε_f increase in comparison with the initial matrix polymer or nanocomposites plasticity enhancement was noted repeatedly. The indicated effect was found out for nanocomposites epoxy polymer/CNT [63], phenylone/CNT [64] and so on. In other words, plasticity increasing effect for nanocomposites indicated class has common enough character. This effect is very important from the practical point of view, since the main deficiency of polymer composites in general is their brittleness, i.e. ε_f reduction at filler contents increasing [62, 65]. Therefore it is important to give theoretical estimation of plasticity enhancement effect for nanocomposites polymer/CNT since in future this gives possibility of production of polymer nanocomposites with operating characteristics unique set: with these materials simultaneous increasing of stiffness, strength and plasticity.

The authors [66, 67] carried out the structural analysis of nanocomposites phenylone/CNT plasticity enhancement within the frameworks of the considered above model. Calculated according to the equation (36) ε_f^T values for the indicated nanocomposites are adduced in table 1 together with necessary for their calculation characteristics C_∞, φ_{cl} and D_{ch}. As it follows from the data of this table, the comparison of experimental ε_f and theoretical ε_f^T values of limiting strain at failure, characterizing nanocomposites phenylone/CNT plasticity, showed their good correspondence – the average discrepancy makes up 8.5 %, that corresponds to the experimental error of this parameter determination.

Let us consider physical principles of the studied plasticity increasing effect for nanocomposites, filled with CNT. CNT distinguishing feature from other nanofillers (disperse particles, organoclay) is their smooth in atomic scale surface [63], that results to matrix polymer macromolecules stretching on this surface and densely-packed interfacial regions polymer matrix – CNT formation [17]. The values of interfacial regions relative fraction φ_{if} for the considered nanocomposites according to the data, of paper [64] are also adduced in table 1. These regions dense packing supposes, that fractal dimension d_f for them (d_f^{if}) achieves the greatest possible value for real solid bodies, namely, d_f^{if}=2.95 [28]. Further the calculation of theoretical dimension d_f (d_f^T) can be fulfilled according to the mixtures rule [66]:

$$d_f^T = d_f^p\left(1-\varphi_{if}\right)+d_f^{if}\varphi_{if}, \tag{39}$$

where d_f^p is fractal dimension of phenylone structure, which is equal to 2.247 (table 1).

Table 1.The molecular, structural and mechanical characteristics of nanocomposite phenylone/CNT [67]

CNT contents, mass %	d_f	C_∞	φ_{cl}	D_{ch}	ε_f, %	ε_f^T, %	φ_{if} [64]	d_f^T
0	2.247	2.33	0.646	1.336	21.8	21.9	-	2.247
3	2.230	2.30	0.667	1.318	22.0	20.2	0.022	2.265
5	2.354	2.55	0.520	1.439	26.3	33.8	0.098	2.316
10	2.280	2.38	0.605	1.376	25.0	25.8	-	2.247

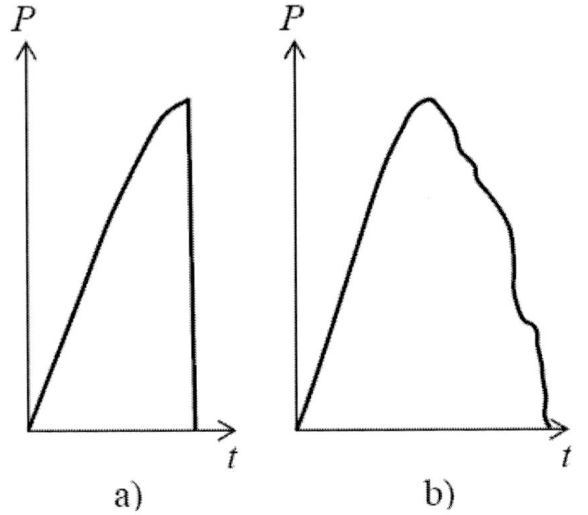

Figure 25. Schematic diagrams load-time (P-t) in instrumented impact tests. The fracture by instable crack (a) and stable one (b) [72].

The calculated according to the equation (39) d_f^T values are also adduced in table 1 and their comparison with d_f values was shown the good correspondence (the average discrepancy of d_f and d_f^T makes up 1.5 %). Taking into account, that the offered above technique of ε_f^T theoretical calculation is founded as a matter of fact on dimension d_f knowledge, one should make the conclusion, that the observed for nanocomposites polymer/CNT plasticity increase effect is due to densely-packed interfacial regions formation on nanotubes surface and, as consequence, to polymer matrix molecular and structural characteristics change (C_∞, d_f, φ_{cl} and D_{ch}, see table 1). Let us also note, that ε_f increase is accompanied by dimension D_{ch} growth (table 1), i.e. molecular mobility intensification. As it has been shown by Kausch [68], such interconnection is common for polymer materials.

And in conclusion let us consider the prediction of properties for nanocomposite phenylone/CNT, contained 30 mass % nonaggregated CNT. In this case the value φ_{if} can be estimated according to the equation (33) and elasticity modulus E_n – according to the equation (8) at the condition E_m=1.25 GPa in tension testing for phenylone. Then the value E_n for the indicated above hypothetical nanocomposite phenylone/CNT contained 30 mass % CNT, makes up 11.85 GPa. The calculation d_f^T according to the equation (39) gives value 2.64 or ν=0.32 according to the formula (16). Hence, yield stress σ_Y value for such nanocomposite according to the relationship (17) makes up 820 MPa. And at last, the calculation according to the equations (3), (18), (36) and (37) gives the value ε_f^T=120 %. To obtain such unique by their mechanical properties totality nanocomposite two technological difficulties hinder: very high nanotubes cost, making up several tens and even hundreds of dollars per gram, and the latter production technology, when nanotubes can be received only in the shape of tangled bundles and available at present dispersion methods work only up to CNT contents of order of several per cents [63, 64, 69].

Thus, the stated above results confirm correctness of structural analysis at limiting strain (plasticity) estimation for nanocomposites phenylone/CNT. The offered model gives the good correspondence of theory and experiment. The physical ground of plasticity increase for nanocomposites, filled with CNT, is densely-packed interfacial regions formation on smooth in atomic scale nanotubes surface. The offered treatment allows to predict nanocomposites properties and to plan their practical realization ways.

The authors of papers [8, 9] found out, that particulate nanofiller $CaCO_3$ introduction in HDPE resulted to nanocomposite HDPE/$CaCO_3$ impact toughness A_p increasing in comparison with the initial polymer approximately by 20 %. The authors [8, 9] carried out a detailed fractographic analysis of this effect and explained A_p observed increase by plastic deformation mechanism change for nanocomposite HDPE/$CaCO_3$ in comparison with the initial HDPE. Without going into details of the indicated analysis, some grounds for doubts in its correctness it is necessary to note. In Figure 25 the schematic diagrams load-time (*P-t*) for polymer materials samples fracture two cases are adduced: by instable (a) and stable (b) cracks. As it is known [70], the value A_p is characterized by area under *P-t* diagram, which gives mechanical energy, spent on samples fracture. Polymer materials macroscopic fracture process, defined by the main crack, begins at the greatest load *P*.

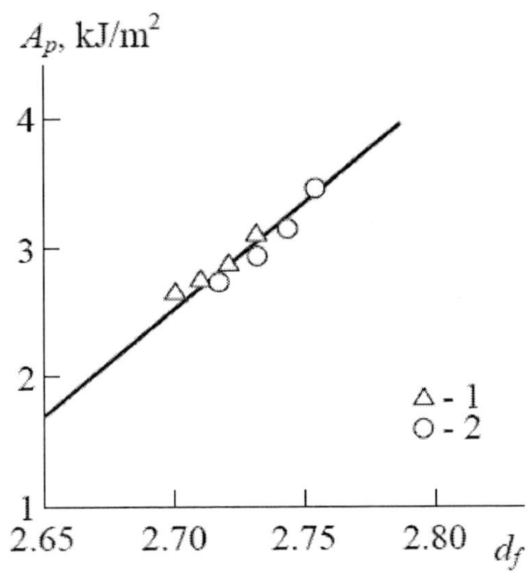

Figure 26. The dependence of impact toughness A_p on structure fractal dimension d_f for HDPE (1) and nanocomposite HDPE/$CaCO_3$ (2) [72].

From the schematic diagrams *P-t* it follows, that actually fracture process does not practically influences on the value A_p in case of crack instable propagation and influences only partly – in case of stable crack. Although the authors [8, 9] carried out impact tests on instrumented apparatus, allowing to obtain *P-t* diagrams, these diagrams were not adduced. Besides, fracture process structural aspect in papers [8, 9] is considered with secondary structures (crazes, shear zones and so on) using. Their interconnection with the initial undeformed material structure is purely speculative. It is obvious, that to obtain quantitative relationships structure-properties (that is the main goal of polymers physics [71]) at such analysis method is not possible. Therefore the authors [72] carried out quantitative structural analysis of HDPE and nanocomposites HDPE/$CaCO_3$ impact tests results within the frameworks of fractal models.

As it is known [73], the fractal dimension d_f is the most general object (in our case – polymer material) structure informant and true structural characteristic, describing structure elements distribution in space. The value d_f can be determined according to the equations (16) and (37). In the second from the indicated equations the value φ_{cl} was determined according to the relationship (4). Estimations according to the equations (16) and (37) gave the following d_f values at testing temperature 293 K: for HDPE 2.68 and 2.73, for nanocomposite HDPE/$CaCO_3$ 2.73 and 2.75, accordingly. As one can see, the good enough correspondence was obtained – the discrepancy by fractional part d_f, which performs the main information amount about structure, does not exceed 7 %.

In Figure 26 the dependence $A_p(d_f)$ is adduced for the considered polymer materials, which turns out to be linear, common for initial HDPE and nanocomposite HDPE/$CaCO_3$ and is described by the following empirical correlation [72]:

$$A_p = 13.5(d_f - 2.5), \quad \text{kj/m}^2. \tag{40}$$

From the equation (40) it follows, that at d_f=2.5 the value A_p=0. The indicated fractal dimension corresponds to the ideal brittle fracture condition [28], that defines the condition A_p=0. For real solid bodies the greatest fractal dimension of their structure is equal to 2.95 [28], that allows to determine the greatest value A_p according to the equation (40), which is equal to ~ 6.1 kj/m^2 [72].

As Kausch showed [68], the energy dissipation at impact grew at polymer material molecular mobility level enhancement. As it was noted above, within the frameworks of fractal analysis this level can be characterized with the aid of the fractal dimension D_{ch} of polymer chain part between its fixation points (chemical cross-linkings, physical entanglement nodes, clusters and so on) [14]. Such analysis method was applied successfully for impact toughness description in case of particulate-filled nanocomposites phenylone/β-sialone [74] and amorphous glassy polymers, treated as natural nanocomposites [75]. The value D_{ch} can be determined with the aid of the following equation [14]:

$$D_{ch} = \frac{\ln N_{cl}}{\ln(4 - d_f) - \ln(3 - d_f)}, \tag{41}$$

where N_{cl} is statistical segments number per chain part between clusters, which is determined as follows. First physical entanglements cluster network density ν_{cl} was determined [14]:

$$\nu_{cl} = \frac{\varphi_{cl}}{C_\infty l_0 S}, \tag{42}$$

where l_0 is the main chain skeletal bond length, which is equal to 1.54 Å for polyethylenes [32].

Then the estimation of polymer chains total length per polymer volume unit L was fulfilled as follows [14]:

$$L = S^{-1}. \tag{43}$$

The chain part between clusters length L_{cl} is determined according to the equation [14]:

$$L_{cl} = \frac{2L}{\nu_{cl}}. \tag{44}$$

And at last, the value N_{cl} can be determined as the ratio [14]:

$$N_{cl} = \frac{L_{cl}}{l_{st}} = \frac{2}{S\nu_{cl} l_{st}}. \tag{45}$$

In Figure 27 the dependence of impact toughness A_p on chain fractal dimension D_{ch} is adduced for the considered materials. As it was expected, A_p growth at D_{ch} increase was observed, described analytically by the following relationship [72]:

$$A_p = 6.75(D_{ch} - 1), \quad \text{kj/m}^2. \tag{46}$$

The equation (46) allows to determine the greatest value A_p for the considered materials at the condition D_{ch}=2.0: this value is equal to 6.75 kj/m^2, that is close to the adduced above estimation according to the equation (40) – the average discrepancy makes up less than 10 %.

Let us consider the condition of zero impact toughness reaching at d_f=2.50, but not at d_f=2.0 ($2.0 \le d_f < 3$ [31]). As it is known [76], at φ_{cl} growth irrespective of its causes structure of polymers in general and HDPE in particular, reaches its quasiequilibrium state, when φ_{cl} growth is balanced by polymer chains entropic stretching and ceases. As it was shown in paper [76], for HDPE structure quasiequilibrium state dimension was equal to 2.50. The calculation according to the adduced above technique (the equations (4) and (41)-(45)) showed, that in this case $D_{ch} \approx 1.0$ and, hence, polymer material became ideally brittle (see the equation (46)).

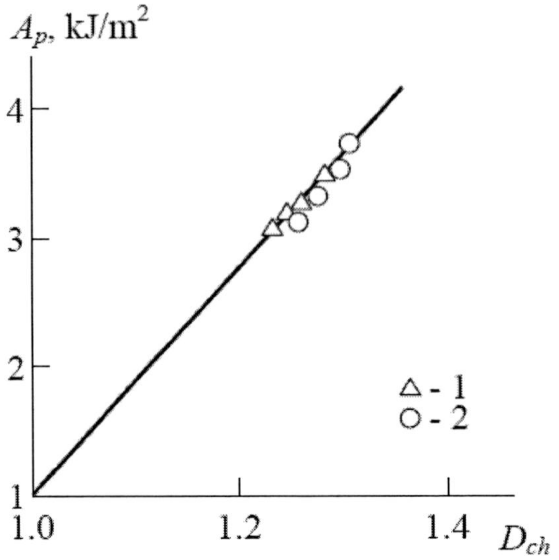

Figure 27. The dependence of impact toughness A_p on fractal dimension D_{ch} of chain part between clusters for HDPE (1) and nanocomposite HDPE/$CaCO_3$ (2) [72].

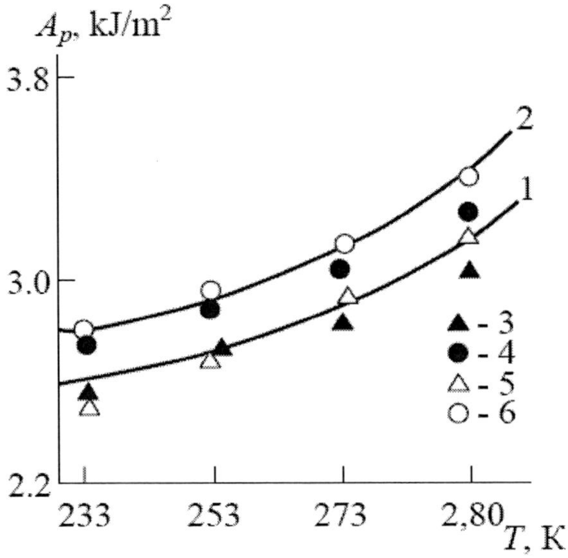

Figure 28. The comparison of experimental (1, 2) and calculated according to the equations (40) (3, 4) and (46) (5, 6) temperature dependences of impact toughness A_p for HDPE (1, 3, 5) and nanocomposite HDPE/$CaCO_3$ (2, 4, 6) [72].

The equations (40) and (46) usage allows to estimate theoretical values A_p (A_p^T) and to compare them with experimental magnitudes of impact toughness. Such comparison is presented in Figure 28, from which a good correspondence of theory and experiment follows (for calculation according to the equation (40) the average discrepancy A_p^T and A_p makes up

3.5 %, according to the equation (46) – 3.9 %, that is essentially less than impact toughness experimental determination usual error, which makes up ~ 10 %).

Let us consider physical grounds of impact toughness increase effect for nanocomposites HDPE/$CaCO_3$ in comparison with matrix polymer HDPE. As it follows from the plots of Figs. 26 and 27, A_p increase is defined by dimensions d_f and D_{ch} enhancement, accordingly. In its turn, as it follows from the equations (37) and (41), the indicated dimensions enhancement is defined by φ_{cl} reduction, since the molecular characteristics C_∞ and S were adduced as constants. As it follows from the equation (4), φ_{cl} reduction depends on crystallinity degree K increase only, since the value T_m for HDPE and nanocomposite HDPE/$CaCO_3$ are practically equal [8, 9]. This means, that A_p increasing effect for nanocomposite HDPE/$CaCO_3$ in comparison with matrix polymer is due to nanofiller $CaCO_3$ action as a nucleator, promoting to crystallinity degree enhancement (see above).

Thus, within the frameworks of fractal analysis the quantitative correlation between impact toughness and structural characteristics is obtained for high density polyethylene and particulate-filled nanocomposites on its basis. An impact toughness increase is due to calcium carbonate action as a nucleator and corresponding change of amorphous phase. The theoretical calculation has shown a good correspondence to the experimental data [72].

The Influence of Polymer Phase Molecular Characteristics on Interfacial Regions Formation in Polymer Nanocomposites

As it is well known [16], the interaction degree or adhesion level for polyphase systems, including nanocomposites polymer/organoclay, influences essentially on their properties. The high level of interfacial adhesion polymer-organoclay is an obligatory condition of necessary for these materials characteristics reaching. The authors [77] carried out computer simulation of interaction energy for nanocomposites PA-6/MMT at 7 quaternary ammonium salts usage as coupling agent. This simulation demonstrated binding energy E_b between PA-6 and MMT growth at the indicated salts molecular volume V_m reduction. So, V_m decrease from 690.8 Å^3 up to 201.2 Å^3, i.e. in 3.4 times, results to E_b growth from 8.22 up to 215.1 kcal/mole, i.e. more than in 26 times. Such essential E_b change should be due to the change of quaternary salts molecules interaction with MMT platelet surface, which is a fractal object with dimension d_{surf}=2.78 [78]. Let us note, that the indicated d_{surf} value was obtained at using as a probe of nitrogen N_2 molecules, whose sizes are essentially smaller than quaternary salts molecules (N_2 molecule volume makes up ~ 17 Å^3), in case of method BET application. Therefore V_m increase from 17 up to 201.2-690.8 Å^3 will result to the appearance on MMT platelet surface a large number of "screened" (inaccessible for quaternary salts) contact sites, that will decrease sharply both effective value of its specific surface S_u and effective dimension of MMT platelet surface d_{surf}^{ef}. This effect results to the reduction of contacts binding agent – MMT number and, as consequence, to their interaction energy E_b decrease. The authors [79] carried out quantitative estimation of the indicated effect within the frameworks of fractal analysis.

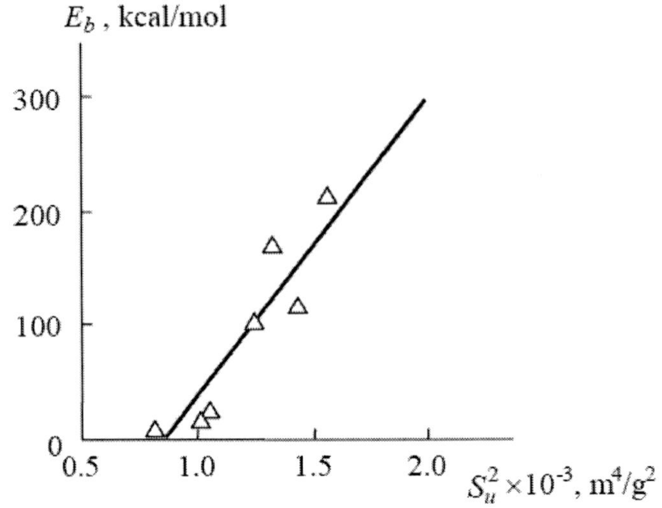

Figure 29. The dependence of PA-6-MMT binding energy E_b on MMT effective specific surface S_u [79].

As it is known [80], the effective specific surface of fractal object S_u depends on the adsorbed molecule cross-section σ_m as follows:

$$S_u \sim \sigma_m^{(2-d_{surf})/2}. \qquad (47)$$

Value σ_m for the considered ammonium salts can be estimated from geometrical considerations, assuming in the first approximation their spherical shape. The proportionality coefficient in the relationship (47) can be determined proceeding from BET results, where N_2 was used as an adsorbent [79]. In this case S_u=73.7 m^2/g, σ_m=8.04 Å^2, d_{surf}=2.78 and then the indicated coefficient is equal to 167.7 m^2/g. The estimations according to the equation (47) at the mentioned above proportionality coefficient showed that the effective value S_u for MMT and seven quaternary ammonium salts is changed within the limits of 28.5-39.4 m^2/g. As it was expected E_b growth at S_u increasing was observed, moreover this growth was stronger than a linear one. Therefore in Figure 29 the dependence of E_b on S_u^2 is adduced (such correlation shape was chosen with the purpose of its linearization). At $S_u \approx 29.6$ m^2/g this dependence supposes zero E_b value and in general case the indicated correlation was described by the following empirical equation [79]:

$$E_b = 0.265\left(S_u^2 - 880\right), \text{ kcal/mole,} \qquad (48)$$

where value S_u, as earlier, is given in m^2/g.

Thus, the equation (48) allows E_b estimation according to the known S_u values. The change of S_u results to the corresponding variation of MMT platelet surface fractal dimension d_{surf}^{ef}, which can be calculated with the aid of the following relationship [16]:

$$S_u = 410L^{(d_{surf}-d)}, \quad (49)$$

where L is characteristic size of MMT platelet, which is accepted here equal to 100 nm [23].

The estimations according to the equation (49) for initial MMT (S_u=73.7 m^2/g) give the value d_{surf}^{ef}=2.61, that is close to this dimension magnitudes, measured by other methods [78]. Reduction S_u up to 28.5-39.4 m^2/g in case of quaternary ammonium salts usage causes d_{surf}^{ef} decrease up to 2.25-2.34, that defines MMT platelets “unscreened” (accessible for contact) surface dimension d_u reduction. This dimension is estimated as follows [81]:

$$d_u = \left(d_{surf}^{ef} - 1\right) + \frac{d - d_{surf}^{ef}}{d_w}, \quad (50)$$

where d_w is salts molecule random walk dimension, which is accepted equal to 2.

The number of ammonium salts molecules with MMT platelets contacts N_u can be estimated according to the relationship [81]:

$$N_u \sim L^{d_u}, \quad (51)$$

where L is accepted equal to half-sum of MMT platelet length and width, i.e. 67.5 nm.

The estimations according to the relationships (50) and (51) show N_u reduction for the smallest S_u value in comparison with the initial MMT in 3 times, that corresponds to N_u reduction for the same conditions approximately in 2.6 times [79].

Although at present theoretical models for elasticity modulus change description for nanocomposites, filled with organoclay, based on the notion of the so-called “effective particle” [23], exist, but these models do not take into consideration two factors, which are the most important for theoretical description of two-phase polymer systems: interfacial adhesion level and polymer matrix molecular characteristics. The first from the indicated factors gains a particular meaning after discovering of the nanoadhesion effect, which supposes availability of interfacial adhesion much higher level in polymer nanocomposites in comparison with traditional composites, filled with particles (fibers) of micron sizes [82]. Therefore the authors [83] carried out the theoretical research of elasticity modulus enhancement at the introduction of montmorillonite in polyimide, accounting for the indicated above factors.

The first series of organically modified MMT (MMT-16C) was prepared via ion exchange reaction in water using concentrated hydrochloric acid, the second MMT series (MMT-OM-m) was obtained also via reaction in water using N-[4-(4’-aminophenoxy)] phenyl phthalimide. Nanocomposites on the basis of polyimide (PI), filled with MMT (PI/MMT-16C and PI/MMT-OM-m) were prepared by polymerization in situ, the details of which are adduced in papers [39, 84].

The reinforcement degree E_n/E_m calculation for the considered nanocomposites on the basis of PI according to the equation (29) gave this parameter understated values:

E_n/E_m=1.15-1.59, whereas reinforcement degree experimental magnitudes are within the range of 1.11-3.65 [39, 84]. The cause of this discrepancy is obvious enough: the equation (29) accounts for nanocomposite polymer chain flexibility with the aid of the parameter l_{st}, but does not account for the interfacial adhesion real level. The last factor accounting can be fulfilled with the aid of the parameter b, which is used at composites thermal expansion description and is determined according to the formula [85]:

$$\alpha_n = \alpha_n^{mix} - b\left(\alpha_n^{mix} - \alpha_n^T\right), \qquad (52)$$

where α_n, α_n^{mix} and α_n^T are nanocomposite thermal expansion coefficient values, obtained experimentally and calculated according to the mixtures rule and Terner equation, accordingly.

Since the calculation according to Terner equation gives lower limiting magnitude of thermal expansion coefficient, then from the equation (52) it follows b=1 at $\alpha_n = \alpha_n^T$, i.e. at perfect adhesion reaching (see the equation (26)). The values b within the range of ~ -0.19-1.39 were obtained for a large number of polymer composites with various polymer matrices and fillers [85]. The nanoadhesion effect supposes much higher values b in nanocomposites case. So, for particulate-filled nanocomposite phenylone/aerosil the value b can reach ~ 12, that corresponds to the condition $\alpha_n < \alpha_n^T$ [86]. The value of parameter b or interfacial adhesion level can be estimated by mechanical tests results as follows. A polymer nanocomposites reinforcement degree can be determined according to the percolation relationship (8). The following relationship between φ_{if} and φ_n was obtained for exfoliated layerend silicates [16]:

$$\varphi_{if} = 1.91\varphi_n b. \qquad (53)$$

Thus, the equation (53) demonstrates, that in the nanoadhesion case the value φ_{if} increases in b times in comparison with the perfect adhesion, for which b=1. The equation (53) substitution in the relationship (8) gives the following formula [83]:

$$\frac{E_n}{E_m} = 1 + 11\left[\varphi_n\left(1 + 1.91b\right)\right]^{1.7}, \qquad (54)$$

which accounts for interfacial adhesion (nanoadhesion) level at reinforcement degree determination and it is true for exfoliated organoclays.

Let us note, returning to the equation (29), that the sole parameter, on which interfacial adhesion level can influence, in this equation is statistical segment length. One can suppose, that the enhancement of interfacial adhesion level, characterized by parameter b growth, or polymer matrix interaction with montmorillonite platelets surface intensification restricts polymer chains mobility in interfacial layer and results to larger rigidity of their chains in comparison with bulk polymer matrix or l_{st} growth, that can be expressed with the aid of the statistical segment effective length l_{st}^{ef} [83]:

$$l_{st}^{ef} = l_0 C_\infty b. \tag{55}$$

Accounting for the equation (55) the relationship (29) can be written as follows [83]:

$$\frac{E_n}{E_m} = 1 + 0.19 W_n^{1/2} l_{st} b. \tag{56}$$

The value b can be calculated with the aid of the equation (54). These estimations showed, that for the considered nanocomposites the parameter b varied within the limits of 4.09-5.31, i.e. for these materials the nanoadhesion was observed (b>1). In Figure 30 the comparison of experimental E_n/E_m and calculated according to the equation (56) $(E_n/E_m)^T$ reinforcement degree values for nanocomposites polyimide/montmorillonite is adduced. As one can see, in this case the good correspondence of theory and experiment is obtained: $(E_n/E_m)^T$ value varies within the limits of 1.56-3.71 and its average discrepancy with experimental values E_n/E_m makes up ~ 16 %.

Hence, the stated above results demonstrated strong influence of interfacial adhesion (nanoadhesion) level on polymer nanocomposites reinforcement degree, which are filled with organoclay. Intensification of interaction nanofiller-polymer matrix, characterized by parameter b, results to essential (in b times) polymer chains flexibility change in interfacial layer in comparison with bulk polymer matrix. The offered modified equation for nanocomposites reinforcement degree determination, accounting for both polymer matrix molecular characteristics and interfacial adhesion level, gives the good correspondence to experiment.

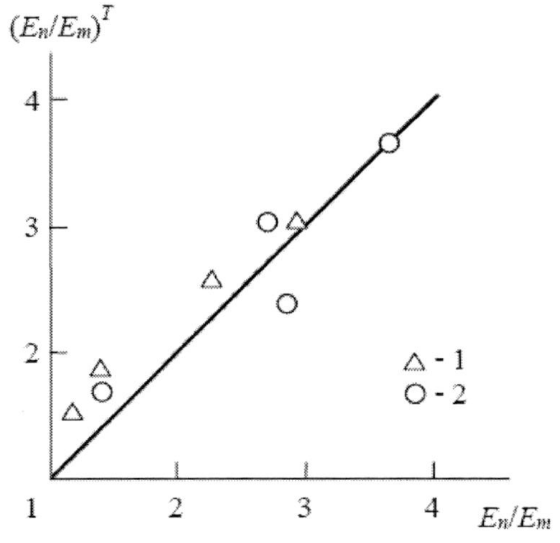

Figure 30. The comparison of experimental E_n/E_m and calculated according to the equation (56) $(E_n/E_m)^T$ reinforcement degree values of nanocomposites PI/MMT-16C (1) and PI/MMT-OM-m (2) [83].

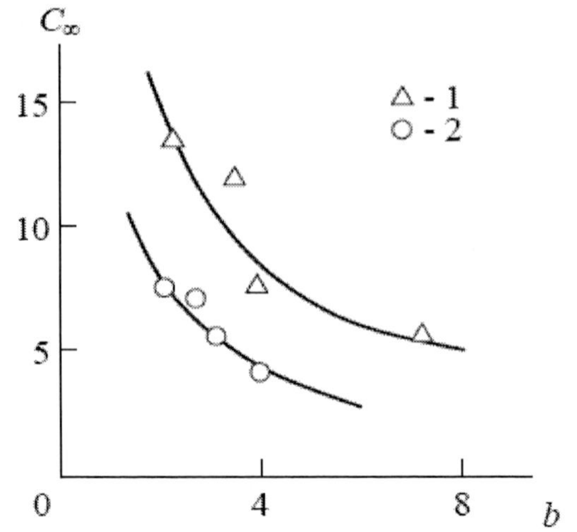

Figure 31. The dependence of characteristic ratio C_∞ on parameter b for nanocomposites LDPE/MMT-1 (1) and LDPE/MMT-2 (2) [87].

At present it is supposed, that interfacial interactions intensification (increase b) facilitates polymer chains arrangement on organoclay platelets surface, increases interfacial regions fraction and rises nanocomposites reinforcement degree [16]. The authors [87] considered more general variant of polymer matrix molecular characteristics (C_∞) and interfacial adhesion level (b) interconnection on the example of nanocomposites LDPE/organoclay [88].

Two kinds of Na^+-montmorillonite were used as nanofiller: the first of them was industrial sample of mark Cloisite 20A (MMT-1) and the second was an experimental sample with application as compatibilized agent of trimethyl hydrogenated-tallow ammonium (MMT-2). In Figure 31 the dependence of characteristic ratio C_∞ on parameter b for two considered series of nanocomposites LDPE/MMT is adduced. As one can see, interfacial adhesion level, characterized by parameter b, reduction results to C_∞ growth, i.e. to l_{st} increase (the equation (27)), since l_0 value is constant and equal to 0.154 nm for polyethylenes [32]. At the first sight such result seems unexpected. Reduction b is due to organoclay contents increase, this effect is common one for nanocomposites polymer/organoclay and is defined by silicate platelets aggregation [16]. In practice this effect is reflected in E_n/E_m growth decelaration or its cessation at W_n increasing within the range of 5-10 mass % [24]. According to the stated above results, interfacial adhesion intensification should be promoted to polymer macromolecular coil "stretching" on montmorillonite platelets surface and, hence, to C_∞ increase. The observed discrepancy can be explained by the obvious fact, that macromolecular coil "stretching" is defined not only by interfacial interaction degree or parameter b, but by nanofiller particles surface area, on which this "stretching" is realized. Since the indicated area is proportional to nanofiller contents W_n, then in Figure 32 the dependence of C_∞ on complex parameter bW_n is adduced, which, as it has been expected, shows C_∞ increase (and, hence, l_{st} enhancement) at bW_n growth. In other words, the increase of both factors (b and W_n, where W_n characterizes contacts nanofiller-polymer number enhancement) promotes to C_∞ growth and nanocomposites reinforcement degree rising (see the equations (27) and (56)). The relation between C_∞ and bW_n, shown in Figure 30, can be expressed analytically as follows [87]:

$$C_\infty = 2 + 0.6bW_n. \tag{57}$$

The characteristic feature of the dependence $C_\infty(bW_n)$ is its extrapolation to C_∞=2 at bW_n=0. Since the condition W_n=0 for nanocomposites has no physical significance, then the indicated extrapolation means, that at interfacial adhesion absence or b=0 macromolecular coil has only tetrahedral valent angles, i.e. limitingly compact macromolecular coil is formed, incapable to “stretching” on silicate platelets surface.

Since bW_n increase results to C_∞ growth, then in this case, interfacial regions relative fraction φ_{if} enhancement, which can be determined with the aid of the equation (8), should be expected. In Figure 33 the dependence of φ_{if} on complex parameter bW_n for the considered nanocomposites is adduced, which is approximated well enough by linear correlation, passing through coordinates origin and described analytically as follows [87]:

$$\varphi_{if} = 1.75 \times 10^{-2} bW_n. \tag{58}$$

The equations (8) and (58) combination demonstrates clearly an interfacial adhesion level role in nanocomposites polymer/organoclay reinforcement process. So, for nanocomposites LDPE/MMT-1 W_n increase from 1 up to 7 mass % results to b reduction from 7.2 up to 2.1. If the value b were not reduced at W_n growth and at W_n=7 mass % were equal to 7.2, then nanocomposite elasticity modulus E_n would reach the value of 2167 MPa, whereas experimental E_n magnitude in this case is equal only to 569 MPa. As it follows from the equation (8), the greatest possible magnitude E_n/E_m=12. At value b=7.2 conservation this magnitude might reach at W_n only 7.6 mass % [87].

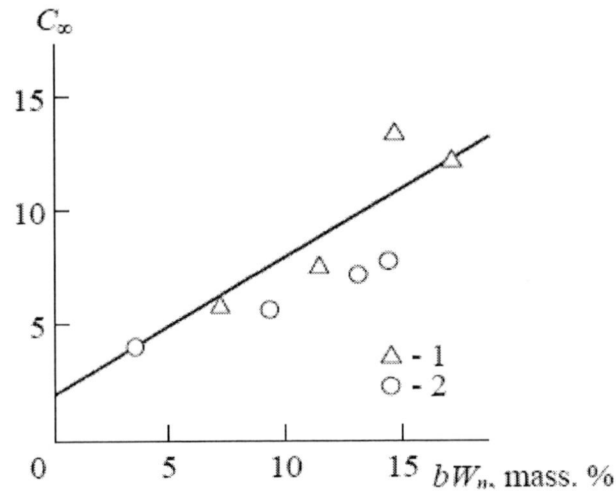

Figure 32. The dependence of characteristic ratio C_∞ on complex parameter bW_n for nanocomposites LDPE/MMT-1 (1) and LDPE/MMT-2 (2) [87].

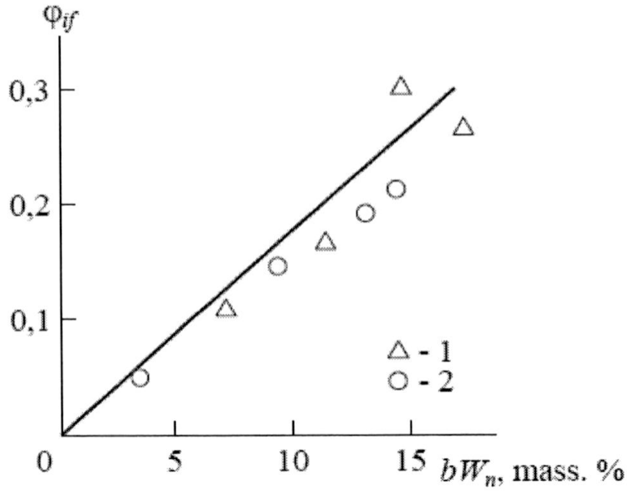

Figure 33. The dependence of interfacial regions relative fraction φ_{if} on complex parameter bW_n for nanocomposites LDPE/MMT-1 (1) and LDPE/MMT-2 (2) [87].

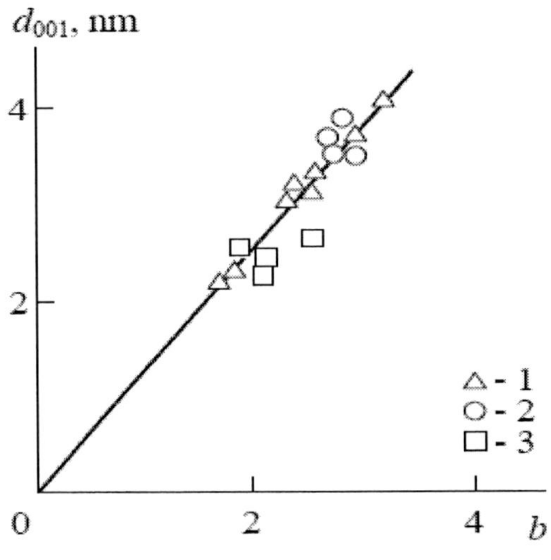

Figure 34. The dependence of interlayer spacing d_{001} on parameter b for nanocomposites HDPE/organoclay (1), PP/organoclay (2) and HDPE/montmorillonite [91] (3) [90].

Thus, the adduced above results demonstrate that characteristic ratio C_∞, which is polymer phase important molecular parameter, depends equally on both interfacial adhesion level and organoclay contents. The minimum value C_∞=2 is reached at interfacial interactions absence irrespective of organoclay contents. The interfacial adhesion level crucial role in nanocomposites polymer/organoclay reinforcement process is shown.

It is well known [23], that interlayer spacing d_{001} is one of the most important characteristics of nanofiller in nanocomposites polymer/organoclay. This parameter characterizes dispersion degree of organoclay in polymer matrix. The value $d_{001}\approx$2-3 nm corresponds to intercalated organoclay and $d_{001}\approx$8-10 nm – to exfoliated one. Layered nanofillers dispersion often (if not always) has the cruiral significance in reinforcement process of nanocomposites polymer/organoclay [16, 23]. It is obvious, that interfacial adhesion level will be influenced on organoclay dispersion degree. An organoclay by no chance is introduced in polymer matrix only with a modifier, that allows to rise interfacial adhesion nanofiller-polymer matrix level [89].

In Figure 34 the dependence $d_{001}(b)$ is adduced, from which interlayer spacing linear increase follows at interfacial adhesion level growth. This dependence is approximated by the following empirical formula [90]:

$$d_{001} = 1.27b, \quad \text{nm.} \tag{59}$$

At b=0 (interfacial adhesion absence) d_{001}=0, i.e. packet (tactoid) of silicate platelets is not split up (not dispersed). Exfoliation, i.e. the condition d_{001}=8-10 nm reaching, is realized at b=6.30-7.90. The adduced results confirm, that nanoadhesion effect realization is necessary for exfoliated organoclay receiving. In Figure 34 the dependence of d_{001} values, obtained experimentally (with the aid of X-raying), on b for nanocomposites HDPE/MMT according to the data of paper [91]. As one can see, these data correspond well to the correlation $d_{001}(b)$, calculated theoretically [90].

CONCLUSIONS

Thus, the obtained in the present review results have demonstrated, that polymer phase in polymer nanocomposites of any class is a complex system, moreover dividing into two structurally distinguishing parts: bulk polymer matrix and interfacial regions. As it has been expected, the characteristic ratio, which characterizes polymer chain statistical flexibility, is the main parameter, defining polymer phase structure changes (crystallinity degree, interfacial regions relative fraction and so on) at nanofiller introduction. The interfacial adhesion nanofiller-polymer matrix level influences strongly on the indicated characteristic. The relationships, obtained with accounting for these factors real values, allow to predict polymer nanocomposites limiting characteristics.

REFERENCES

[1] Ahmed S., Jones F.R. *J. Mater. Sci.,* 1990, v. 25, № 12, p. 4933-4942.

[2] Bobryshev A.N., Kozomazov V.N., Babin L.O., Solomatov V.I. Synergetics of Composite Materials. Lipetsk, NPO ORIUS, 1994, 154 p.

[3] Novikov V.U., Kozlov G.V. *Mekhanika Kompozitnykh Materialov*, 1999, v. 35, № 3, p. 269-290.

[4] Yanovskii Yu.G. Nanomechanics and Composite Materials Strength. Moscow, Publishers IPRIM RAS, 2008, 180 p.

[5] Solomko V.P. A Filled Crystallizing Polymers. Kiev, Naukova Dumka, 1980, 264 p.

[6] Aloev V.Z., Kozlov G.V. The Physics of Orientation Phenomena in Polymer Materials. Nal'chik, Polygraphservice and T, 2002, 288 p.

[7] Kozlov G.V., Zaikov G.E. *Izvestiya KBSC RAS*, 2003, № 1, p. 54-57.

[8] Tanniru M., Misra R.D.K. *Mater. Sci. Engng*., 2005, v. 405A, № 1, p. 178-193.
[9] Deshmane C., Yuan Q., Misra R.D.K. *Mater. Sci. Engng*., 2007, v. 452-453A, № 3, p. 592-601.
[10] Dolbin I.V., Burya A.I., Kozlov G.V. *Kompozitnye Materialy*, 2008, v. 2, № 1, p. 3-7.
[11] Aphashagova Z.Kh., Kozlov G.V., Malamatov A.Kh. *Nanotekhnika*, 2009, № 1, p. 40-44.
[12] Wunderlich B. Macromolecular Physics. Vol. 2. Crystal Nucleation, Growth, Annealing. New York, San Francisco, London, Academic Press, 1976, 556 p.
[13] Aloev V.Z., Kozlov G.V., Zaikov G.E. *Russian Polymer News*, 2001, v. 6, № 4, p. 63-66.
[14] Kozlov G.V., Zaikov G.E. Structure of the Polymer Amorphous State. Leiden-Boston, Brill Academic Publishers, 2004, 465 p.
[15] Wu S. *J. Appl. Polymer Sci*., 1992, v. 46, № 4, p. 619-624.
[16] Mikitaev A.K., Kozlov G.V., Zaikov G.E. Polymer Nanocomposites: the Variety of Structural Forms and Applications. New York, Nova Science Publishers, Inc., 2009, 319 p.
[17] Pfeifer P. In book: Fractals in Physics. Ed. Pietronero L., Tosatti E. Amsterdam, Oxford, New York, Tokyo, North-Holland, 1986, p. 72-81.
[18] Privalko V.P., Kawai T., Lipatov Yu.S. *Polymer J*., 1979, v. 11, № 9, p. 699-709.
[19] Kozlov G.V., Zaikov G.E. *Materialovedenie,* 2002, № 12, p. 13-17.
[20] Shklovskii B.I., Efros A.L. *Uspekhi Fizicheskikh Nauk*, 1975, v. 117, № 3, p. 401-436.
[21] Gopakumar T.G., Lee J.A., Kontopoulou M., Parent J.S. *Polymer*, 2002, v. 43, № 16, p. 5483-5491.
[22] Dzhangurazov B.Zh., Kozlov G.V., Ovcharenko E.N., Mikitaev A.K. *Prikladnaya Fizika,* 2011 (in press).
[23] Sheng N., Boyce M.C., Parks D.M., Rutledge G.C., Abes J.I., Cohen R.E. *Polymer,* 2004, v. 45, № 2, p. 487-506.
[24] Antipov E.M., Barannikov A.A., Gerasin V.A., Shklyaruk B.F., Tsamalashvili L.A., Fischer H.R., Razumovskaya I.V. *Vysokomolek. Soed.* A, 2003, v. 45, № 11, p. 1885-1899.
[25] Family F. J. *Stat. Phys.,* 1984, v. 36, № 5/6, p. 881-896.
[26] Kozlov G.V., Aphashagova Z.Kh., Malamatov A.Kh. *Nanotekhnika*, 2011 (in press).
[27] Budtov V.P. Physical Chemistry of Polymer Solutions. Sankt-Peterburg, *Khimiya,* 1992, 384 p.
[28] Balankin A.S. Synergetics of Deformable Body. Moscow, Publishers Ministry of Defence SSSR, 1991, 404 p.
[29] Kozlov G.V., Sanditov D.S. Anharmonic Effects and Physical-Mechanical Properties of Polymers. Novosibirsk, *Nauka*, 1994, 261 p.
[30] Wu S. J. *Polymer Sci.: Part B: Polymer Phys*., 1989, v. 27, № 4, p. 723-741.
[31] Feder F. Fractals. New York, Plenum Press, 1989, 252 p.
[32] Aharoni S.M. *Macromolecules*, 1983, v. 16, № 9, p. 1722-1728.
[33] Kozlov G.V., Sultonov N.Zh., Mikitaev A.K. Proceedings of VI Intern. Sci.-Techn. Conf. "Machines Study and Planning Problems". Penza, PSU, 2010, p. 87-93.
[34] Sultonov N.Zh., Mikitaev A.K. Proceedings of VI Intern. Sci.-Pract. Conf. "New Polymer Composite Materials". Nal'chik, KBSU, 2010, p. 392-398.

[35] Kalinchev E.L., Sakovtseva M.B. Properties and Processing of Thermoplastics. Leningrad, *Khimiya*, 1983, 288 p.
[36] Aharoni S.M. *Macromolecules*, 1985, v. 18, № 12, p. 2624-2630.
[37] Chang J.-H., An Y.U., Kim S.J., *Im S. Polymer*, 2003, v. 44, № 15, p. 5655-5661.
[38] Chang J.-H., Seo B.-S., Wang D.-H. *Polymer*, 2002, v. 43, № 7, p. 2969-2974.
[39] Liang Z.-M., Yin J., Xu H.-J. *Polymer*, 2003, v. 44, № 4, p. 1391-1399.
[40] Chen J.-S., Poliks M.D., Ober C.K., Zhang Y., Wiesner U., Giannelis E. *Polymer*, 2002, v. 43, № 13, p. 4895-4904.
[41] Huang J., He C., Xiao Y., Mya K.Y., Dai J., Siow Y.P. *Polymer*, 2003, v. 44, № 12, p. 4491-4499.
[42] Tugov I.I., Shaulov A.Yu. *Vysokomolek. Soed.* B, 1990, v. 32, № 7, p. 527-529.
[43] Kozlov G.V., Yanovskii Yu.G., Zaikov G.E. Structure and Properties of Particulate-Filled Composites: the Fractal Analysis. New York, Nova Science Publishers, Inc., 2010, 349 p.
[44] Malamatov A.Kh., Burya A.I., Kozlov G.V. *Sovremennye Naukoemkie Tekhnologii*, 2005, № 11, p. 16-18.
[45] Kozlov G.V., Burya A.I., Lipatov Yu.S. *Doklady NAN Ukraine*, 2008, № 1, p. 132-136.
[46] Kozlov G.V., Burya A.I., Zaikov G.E. *J. Balkan Tribologic. Assoc.*, 2007, v. 13, № 4, p. 475-479.
[47] Kozlov G.V., Yanovskii Yu.G., Zaikov G.E. Synergetics and Fractal Analysis of the Polymer Composites Filled with Short Fibers. New York, Nova Science Publishers, Inc., 2010, 248 p.
[48] Yoon P.J., Hunter D.L., Paul D.R. *Polymer*, 2003, v. 44, № 14, p. 5323-5339.
[49] Kozlov G.V., Yanovskii Yu.G., Lipatov Yu.S. Mekhanika Kompozitsionnykh Materialov i Konstruktsii, 2002, v. 8, № 1, p. 111-149.
[50] Hentschel H.G.E., Deutch J.M. *Phys. Rev.* A, 1984, v. 29, № 12, p. 1609-1611.
[51] Malamatov A.Kh., Kozlov G.V., Antipov E.M., Mikitaev A.K. *Perspektivnye Materialy*, 2006, № 5, p. 54-58.
[52] Chen T.K., Tien Y.I., Wei K.H. *Polymer,* 2000, v. 41, № 3, p. 1345-1353.
[53] Kim B.K., Seo J.W., Jeong H.M. *Eur. Polymer J.*, 2003, v. 39, № 1, p. 85-91.
[54] Dzhangurazov B.Zh., Kozlov G.V., Mikitaev A.K. Proceedings of 29-th Intern. Conf. “Composite Materials in Industry”. Kiev, Nauka, Tekhnika, Tekhnologiya, 2009, p. 325-327.
[55] Aphashagova Z.Kh., Kozlov G.V., Burya A.I., Mikitaev A.K. *Materialovedenie*, 2007, № 9, p. 10-13.
[56] Edwards D.C. *J. Mater. Sci.*, 1990, v. 25, № 12, p. 4175-4185.
[57] Kozlov G.V., Burya A.I., Lipatov Yu.S. *Mekhanika Kompozitnykh Materialov*, 2006, v. 42, № 6, p. 797-802.
[58] Kozlov G.V., Yanovskii Yu.G., Karnet Yu.N. *Mekhanika Kompozitnykh Materialov i Konstruktsii*, 2005, v. 11, № 3, p. 446-450.
[59] Dzhangurazov B.Zh., Kozlov G.V., Mikitaev A.K. Proceedings of the Third Intern. Conf. “Deformation and Fracture of Materials and Nanomaterials – DFMN-2009”, Moscow, 2009, p. 142-143.
[60] Kozlov G.V., Serdyuk V.D., Dolbin I.V. *Materialovedenie*, 2000, № 12, p. 2-5.
[61] Aphashagova Z.Kh., Kozlov G.V., Malamatov A.Kh. *International Journal of Applied and Fundamental Researches*, 2009, № 3, p. 38-41.

[62] Dzhangurazov B.Zh., Kozlov G.V., Zaikov G.E., Mikitaev A.K. *Nanomechanics Sci. and Techn.*, 2010, v. 1, № 1, p. 31-48.

[63] Khabashesku V.N., Barrera E.V., Lobo R.F.M. *Current Research on Nanotechnology*, 2007, v. 1, № 2, p. 165-182.

[64] Yanovskii Yu.G., Kozlov G.V., Burya A.I., Lipatov Yu.S. *Fizicheskaya Mezomekhanika*, 2007, v. 10, № 6, p. 63-67.

[65] Wu S.-H., Wang F.-Y., Ma C.-C.M., Chang W.-C., Kuo C.-T., Kuan H.-C., Chen W.-J. *Mater. Lett.*, 2001, v. 49, № 2, p. 327-333.

[66] Kozlov G.V., Burya A.I., Aphashagova Z.Kh., Mikitaev A.K. *Nanotekhnika*, 2008, № 2, p. 33-36.

[67] Zaikov G.E., Kozlov G.V., Aphashagova Z.Kh., Mikitaev A.K. *J. Appl. Polymer Sci.*, 2009, v. 111, № 5, p. 2621-2624.

[68] Kausch H.H. Polymer Fracture. Berlin, Heidelberg, New York, Springer-Verlag, 1978, 423 p.

[69] Ryabenko A.G. Proceedings of Intern. Conf. "Deformation and Fracture of Materials and Nanomaterials", DFMN-2007, Moscow, 2007, p. 411-412.

[70] Bucknall C.B. Toughened Plastics. London, *Applied Science*, 1977, 328 p.

[71] Bartenev G.M., Frenkel S.Ya. Physics of Polymers. Leningrad, *Khimiya*, 1990, 432 p.

[72] Kozlov G.V., Misra R.D.K., Aphashagova Z.Kh. *Nanotekhnika*, 2009, № 2, p. 71-74.

[73] Kuzeev I.R., Samigullin G.Kh., Kulikov D.V., Zakirnichnaya M.M. A Complex Systems in Nature and Engineering. Ufa, Publishers USNTU, 1997, 225 p.

[74] Aphashagova Z.Kh., Kozlov G.V., Burya A.I., Malamatov A.Kh. Proceedings of Khar'kovskoi Nanotekhnologicheskoi Assamblei, Khar'kov, KhSU, 2007, p. 204-205.

[75] Bashorov M.T., Kozlov G.V., Mikitaev A.K. *Deformatsiya i Razrushenie Materialov*, 2009, № 11, p. 37-39.

[76] Kozlov G.V., Zaikov G.E. In book: Fractals and Local Order in Polymeric Materials. Ed. Kozlov G., Zaikov G. New Yprk, Nova Science Publishers, Inc., 2001, p. 55-63.

[77] Fermeglia M., Ferrone M., *Pricl S. Fluid Phase Equilibrium*, 2003, v. 212, № 2, p. 315-329.

[78] Pernyeszi T., Dekany I. *Colloid Polymer Sci.*, 2003, v. 281, № 1, p. 73-78.

[79] Dzhangurazov B.Zh., Kozlov G.V., Mikitaev A.K. Proceedings of V Intern. Sci.-Techn. Conf. "Progressive Technologies in Modern Mechanical Engineering". Penza, PSU, 2009, p. 35-38.

[80] Avnir D., Farin D., Pfeifer P. *Nature,* 1984, v. 308, № 5959, p. 261-263.

[81] Stanley H.E. In book: Fractals in Physics. Ed. Pietronero L., Tosatti E. Amsterdam, Oxford, New York, Tokyo, North-Holland, 1986, p. 463-477.

[82] Kozlov G.V., Burya A.I., Yanovskii Yu.G., Aphashagova Z.Kh. *Nanotekhnika*, 2008, № 1, p. 81-85.

[83] Dzhangurazov B.Zh., Kozlov G.V., Mikitaev A.K. Proceedings of Inter. Meeting "Low Dimensional Systems". Rostov-na-Donu, SKHTs VSh YuFU, 2010, p. 126-129.

[84] Liang Z.-M., Yin J., Wu J.-H., Qiu Z.-X., He F.F. *Eur. Polymer J.*, 2004, v. 40, № 2, p. 307-314.

[85] Holliday L., Robinson J.D. In book: Polymer Engineering Composites. Ed. Richardson M.O.W. London, Applied Science Publishers LTD, 1978, p. 241-283.

[86] Kozlov G.V., Burya A.I., Yanovskii Yu.G., Zaikov G.E. In book: Progress in Monomers, Oligomers, Polymers, Composites and Nanocomposites. Ed. Pethrick R., Zaikov G., Pielichowski J. New York, Nova Science Publishers, Inc., 2009, p. 343-359.
[87] Dzhangurazov B.Zh., Kozlov G.V., Mikitaev A.K. *Kondensirovannye Sredy i Mezhfaznye Granitsy,* 2010, v. 12, № 2, p. 119-122.
[88] Hotta S., Paul D.R. *Polymer,* 2004, v. 45, № 22, p. 7639-7654.
[89] Ranage A., Nayak K., Fairbrother D., D'Souza N.A. *Polymer*, 2005, v. 46, № 21, p. 7323-7333.
[90] Dzhangurazov B.Zh., Kozlov G.V., Mikitaev A.K. *Poverkhnost',* 2011 (in press).
[91] Gianelli W., Ferrara G., Camino G., Pellegatti G., Rosenthal J., Trombini R.C. *Polymer,* 2005, v. 46, № 21, p. 7037-7046.

In: Polymer Phase Behavior
Editors: T. P. Ehlers and J. K. Wilhelm, pp. 171-203
ISBN: 978-1-61324-336-7
© 2011 Nova Science Publishers, Inc.

Chapter 4

PHASE INVERTING POLYMER SYSTEMS IN DRUG DELIVERY AND MEDICINE

Luis Solorio,* Loran D. Solorio,# Sarah Gleeson,† Alexander M. Olear,‡ Angela N. Carlson¶ and Agata A. Exner•

Department of Biomedical Engineering,
Case Western Reserve University,
Cleveland, Ohio, USA

ABSTRACT

Phase inverting polymer systems are primarily utilized in industrial applications such as the microfiltration of bacteria and reverse osmosis, but their use has been rapidly expanding in other areas. In the medical field the predominant role of these systems has been in development of new biomaterial matrixes for drug delivery and tissue engineering. The use of phase inverting systems for the controlled release of therapeutic agents is of interest due to the injectable nature of the implants, which provides a less invasive means of physically placing the implant at or near the site of action. The goals of this chapter are to: provide a basic description of the phase inversion process involved in medical implants, describe factors that affect the phase inversion and drug release processes, overview the techniques used to characterize these systems, and provide insight into the *in vivo* behavior to include biocompatibility and deviations from *in vitro* behavior. *In situ* forming implant systems are an exciting field of study, and have been successfully used to treat diseases that range in severity from prostate cancer to periodontitis. These systems provide a compelling alternative to preformed polymer

* Phone: 216-844-0077. fax: 216-844-5922. e-mail: luis.solorio@case.edu.

Phone: 216-502-1537, fax: 216-844-5922. e-mail: loran.vieregge@case.edu.

† Phone: 216-983-3011, fax: 216-844-5922. e-mail: sxg373@case.edu.

‡ Phone: 216-983-3011 fax: 216-844-5922. e-mail: amo29@case.edu.

¶ Phone: 216-983-3011, fax: 216-844-5922. e-mail: anc8@case.edu.

• Phone: 216-844-3544. e-mail: agata.exner@case.edu.

implants, and may prove to be paramount in overcoming the intrinsic obstacles of the physical targeting of polymer implants for the local delivery of therapeutic agents.

Keywords: *in situ forming implants, drug delivery, phase inversion, ultrasound, in vivo*

1. INTRODUCTION

Various polymeric biomaterials play a significant role in the fields of medicine and biomedical engineering. One particularly prolific area of biomaterials research is the development of new functional delivery systems for the administration of drugs. In phase inverting drug delivery systems, a therapeutic agent is packaged into a protective carrier typically comprised of a biocompatible, biodegradable polymer that facilitates its diffusion in the body and activity at the target tissue. These *in situ* forming drug depots are uniquely advantageous because they are injectable, providing a noninvasive method for the localized delivery of therapeutic agents and potentially protecting their bioactivity. This chapter will provide a basic description of the phase inversion process that occurs in *in situ* forming implants as well as a discussion of implant and environmental factors affecting phase inversion and drug release. Additionally, the characterization techniques for these systems will be overviewed, and there *in vitro* behavior will be considered.

Drugs can be administered in a variety of ways, ranging from swallowing a pill or applying a dermal patch, to surgically implanting a controlled drug delivery device [1, 2]. While the vast majority of drugs are given orally, factors such as rapid metabolism, poor oral bioavailability, narrow therapeutic index, and nonspecific targeting can render oral delivery methods undesirable or impossible [1, 2]. The development of more sophisticated parenteral methods of drug delivery has become critical as treatment options advance and become more complex. One example that highlights the immediate need for more advanced delivery techniques is the administration of drugs for the treatment of cancer. For many chemotherapeutic treatments, highly toxic drugs must be ingested, traveling through the gastrointestinal (GI) tract before entering the circulatory system. Once the drug is absorbed into the circulatory system, or following intravenous injection which bypasses the GI phase, it can be metabolized, eliminated, or passed into the interstitial space of both diseased and healthy tissues. If the drug reaches the cancerous lesion, it must then diffuse against pressure gradients induced by poor lymphatic drainage in order to reach the target, a single cancer cell. Since many chemotherapeutic drugs act on cancer cell by targeting their propensity for cell proliferation, systemic exposure to these toxic drugs affects healthy, mitotically active tissues such as hair follicles, bone marrow and the lining of the GI tract. As a result, these patients often experience hair loss, nausea, anemia, and infection due to immune suppression. More advanced methodologies of drug administration could enable local rather than systemic chemotherapeutic drug delivery, potentially limiting the adverse effects on healthy tissues.

Pre-formed polymer implants have been developed as a method of protecting drugs from harsh physiological conditions as well as providing a means for localizing a therapeutic agent to a target area. In these implants, a drug is encapsulated in an implantable polymer matrix of fixed geometry then surgically inserted directly into the site of action. Utilizing this system, high local levels of drug can be delivered without systemic involvement, ultimately reducing side effects and protecting drug activity. One of the most commercially successful pre-formed depots for local drug delivery is the Gliadel® wafer, used to treat malignant glioma as an adjunct to surgical resection and radiation. Gliadel® wafers, composed of poly(bis[p-carboxyphenoxy]) propane:sebacic acid (PCPP:SA), degrade by surface erosion to locally release carmustine. When placed directly into voids left after tumor resection, the wafers release carmustine into the adjacent tissue space [3-10], treating residual tumor cells along the periphery of the resection zone. These implants have shown modest success in both extending the patient survival time (by as much as two months), as well as increasing the patient's chances for survival (by as much as 25%) [10].

While some drugs may require local delivery to provide the most effective treatment, other situations may necessitate a sustained, system-wide drug administration. Percutaneous placement of pre-formed drug eluting implants has provided a means by which constant plasma levels of drug can be achieved. In the treatment of prostate cancer, one form of therapy was daily injections of leuprolide acetate, a potent gonadotropin hormone-releasing hormone agonist, which reduces circulating levels of androgens. As an alternative to daily injections, the Leupron® Depot was developed and launched in 1989 [11-14]. The Leupron® Depot is a polymer microsphere formulation injected subcutaneously to provide sustained delivery of leuprolide acetate for up to 4 months [14]. With this treatment, patient compliance improved due to a single quarterly injection replacing inconvenient and potentially painful daily injections [5, 9, 11, 15-23]. The Leupron® Depot has since been used as a treatment for a number of other hormonal disorders as well, including endometriosis and precocious puberty[13].

Despite the benefits of drug encapsulation in polymers, both pre-formed implants and injectable microsphere suspensions have disadvantages. For example, one major drawback of the Leupron® Depot is the complicated and expensive multistep process necessary for microsphere fabrication. Additionally, if a complication arises due to the treatment, removal of the microspheres is a difficult process [12, 14]. Preformed implants such as the Gliadel® Wafer have a more favorable manufacturing cost, since a large number of these implants are formed through compression molding and extrusion techniques, but they have limited application because they require surgical placement and simple modifications in formulation to customize the system can be difficult to implement [5, 9].

An alternative class of drug eluting implants was first described by Dunn et al. in a series of patents that described a system utilizing biocompatible, biodegradable polymers dissolved in a biocompatible solvent that could be mixed with an active therapeutic agent [24, 25]. This system provided a means by which many of the complications involved in drug eluting depots could be circumvented [12-14, 19, 23-26]. Upon injection of the polymer solution into the body via a syringe, a solid drug eluting depot forms *in situ* through a process known as phase inversion [9]. Phase inversion refers to the process by which a polymer precipitate is formed through the immersion of the polymer implant solution into a bath that is miscible with the implant solution solvent but acts as a nonsolvent for the polymer phase [27, 28]. Once the implant solution is in contact with the nonsolvent bath, mass transfer of solvent and nonsolvent occurs, resulting in an unstable ternary system [28, 29]. With continued mass transfer of solvent and nonsolvent, stability of the system is restored. The mass transfer results in liquid-liquid demixing that facilitates precipitation of the polymer from the solvent/nonsolvent milieu [22, 28-32]. While the use of phase-inverting polymer systems is a well established area of research for the fabrication of asymmetric and symmetric membranes, the use of phase inversion to design drug eluting polymer depots was not established until 1990 [11-14, 16, 19, 22, 24, 26, 28, 29].

Phase-inverting drug depot systems have a number of advantages over traditional preformed implants. *In situ* forming implants (ISFIs) provide a means by which a drug eluting depot can be delivered in a minimally invasive manner directly to the site of action through injection of the polymer solution. The placement of the implant at the site of action ultimately reduces systemic exposure to therapeutic agents while maintaining elevated local levels of the drug [5, 9]. In addition to local delivery of therapeutic agents, these implants can be injected subcutaneously to achieve sustained delivery of therapeutic levels of quickly eliminated hormones or growth factors for the treatment of diseases such as diabetes and prostate cancer [5, 9, 11-15, 17, 21, 22, 26, 33-46].

A large body of literature outlines many of the parameters affecting drug release rates from these phase-inverting implants. The goals of this chapter are to discuss the methods used to characterize ISFIs both *in vivo* and *in vitro*, outline factors that affect drug release profiles from these implants, highlight the issues of biocompatibility, and finally present an overview on the effects of injection site on the release behavior of these implants.

2. Techniques for Evaluating the Phase Inversion Process

There are several commonly used methods for the characterization of *in situ* forming polymer implants, and each provides distinct information about the structure and dynamics of the phase inversion process. In this section, three specific modes of analysis are considered: dark ground optics, electron paramagnetic resonance (EPR), and ultrasound imaging. The fundamental principles behind each technique are discussed as well as their advantages and limitations in terms of describing the behavior of phase-inverting systems. The data extracted using these various modalities includes information used to determine water diffusion, drug release, gel formation, polymer erosion, and implant geometry. Only EPR and ultrasound can be used for continuous *in vivo* implant analysis, which gives insight into differences in behavior between *in vivo* and *in vitro* implants.

2.1. Dark Ground Optics

Dark ground optics is an imaging method developed by McHugh and colleagues which provides information regarding the mechanisms of polymer phase inversion [16, 19, 22, 31, 32, 47-49]. This characterization technique is employed when a polymer solution is exposed to a nonsolvent bath, creating both dark ground and reflected light images [31]. The process is simple and quick, typically requiring less than three minutes of total run time and an uncomplicated data analysis [49]. From the dark ground video images, information about diffusion coefficients, morphology, liquid-liquid phase separation and gel formation can be obtained, enabling the characterization of phase inversion behavior. Using this technique Brodbeck et al. determined the protein release kinetics of a PLGA solution by monitoring the propagation rate of the liquid demixing region [16]. Graham et al. also studied a PLGA system and used dark ground techniques to characterize the effect of formulation changes on drug release mechanisms *in vitro* [19].

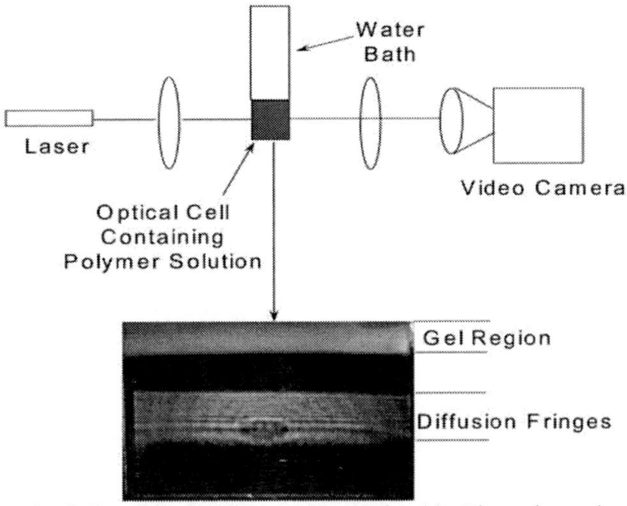

"Reprinted from Journal of Controlled Release, 58, Author(s), Phase inversion dynamics of PLGA solutions related to drug delivery, 233-245, Copyright (1999) with permission from Elsevier."

Figure 1. Schematic of the dark ground imaging system, and a representative image showing the gel region and the diffusion fringes.

Dark ground imaging is used to track the motion of fronts during phase inversion, and can also be used to monitor changes in solvent concentration at the polymer/nonsolvent interface [50]. Figure 1 shows the arrangement of the imaging system. A polymer solution is situated within an optical quality quartz cell and placed in a nonsolvent bath. A thin metal foil prevents any initial interaction between the solvent and bath-side, and the phase inversion process begins following the removal of the foil when the two liquids come into contact [19]. Collimating lenses focus a filtered laser onto the quartz cell. After passing through the quartz cell, the diffraction image of the refracted laser beam is filtered onto the back focal plane of the transform lens with a circular stop. This ensures that the unrefracted light is eliminated before the image is focused through a camera. The fringes are recorded by the camera and overlaid on the cell to visualize the relative fringe locations [31]. As water diffuses into the polymer system, interference fringes are created and dark ground video imaging records the concentration gradient that forms [51]. The interference fringes indicate the depth of water diffusion into the solution, and appear as striations in the image (Figure 1). Using reflected light, the region of the polymer solution which becomes phase-separated can be visualized. By monitoring reflected light and refractive index distributions, diffusion and liquid-liquid gelation fronts are identified relative to the initial thin film [31]. Both fronts move farther into the polymer region as more nonsolvent comes in contact with the solution. Figure 1 shows a dark ground image where both diffusion and gelation fronts can be seen. By tracking the position of the two fronts over time, a plot of position squared vs. time can be created for each. The diffusion fringe indicates the rate at which water enters the polymer system, and the liquid-liquid fringe reveals the rate of gelation of the system [19]. Transmitted light images can be analyzed to obtain the precipitation times [31].

Through the use of dark ground imaging, the relationship between phase inversion and implant morphology and the resultant effects on drug release have been established [19]. Graham et al. found that an increase in the gelation rate of a solution correlates to a greater initial drug burst, which suggests that drugs diffuse more quickly through a gelled system than a single-phase solution [19]. McHugh et al. also acquired data which indicates that the nonsolvent diffusion rates and gel formation kinetics ultimately affect drug diffusion through the phase-inverted polymer system [31]. This information provides a basis for the manipulation of system properties in order to control drug release kinetics. In addition, dark ground optical analysis gives insight into the influence of solvent and nonsolvent compositions on the dynamics and release characteristics of phase-inverting systems.

One limitation of the dark ground optics system is that it can only image phase-inverting systems comprised of thin polymer films. Additionally, this technique only applies to shorter time courses ranging from a few minutes to several hours [52]. The use of other imaging modalities is necessary for longer-term studies such as implant degradation analyses. *In vitro* phase inversion systems can be characterized with dark ground optics, but no *in vivo* images can be obtained.

2.2. Electron Paramagnetic Resonance Spectroscopy

The exploration of *in vivo* behavior of phase-inverting implants is crucial to the design of effective drug delivery systems. Common *in vivo* characterization methods including IR spectroscopy, chromatography, and calorimetry require implant extraction from the tissue and cannot provide continuous serial data [53]. One technique which provides noninvasive analysis of phase inversion is Electron Paramagnetic Resonance (EPR) spectroscopy, also called Electron Spin Resonance (ESR), and this method has recently been employed to study ISFIs. EPR is based upon the interaction of electrons with magnetic fields due to the magnetic moments of the electrons. These magnetic moments cause electrons to align with the applied magnetic field within the resonator. Electromagnetic radiation is applied to a sample, exciting the unpaired electrons at a recognizable resonance frequency [54]. *In vivo* frequencies range from 9000 MHz to 250 MHz [55].

Continuous-wave EPR (CW-EPR) is the most effective and commonly used EPR method for *in vivo* characterization. This machine has a resonator chamber into which the sample is placed. The spectrometer is tuned to an optimal frequency and matched so that electromagnetic waves are not reflected back from the sample. The CW-EPR apparatus directs microwaves at the resonator cavity at the resonance frequency. The spectrometer applies a magnetic field at a value below the resonance level and slowly increases the field above the resonant frequency. This magnetic field sweep causes all unpaired electrons to switch their alignment in the field, which in turn changes the matching of the resonator cavity. Paramagnetic species are recognized because the microwaves they reflect back are detected and compiled into an absorption EPR spectrum. In order to receive a sufficiently strong signal, magnetic field modulation is used. An oscillating magnetic field with low amplitude is applied during the initial magnetic field sweep and modulates the amplitude of the signal at the resonance frequency [54]. Another frequently used EPR method is Spectral Spatial EPR imaging, which is able to make a localized assessment to determine the layer of the polymer system which corresponds to the EPR spectra [56].

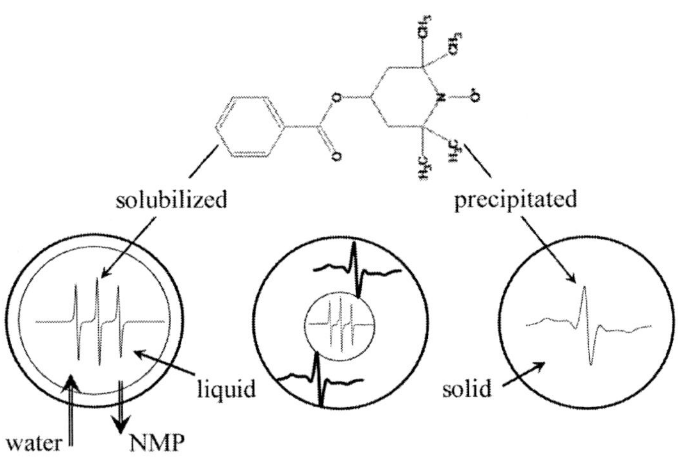

"Journal of Controlled Release, 130, S. Kempe, H. Metz, K. Mäder, Do in situ forming PLG/NMP implants behave similar in vitro and in vivo? A non-invasive and quantitative EPR investigation on the mechanisms of the implant formation process, 220–225, Copyright (2008), with permission from Elsevier."

Figure 2. Schematic of the effect of polymer precipitation on the EPR signal using a tempolbenzoate (TB) probe.

The EPR spectrometer identifies paramagnetic compounds such as free radicals and is able to characterize samples based on the interaction of these species with their environment [52]. It is non-invasive, non-destructive, and continuous and therefore a time course of data can be taken from a single implant in a live animal. Discrepancies between *in vitro* and *in vivo* data can be examined to determine how drug release dynamics and implant formation are affected by the physiological conditions [57]. Mäder et al. found that for P(CPP-SA)-polymers, the drug release rate is 1.5 times faster *in vitro* than *in vivo* [53]. Other studies have shown that *in vitro* systems undergo a more complete drug release than the same systems *in vivo* [58]. By understanding the differences in release mechanisms inside the body, ideal drug delivery systems can be designed while taking into account variable *in vivo* factors such as pH, water content, and enzyme activity [54].

Many properties of polymer phase inversion systems can be studied using EPR. EPR spectra indicate several characteristics of the sample including the spatial distribution of micropolarity, microacidity, microviscosity, and nitroxide concentration [54]. These parameters provide information on the internal polymer environment from which the drug is released, and they can be tailored to modify the release behavior [57]. The drug release and polymer erosion mechanisms can be profiled by continuously measuring drug concentration and mobility through the system over time [58]. Additionally, EPR has provided a means for the *in vivo* measurement of polymer degradation and water penetration. Solvent/nonsolvent interaction dynamics and exchange can also be monitored due to the sensitivity of EPR to motion and polarity within an implant. Kempe et al. traced the kinetics of the exchange between the solvent NMP and water within PLGA implants both *in vitro* and *in vivo* using EPR [57]. The presence of paramagnetic species and their relative amounts can be concluded from EPR information, as well as their physical and chemical properties. EPR is effective for the characterization of multiphase systems, non-transparent samples and small-scale measurements [54]. Non-invasive and continuous *in vivo* monitoring of polymer systems is also possible through other means such as radioactive labeling, but EPR has the advantage of providing valuable structural information [59].

The results of EPR spectroscopy are used to determine parameters such as the hyperfine coupling parameter and rotational correlation time [52]. The hyperfine coupling parameter is calculated as the distance between the high and low field lines on EPR spectra, and is directly related to the polarity of the environment. The rotational correlation time designates the microviscosity of the environment and measures rotational movement of spin markers [57].The presence of magnetic nuclei in a sample can cause more than one resonance and split the EPR spectrum into three lines when they interact with paramagnetic species, as seen in Figure 2 [54]. EPR has had many pharmaceutical applications as well, such as *in vivo* studies on the behavior and effects of antacids in mice [56].

Most drug delivery polymer systems are diamagnetic instead of paramagnetic and therefore are undetectable using EPR. The addition of paramagnetic compounds to the sample is often necessary for EPR detection. These species, such as free radicals, act as spin probes and are identified by the spectrometer at the resonance frequency [52]. Stable nitroxide free radicals are common additives to polymer systems, and the concentrations of these paramagnetic materials are discovered using EPR [60]. Through varying the properties of the nitroxide used, optimal EPR results can be obtained [53]. Copper and manganese are other examples of substances which are paramagnetic and can be added to polymer systems [54]. Paramagnetic species which are naturally occurring in living organisms include: molecular oxygen, free radical intermediates of metabolism and drugs, and metal ions in paramagnetic states. These compounds are useful for *in vivo* detection but are typically found in low concentrations and require a supplemental paramagnetic compound [55]. Certain nitroxides are conducive to measuring proton activity within a polymer sample using EPR, which provides information regarding pH distribution and microacidity [52]. In dry environments, irradiation can form free radicals that are stable yet disappear upon contact with water. In this manner, paramagnetic species can measure the kinetics of water penetration into an implant [54].

Mäder and coworkers were the first to utilize EPR for the purpose of characterizing *in vivo* biodegradable implants, and this technique has proven to be effective [59]. EPR is a desirable analytical method since low frequency spectrometers enable monitoring of micro-viscosity which is can be related to the phase inversion (Figure 2). This system can monitor implants *in vivo*, making EPR one of the only non-invasive *in vivo* characterization techniques currently in use [51].

EPR has limitations on the type and accuracy of the data collected. EPR cannot produce images of implant systems, and instead the data from the spectra are analyzed for characterization. Additionally, small molecular weight (Mw) spin probes are required for EPR measurements [51]. Another complication for *in vivo* studies is that levels of detectable paramagnetic species in animals is often low, requiring the addition of nitroxides or other compounds. *In vivo* environments can absorb waves at non-resonant frequencies due to their dielectric properties, necessitating the use of low ESR frequencies to prevent this. Moreover, the EPR resonator chamber must have dimensions to accommodate the animal (usually a mouse) for *in vivo* studies [55]. EPR does not indicate how the *in vivo* environment is affected by the polymer implant, which is important to know in order to reduce negative physiological reactions in the body. Also, the size and shape of the injected polymer cannot be determined through EPR alone, so this technique is commonly paired with non-invasive MRI to give a more complete picture [57].

Since this procedure is used to track the mechanisms of polymer systems implanted into live animals, problems can arise from animal subject movement such as breathing or circulation. This additional signal noise can be as strong as the signal from paramagnetic species, and causes the spectrometer to read incorrectly. This motion-induced data can be diminished by automatic frequency control (AFC) and automatic matching control (AMC) systems. Besides changing the feedback control, an additional option to decrease physiological noise is to use longitudinally detected EPR (LODEPR). This detector does not take into account tuning or matching changes in the resonator. In LODEPR, the magnetic field is not modulated as in CW-EPR but instead the amplitude of the electromagnetic waves is altered. A solenoidal receiver coil is tuned to a frequency that is double the modulation frequency, and is used to directly detect oscillating paramagnetic species. Since detuning and dematching of the signal are not used in LODEPR detection, physiological motion has little effect on the resultant spectrum. Despite these challenges, EPR is an effective *in vivo* technique [54].

2.3. Diagnostic Ultrasound Imaging

In addition to EPR spectroscopy, ultrasound has recently been proposed as a noninvasive *in vivo* characterization technique for polymer implants [51]. Ultrasonic waves have frequencies exceeding the audible range for humans (above 20 kHz) and can travel through a substance as a continuous wave or in brief pulses [61]. A transducer linearly emits short bursts of ultrasound waves into a medium and records the echoes that result from differences in elastic properties of the composite material [62]. The backscattered signal can be characterized as a function of acoustic impedance where acoustic impedance is defined as a function of density and the propagation speed of sound [62] such that:

$$Z=\rho C$$

Z is the acoustic impedance, ρ is the density of the material, and C is the propagation speed of sound in the material [62]. Subsequently, the portion of the signal reflected back to the transducer is a function of impedance difference between two materials [62]:

$$R=\left[\frac{Z_1-Z_2}{Z_1+Z_2}\right]^2$$

where R is the intensity of the backscattered signal, Z_1 is the impedance of material 1, and Z_2 is material 2 [62]. Therefore as the polymer solution transitions from liquid to solid, the acoustic impedance changes, resulting in the development of a backscattered signal [51]. If the material were to phase invert into a perfectly homogenous material, then one would expect to see only the edges of the implant. However, because the gelation of the polymer results in the formation of polymer lean domains composed predominantly of the nucleation of solvent which leads to pore formation intermixed with polymer rich domains, the precipitation can be imaged (Figure 3). Consequently, regions of high solvent concentration would lead to low signal intensity due to the homogenous nature of the polymer and solvent domains.

For phase-inverting systems, the development of a signal due to changes in the acoustic resistance is caused by compositional changes of the polymer matrix. The pulse echo information is compiled into a two or three dimensional grey-scale image [62]. Within a polymer implant, the phase inverted portion reflects ultrasound waves back strongly, while the polymer solution region does not (Figure 4) [51]. Therefore, the ultrasound image that is generated depicts a real time picture of phase inversion and analysis of these images can be used to trace changes in the shape and composition of the implant over time.

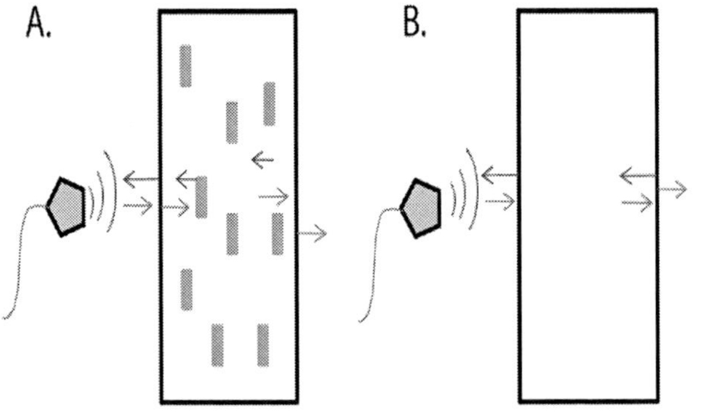

Figure 3. Representative illustration highlighting the effect of microdomains (A) and a homogenous structure (B) on ultrasound reflection.

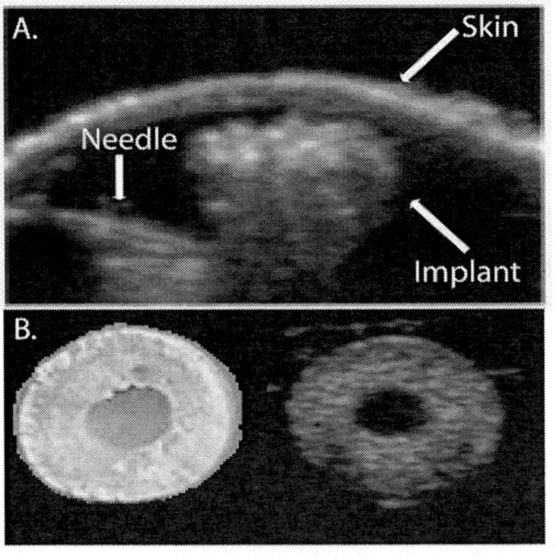

Figure 4. Implant formation in vitro and in vivo. (A) Representative ultrasound image an implant injection into a subcutaneous tumor, the arrows highlight the needle used for injection, the implant, and the skin. (B) Representative ultrasound image of an implant formed in a PBS bath along with a photo of the sectioned implant.

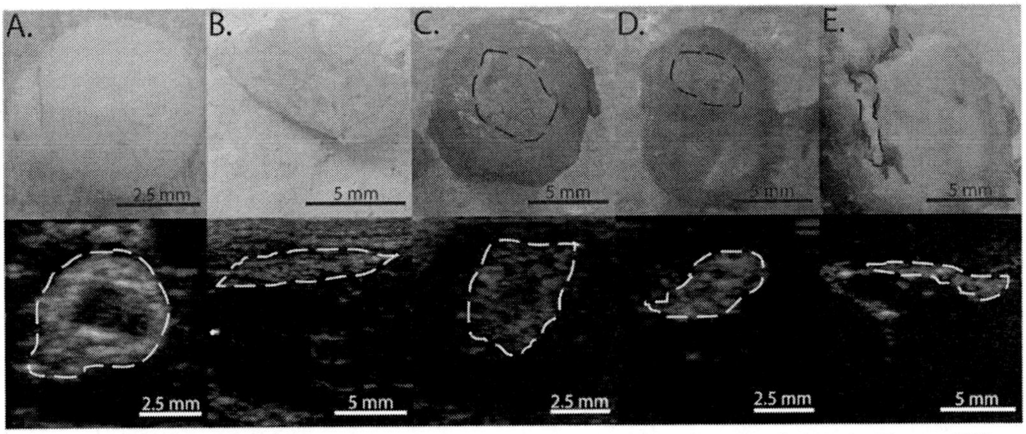

"Journal of Controlled Release, 147, R.B. Patel, L. Solorio, H. Wu, T. Krupka, A.A. Exner, Effect of injection site on in situ implant formation and drug release *in vivo*, 350–358, Copyright (2010), with permission from Elsevier."

Figure 5. Representative images of implants formed in different environments 7 days after implant formation in order of increasing stiffness of the injection environment *in vitro*, subcutaneous, non-necrotic tumor, necrotic tumor, and ablated tumor (A–E) with the corresponding ultrasound images (F-J).

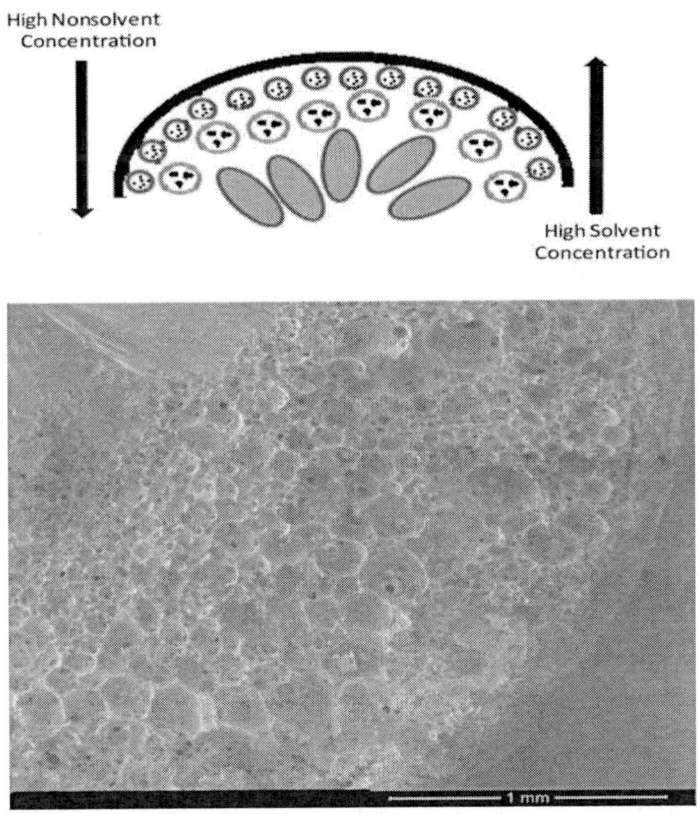

Figure 6. A schematic illustration summarizing the effects of opposing concentration gradients on macrovoid formation, with a corresponding SEM image highlighting these features.

Table 1. Characterization Techniques

Technique	Benefits	Limitations
Dark Ground Optics	• Solvent-nonsolvent exchange can be monitored • Simple and fast characterization	• Only thin film polymer systems can be studied • Studies over short time courses only • No *in vivo* imaging is possible
EPR	• Noninvasive, nondestructive • Can collect *in vivo* data • Provides structural information about the implant • Continuous data can be collected • Solvent-nonsolvent exchange can be monitored • Small scale measurements are possible • Can be used to characterize multiphase systems	• Additional signal noise can arise from animal subject movement • No images are produced directly • Cannot determine implant geometry • Large resonator chamber and small MW spin probes are required • Low EPR frequencies must be used • No information on the *in vivo* environment is found • Paramagnetic species must be added for detection
Ultrasound	• Noninvasive, nondestructive • Can collect *in vivo* data • Does not require a probe • Studies can be followed over long time courses	• Cannot directly detect solvent-nonsolvent exchange • Low resolution prevents small scale imaging

Ultrasound is an attractive characterization technique in many respects. It is the only noninvasive imaging process that is able to collect data describing the polymer behavior *in vivo* as well as *in vitro*. Similar to EPR, it is noninvasive and nondestructive, allowing an entire set of *in vivo* time course data to be collected using the same implant and animal and thereby reducing environmental variability and the number of animals used. Through compiling and analyzing multiple ultrasound images, data for a single implant can be extrapolated over time and quantified. While EPR requires the addition of a tracer element to collect data, ultrasound does not and it is thus a direct visualization method [51].

One shortcoming of ultrasound characterization is that it is unable to directly detect the solvent/nonsolvent exchange that occurs during phase inversion, as both dark ground optics and EPR can. Instead, properties such as polymer precipitation and swelling can be quantified to determine the inversion dynamics. Additionally, ultrasound imaging resolution is dependent on the frequency of the transducer used, which may limit small-scale examinations of polymer implants. Combining ultrasound data with a higher resolution imaging method provides more detailed image information [51].

Solorio et al. utilized ultrasound imaging to relate implant formation to initial drug release kinetics, as well as to determine the effect of polymer Mw on implant formation and swelling over time [51]. This study also demonstrated that ultrasound is a functional and effective characterization technique for both *in vitro* and *in vivo* studies (Figure 3). The *in vitro* study was conducted using 1% agarose as a mock tissue environment and fluorescein as a model drug. PLGA/NMP implants were implanted into a cavity in the agarose, then the phantom was aligned in a holder over an upright ultrasound transducer and imaged at time points spanning five days. For *in vivo* experiments, polymer was injected subcutaneously into rats and imaged over the course of seven days (figure 5) [51, 63].The authors created grey-scale arrays to analyze the ultrasound images for both *in vivo* and *in vitro* studies (figure 5). The impedance of each array was calculated in reference to the grey-scale, and the change in area of the implant images, measured in pixels, quantified implant swelling over time. They found, through ultrasound characterization, that as an implant begins to form, an outer shell encapsulates it and grows in proportion to polymer precipitation. Ultrasound is an effective noninvasive tool for phase inversion analysis, and offers a means by which to correlate *in vitro* and *in vivo* data [51]. An overview of the benefits and limitations of each characterization technique can be found in Table 1.

3. Factors Affecting *In Vitro* Release and Phase Inversion

The phase inversion dynamics within in situ forming implant systems are governed by a number of factors including solvent type, polymer composition, and various additives. Within these systems, the drug release kinetics are related to both the phase inversion behavior and the resulting implant morphology. This section is subdivided into descriptions of fast and slow phase inverting systems, cosolvent mixtures, polymer types, and various additives.

3.1. Fast Phase-Inverting Systems

Implants that form via phase inversion can be classified into two morphological classes: fast phase inverting (FPI) and slow phase inverting (SPI) systems. FPI systems occur when a polymer is dissolved in a strong, highly water miscible solvent such as NMP (Table 3), and the resultant implant morphology has a vast network of interconnected pores and macrovoids (Figure8) [16, 17, 19, 22, 47, 48, 64]. These FPI implant formulations often have much lower viscosities than the SPI systems, and begin to precipitate within seconds after they have been exposed to the aqueous environment. For ternary systems (consisting of solvent, polymer, and nonsolvent phases) the internal pore formation is hypothesized to be a result of changes in nonsolvent diffusivity [28]. For FPI systems, instantaneous demixing occurs when the polymer solution comes into contact with the bath-side (nonsolvent). This instantaneous demixing process happens as a consequence of the high miscibility between the solvent and nonsolvent and results in the formation of a thin, dense polymer shell [28]. Once formed, the polymer shell acts as a diffusional barrier, causing the development of opposing concentration gradients of nonsolvent and solvent within the implant [28]. If one were to consider a spherical implant, the shell would exist at the surface exposed to the nonsolvent, with the nonsolvent concentration highest near the polymer shell. The solvent would have a concentration gradient in the opposite direction, with the solvent concentration highest at the center of the implant (Figure 6). The polymer precipitation at the implant surface is instantaneous, and polymer precipitation is delayed towards the center, again as a result of the opposing concentration gradients.

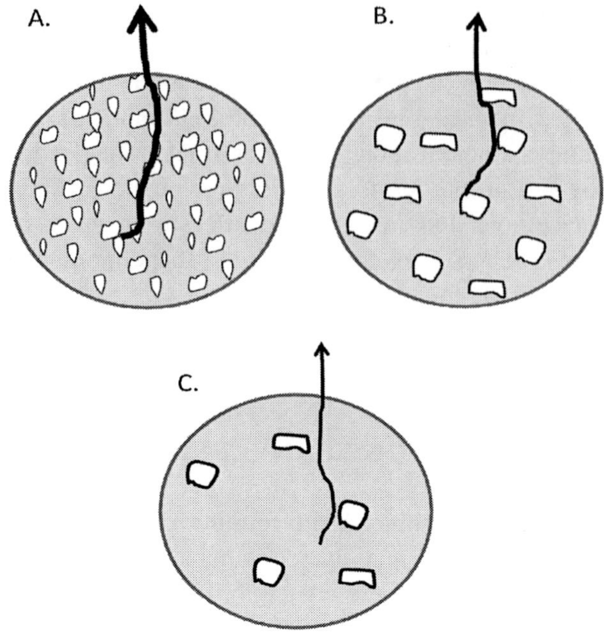

Figure 7. A schematic illustration of the effects of the interconnected porous network on the release of drug through spherical implants composed of different solvents (A) NMP, (B) Triacetin, and (C) ethyl benzoate. The arrow thickness indicates the rate of diffusion through the matrix.

Just inside the interfacial polymer shell, droplets consisting predominantly of a mixture of solvent and nonsolvent form. These droplets are referred to as polymer-lean domains [28]. The polymer-lean domains are stabilized if diffusional flow of nonsolvent into the polymer phase is sufficiently large, so that polymer precipitation occurs and droplet expansion is arrested [28]. Since the concentration of solvent is sufficiently high towards the implant interior, the nucleated polymer-lean droplets begin to expand and aggregate as a result of insufficient nonsolvent exchange from the droplets into the surrounding polymer solution, ultimately forming macrovoids [28]. Since the development of macrovoids results from a transition from instantaneous demixing to delayed demixing, implants that have a sufficiently thin cross-section will not form macrovoids, but simply form a dense polymer shell [28, 29, 65]. One hallmark trait of FPI systems is the initial burst release of hydrophilic drugs followed by a plateau, due to the porous networks that are formed within these implants [16, 19, 22, 39].

Therefore, the phase inversion dynamics directly affect the implant morphology, and consequently the drug release behavior. The initial burst of drug characteristic of FPI systems has been hypothesized by McHugh et al. to occur as a result of drug dissolution into the polymer-lean phase. As a result of the lower viscosity and improved diffusion of drug through these polymer-lean phase regions, a period of burst release occurs until the drug is depleted from the porous network. After the drug has been depleted from the porous network, cumulative drug release is significantly reduced. Prolonged release is a function of polymer degradation as well as diffusion of drug through the polymer-dense regions (Figure 7) [16, 19, 22, 66]. A table of various solvents typically used in ISFIs and their solubility characteristics is shown below:

Table 2. Solvent Characteristics

Solvent	Miscibility (mg/mL)	Phase Inversion System
N-Methyl Pyrollidone	150[67]	Fast Phase Inversion
Dimethyl Sulfoxide	>95[67]	Fast Phase Inversion
Triacetin	77.8[68]	Intermediate Phase Inversion
Ethyl Benzoate	Insoluble[69]	Slow Phase Inversion
Benzyl Benzoate	3.4 in hot water[69]	Slow Phase Inversion
Benzoic Acid	40[69]	Slow Phase Inversion
PEG-400	Soluble	Fast Phase Inversion

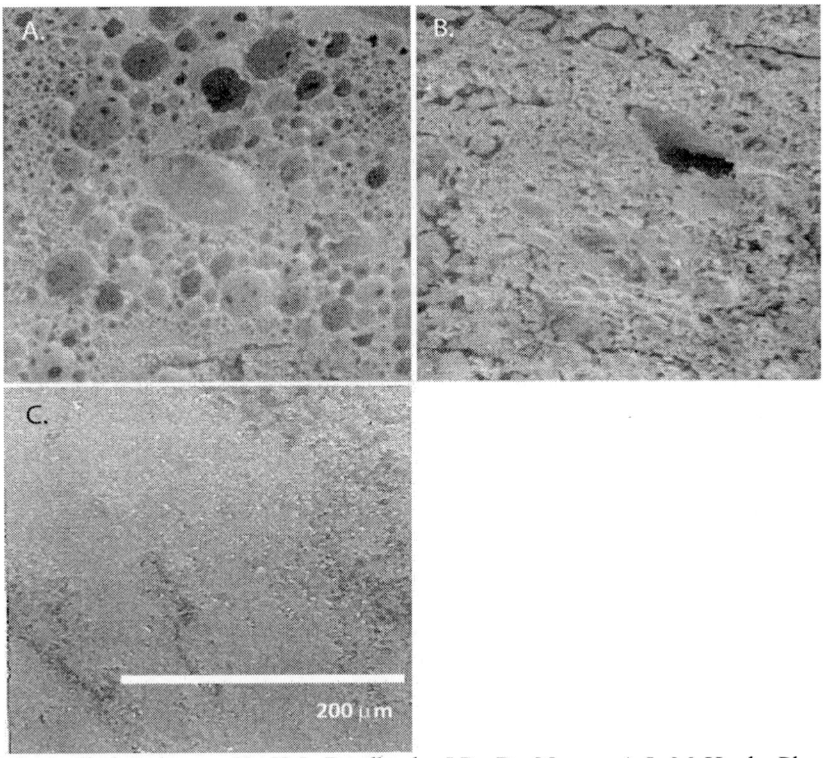

"Journal of Controlled Release, 62, K.J. Brodbeck, J.R. DesNoyer, A.J. McHugh, Phase inversion dynamics of PLGA solutions related to drug delivery Part II. The role of solution thermodynamics and bath-side mass transfer, 333-344, Copyright (1999), with permission from Elsevier."

Figure 8. SEM micrographs of drug eluting depots made using 3 different solvents (A) NMP, (B) triacetin, and (C) ethyl benzoate.

3.2. Slow Phase-Inverting Systems

In contrast to FPI systems, SPI implant systems incorporate a solvent with a weak affinity for the nonsolvent, such as ethyl benzoate which is poorly miscible in water (Table 3). Often times these water-immiscible solvents are also weak solvents for the polymer used to make the implants, which results in a high viscosity solution [11, 22, 47, 64]. When amorphous polymers are used, delayed demixing occurs without a period of instantaneous demixing due to the low solvent miscibility in SPI systems. As a result of the slow precipitation, the composition throughout the implant is maintained as a constant, and no macrovoid formation occurs [16, 22, 28, 32, 47, 49]. Due to the absence of an internal porous network, drug is released from the implant via diffusion through the viscous polymer phase (Figure 8) [16, 19, 22, 66]. One advantage of these SPI systems is that they have a nearly zero-order release, with almost no initial burst. However, these implants are more difficult to inject than FPI implants due to their relatively high viscosities, and thin film precipitation requires several hours or even days. Due to the morphological differences that arise during polymer precipitation, a high correlation exists between the phase inversion and drug release [16, 19, 20, 22, 51, 63].

A unique feature of SPI systems is that the drug release from these implants can be significantly enhanced by increasing the crystallinity of the polymer used [47]. For these systems, the presence of water can induce polymer crystallization as early as four days after implant formation [47]. As water continues to diffuse into the implant, a solid-liquid separation leads to the development of a semi-crystalline matrix, which excludes the drug into the polymer-lean domains. The resultant morphology of SPI systems utilizing crystalline polymers is a highly interconnected network of small pores (substantially smaller than the macrovoids seen in FPI systems). As in FPI systems, the development of a porous network leads to a substantial burst in drug release [16, 19, 47]. Unlike the burst release seen with FPI systems, burst release from SPI implants made with crystalline polymers begins four to five days after implant formation and continues for about one week. After this burst period, there is little change in the structure, and most of the drug has been depleted indicating that very little drug is trapped within the polymer dense regions [47]. The burst can be modified adding varying concentrations of an amorphous polymer, which will result in prolonged drug release [47]. Thermoreversible crystallization can also occur for amorphous polymers dissolved in benzyl benzoate [70]. Crystallization begins to occur as soon as 6 h after precipitation during polymer storage at 25°C, and can be reversed by heating the sample to 65°C. If the samples are not properly heated before use, there is a significant increase in drug release from the implants.

Within SPI systems, the composition of the injection environment plays a significant role in determining the drug release performance of an ISFI implant. In general, as the polymer solution becomes less miscible with water, the bath-side composition becomes increasingly more important to the drug release behavior [16]. The effects of bath-side additives such as triglycerides and organic salts on SPI systems have been studied, as both are found in the subcutaneous environment where these depots would likely be injected. The bath-side addition of triglycerides and salts both result in a significant increase in overall protein release rate, but these effects are governed by different phenomena. The first case is demonstrated by adding triacetin, a short-chain triglyceride, to the bath-side of implant systems using the weakly miscible solvent ethyl benzoate [16]. In this system, triacetin added to the bath-side solution is relatively more hydrophilic than the polymer-solvent depot. As triacetin diffuses into the implant, the overall implant viscosity is decreased and its capacity for water absorption is increased. As a result, the liquid demixing process is accelerated, triggering an increase in protein release rate. In the case of adding salt to the bath-side solution, no changes are observed in the liquid demixing rate [16]. Instead, surface erosion of the polymer depot occurs, and the increased protein release is attributed to polymer surface degradation. As a whole, these studies demonstrate that the specific composition of the aqueous implant environment is very important in SPI systems, and one must be aware that components entering the system may influence the drug release behavior.

3.3. Effects of Cosolvent Composition

While the classical systems described above demonstrate the effect of a pure solvent on drug release, phase inversion, and implant morphology, systems which combine solvents have been developed as a means for achieving intermediate release profiles [18, 19, 34, 35, 38, 42, 71-77]. The combination of NMP and triacetin reduces the affinity of the polymer depot for water, and subsequently reduces both the rate of phase inversion and drug release [19, 38]. Morphological analysis shows that the addition of 10wt% of triacetin into the implants causes a transition from a vastly porous network with a large number of macrovoids, to a more spongy structure, with additional increases in triacetin leading to a more dense structure [19]. Additional studies have shown that the introduction of solvents with low water miscibility has resulted in a reduction in burst release from the FPI implants [38, 72]. In some instances a limited burst is desired, and SPI systems can be modified by the addition of water miscible solvents [35, 76, 77].

Both fast and slow phase inverting systems have benefits and drawbacks, but because multiple parameters can be modified in order to produce the desired drug release profile, the desired release profile often dictates which implant formulation is used. Disorders that required near constant drug plasma levels, such as maintenance of basal insulin levels for the treatment of diabetes or the treatment of Turner syndrome by delivery of human growth hormone (hGH), may require the use of a SPI system to achieve optimal therapeutic results [16, 17, 42, 43, 63, 78]. In the case of diseases that require a high initial dose of drug followed by a small maintenance dose, such as the treatment of hormone imbalances, where the target organ needs to receive a saturation of signal in order to induce down regulation of the receptors, or cancer chemotherapy where one wants as much drug as possible to be taken up by the cells, so that drug resistance can be avoided, FPIs are a better choice [12-14, 63, 78-80].

3.4. Polymer Type

PLGA is among the most common polymer used for ISFIs, but implants have also been formulated using a number of different polymers including PCL, PLA, Poly(ethylene carbonate), sucrose acetate isobutyrate, and fluoroalkyl-ended poly(ethylene glycol) [11, 23, 40, 43, 44, 47, 71, 81-84]. The chemical composition of a polymer (including but not limited to the Mw, functional groups, and degree of hydrophobicity) affects how the polymer will interact with other components in the ISFI system. Therefore the choice of polymer provides another tool by which researchers can tune the rate of phase inversion and drug release in order to fit their system demands. For example, implants formed using hydrophobic polymers have been demonstrated to release less drug than implants formed from their hydrophilic counterparts [82, 85, 86].

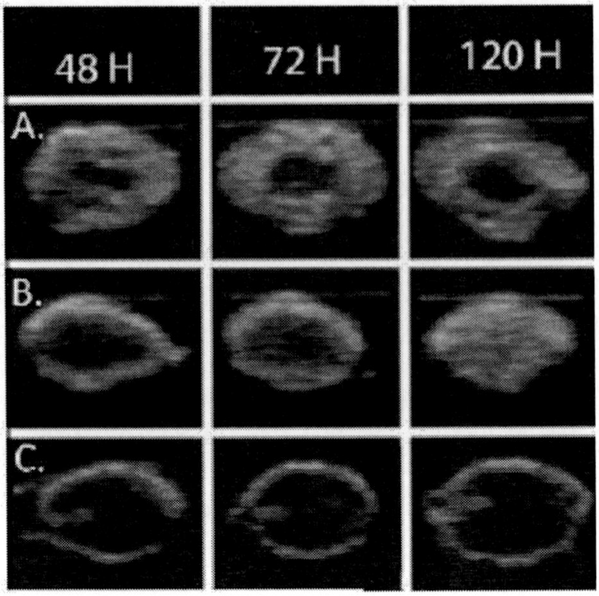

Figure 9. Representative ultrasound images of 15kDa (A), 29kDa (B), and 53kDa (C) implants formed in PBS illustrating the polymer precipitation of the implants over time. The 15kDa implants have already undergone phase inversion, and the central black region is result of macrovoid formation. (B) illustrates the gradual precipitation of the polymer before macrovoid formation occurs after 120 h. (C) illustrates delayed demixing after shell formation.

Additionally, changes in polymer Mw have been shown to alter drug release from ISFIs [38, 51, 63, 78, 82, 87-89]. Interestingly, several studies have demonstrated specific release behaviors for polymers of specific Mw. For implants formulated using 12 kDa PLGA, there is a significant burst of drug during the first 24 h (for leuprolide acetate, as much as a 36% burst release of drug), followed by a short period of diffusion-facilitated release (2-3 days), which is followed by a period of increased drug release due to polymer degradation[14, 87-89]. For implants formulated using 30 kDa PLGA, the initial burst is higher than that from 12 kDa ISFIs, but the rate of release quickly reaches a plateau and no significant release occurs for an extended period of time (2 weeks), until polymer degradation results in elevated drug release [38, 85, 88, 89]. When the Mw is increased to 48 kDa, the burst release of drug is significantly reduced, with elevated release occurring due to polymer degradation occurring 4-5 days after implant formation [38, 85, 87, 88]. The general trend described for implants made using 12 kDa, 34 kDa, or 48 kDa Mw polymers have been seen in a number of studies, even when the solvent or drug differs [38, 72, 85, 87-90].

The effects of polymer Mw on phase inversion varies based on a number of factors. Intuitively, the rate of phase inversion would appear to be inversely related to the critical water concentration of the polymer solution. Due to the lower water affinity, one may assume that the formation of implants with small Mw would occur more slowly than implants made from polymers with larger Mw. However, the order of phase inversion is reversed. It has been demonstrated that the phase inversion of polymer films formed from polyethersulfone occurred more slowly with an increase in Mw [19, 31]. This trend has also been observed in our laboratory, with implants made from smaller Mw PLGA phase invert faster than implants made from larger Mw PLGA (Figure 9). This change may be attributed to many factors including a sufficient increase in opposing concentration gradients due to the formation of the dense polymer shell, a change in the hydrophobicity resulting in a decrease in water absorption, and changes in diffusivity throughout the ISFI [19, 31, 91].

In addition to varying the polymer Mw, another method of modifying system characteristics is through changing the polymer concentration. Increasing the polymer concentration has been shown to decrease overall drug release as well as the rate of phase inversion in ISFIs [14, 15, 19, 28, 31, 35, 38, 39, 71, 85, 89, 92]. In FPI systems, the use of a more concentrated polymer solution influences the implant morphology, changing the microstructure from a highly interconnected porous network into a less porous spongy structure [19, 28, 30, 31]. The change in implant behavior can attributed to a number of factors including lower water uptake, thicker skin formation, increased hydrophobicity of the polymer solution, as well as decreased diffusivity [19].

3.5. Polymer Formulation Additives

Changes in either the solvent or polymer used for ISFI fabrication provide a means by which the drug release and phase inversion profiles of ISFIs can be manipulated [16, 19, 20, 38, 51, 63, 76, 78, 81]. Due to the high viscosity of SPI implant solutions, injection is difficult or impossible without preheating the polymer solution [48]. Thus, while the ideal delivery profile may be achievable utilizing SPI systems, their use in a clinical setting is sometimes impractical. Therefore, methods by which the phase inversion and drug release rate of FPI systems can be altered through the use of additives, is an ongoing area of research [19, 22, 46, 48, 76, 78, 93]. Since burst release is often times an undesirable affect of the phase inversion process, additives have most often been used to reduce the burst release of drug from implants [22, 47, 78, 79].

The use of Pluronic as an additive (which is an amphiphilic triblock copolymer of polyethylene oxide (PEO) and polypropylene oxide (PPO), shown to increase a cell's sensitivity to cancer drugs and heat [21, 94, 95]), provides a means by which the drug release can be reduced [48, 78]. Despite the reduction in drug release, the rate of phase inversion and water absorption of implants formulated with Pluronic were shown to increase due to the presence of the hydrophilic PEO blocks [48]. While morphological changes may occur within the ISFI (depending on the choice of Pluronic used), the changes do not elicit the transition into a more dense spongy microstructure [48, 78]. Instead the reduction in drug release is hypothesized to occur as a result of the orientation of the Pluronic molecule in the polymer matrix [48]. Since the PPO block is hydrophobic, it is reasonable to assume that the PPO block would readily associate with the polymer, while the PEO block would extend into the polymer-lean pores [48]. If the concentration is raised significantly, the Pluronic begins to fill the polymer-lean domains, resulting in a diffusivity barrier for the drug [48]. While increased Pluronic concentrations have been shown to decrease drug release from implants, this effect can be lost if the concentration of the additive exceeds the percolation threshold leading to increased diffusivity through the implant [78]. Other additives for reducing the rate of drug release include ethyl heptanoate and glycerol [79].

Table 3. Additives

Additive	Effect
Pluronic P85	Reduced Burst
Pluronic L101	Reduced Burst and overall release
Glycerol	Reduced Burst
Ethyl Heptanoate	Reduced Burst

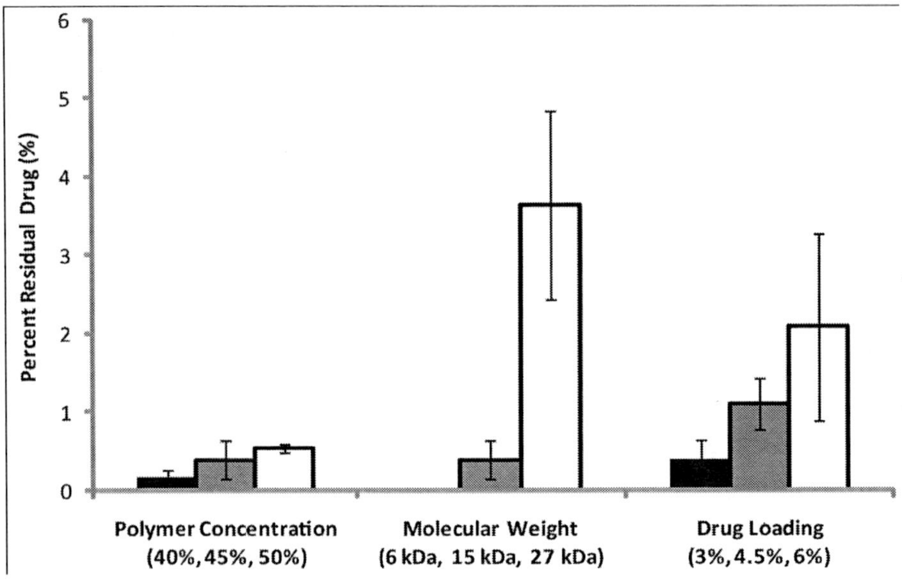

Figure 10. Bar graph generated using the percent of residual drug content data from: International Journal of Pharmaceutics 194 (2000) 181–191, with implants explanted 105 days after implant formation in male Sprague–Dawley rats.

While the majority of additives studied are used to reduce drug release, some additives have the opposite effect on release [19, 96]. The use of hydrophilic polymers (such as polyvinylpyrrolidone (PVP)) has been shown to eliminate macrovoids if a large enough mass is included in the polymer solution [19, 27, 28]. However, when only a small mass of PVP is added to the polymer solution, an increase in the rate of polymer gelation and drug release occurs without increasing the rate of water absorption [19]. Another additive that can be used to increase release from FPI systems include the use of aliphatic esters which have been shown to increase the water absorption, solvent release, and enhance polymer degradation [96]. A table of additives and their effects is provided below.

4. The Role of Injection Site on *In Vivo* Implant Behavior

In the twenty years since the development of phase sensitive ISFIs, an extensive library has been developed describing the behavior of these implants *in vitro*. Techniques such as dark ground imaging and scanning electron microscopy have provided insight into the phase inversion and drug release dynamics of several polymer/solvent systems [16, 17, 19, 22, 31, 32, 47-49, 66]. While investigating the *in vitro* behavior of ISFIs is not a trivial task, understanding the behavior of these implant systems *in vivo* has been a complicated endeavor. One consequence resulting from the destructive nature of analytical techniques used to evaluate implants *in vivo* is that often times direct comparisons between the different implant environments is difficult or impossible [63]. Through the use of ultrasound imaging, EPR spectroscopy, and MR imaging, the phase inversion, swelling, rheology, and solvent exchange behavior of implants can now be evaluated nondestructively *in situ* [51, 52, 57, 63]. Through the use of these nondestructive techniques, the factors that lead to the deviation of behavior between *in vivo* and *in vitro* implants can be better understood. This section will provide an overview of the parameters that affect *in vivo* phase inversion, drug release, and implant swelling.

4.1. *In Vivo* Release

Early *in vivo* ISFI studies focused predominately on evaluating drug release through analysis of serum plasma levels which screened for released drug, proteins, or specific drug efficacy markers such as testosterone [12-15, 17, 21, 34, 35, 37, 41, 42, 97, 98]. While the release data obtained from these studies cannot be directly compared to that obtained from *in vitro* release studies (due to metabolism, adsorption, and drug clearance occurring in the body), the effect of implant formulation on release could still be evaluated [63]. In a series of studies performed in both rat and dog models, Dunn et al. measured testosterone levels as an efficacy marker for leuprolide acetate release from ISFIs. Testosterone can be used to determine leuprolide acetate efficacy, because continuous delivery of leuprolide acetate causes a reduction in circulating sex hormones due to pituitary desensitization [13, 14]. FPI systems are optimal for this application due to their characteristic burst release, which can rapidly increase pituitary signaling to desensitize the gland more rapidly. The role of polymer Mw, polymer concentration, polymer hydrophobicity, and drug loading were evaluated for FPI implants consisting of NMP and 75:25 PLGA [13, 14]. From these studies, it was shown that changes in polymer concentration and drug loading had a negligible effect in reducing serum testosterone levels, while changes in polymer Mw and hydrophobicity were more significant factors for controlling drug release in these *in vivo* FPI systems. The average percentage of residual leuprolide acetate followed the expected trend of increased drug concentration with increases in polymer concentration, loading drug concentration, and polymer Mw (Figure 10). Additionally, changing the polymer hydrophobicity resulted in a prolonged lag time before desirable levels of testosterone were achieved, taking 35 days compared to 14 days for the more hydrophilic counterpart [14].

Another common formulation parameter that has been varied in order to evaluate its effects on *in vivo* release is the organic solvent used to dissolve the polymer [17]. The sustained release of human growth hormone (hGH) from 50:50 PLGA was evaluated using polymer formulations comprised of solvents with increasing nonsolvent miscibility [17]. Serum hGH levels were monitored after subcutaneous injection of implants made with NMP, triacetin, ethyl benzoate, or benzyl benzoate. The solvent affinity for water played a significant role in altering the release profile of implants. It was shown that ISFIs made with benzyl benzoate maintained a relatively flat serum hGH profile when compared with the other solvent systems, indicating a more constant rate of growth factor release, which is indicative of a SPI system [17].

4.2. Comparison of *In Vitro* and *In Vivo* Release and Phase Inversion

Early studies demonstrated that implants formed *in vivo* and *in vitro* follow the same fundamental principles of phase inversion, such that implants made from a solvent with higher water miscibility will phase invert more rapidly. These studies did not directly compare the release profiles found from *in vitro studies* with those found from *in vivo* studies; therefore differences in release due to the injection site were not well understood. In a study by Solorio et al., ultrasound was used to noninvasively monitor the phase inversion behavior of FPI implants both *in vitro* and *in vivo*. To mimic the injection site environment, implants were imbedded in 1% agarose phantoms. It was noted that implant swelling was inversely related to the Mw of the polymer used to fabricate the implants. This was hypothesized to be a result of increased osmolarity of the smaller Mw implants due to the larger number of moles contained within the polymer solution. The higher solvent affinity of the smaller Mw polymers would result in an increased mass of residual solvent, maintaining the osmotic drive on nonsolvent influx towards the center of the implant. As a consequence, the movement of nonsolvent into the implants would be greater than the movement of solvent out, resulting in an overall increase in volume. After the implants were imbedded in agarose for a sufficient period of time, the residual non-precipitated polymer solution ruptured the phantoms and leaked into the surrounding agarose. Leakage of polymer solution was dependent on the polymer Mw; occurring more slowly as the polymer Mw increased due to decreased osmotic drive (Figure 11). The results suggested that implant swelling causes a significant amount of pressure at the injection site, ultimately causing the phantom to crack, leading to the leakage of residual polymer/solvent into the phantom (Figure 11).

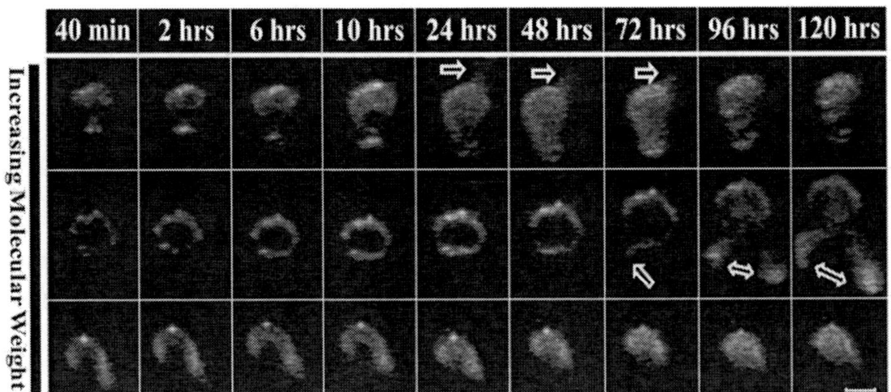

"Journal of Controlled Release, 143, L. Solorio, B.M. Babin, R.B. Patel, J. Mach, N. Azar, A.A. Exner, Noninvasive characterization of in situ forming implants using diagnostic ultrasound, 183–190, Copyright (2010), with permission from Elsevier."

Figure 11. Representative ultrasound images of the implants formed in agarose over time, with implants formulated with increasing Mw Polymer (15kDa, 29kDa, and 64kDa PLGA). The scale bar represents 2.5 mm. Leakage is indicated by the arrows.

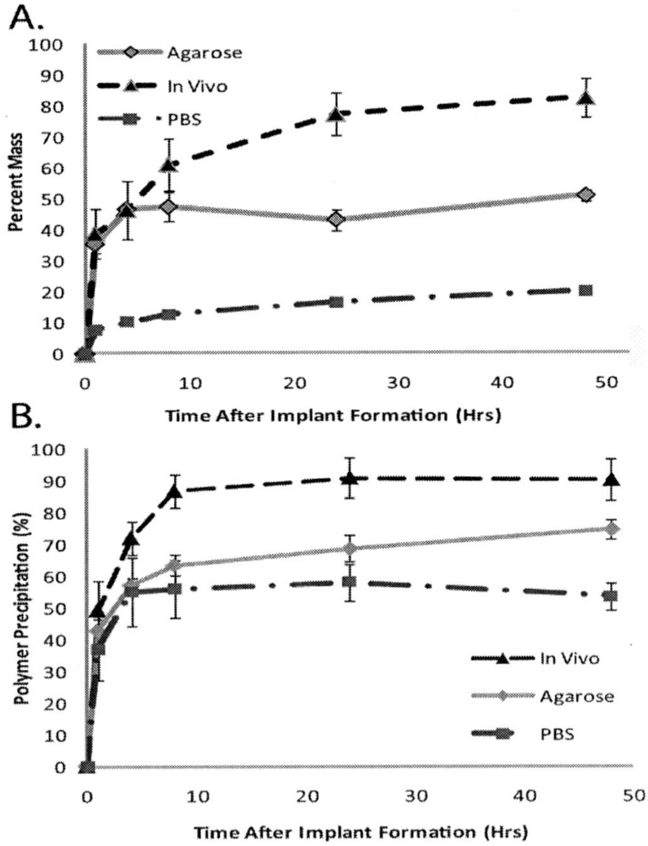

Figure 12. The effect of injection site on drug release (A), and on phase inversion (B). Implants were formed in 3 different environments: PBS bath, agarose, and subcutaneous injection under the dorsal skin flap of BDIX rats.

While it was clear that implants were affected by the injection environment, the roles of the injection site on phase inversion and drug release have only recently been evaluated. The phase inversion and drug release profiles of implants formed subcutaneously under the dorsal skin flap of rats were compared to implants formed in agarose or PBS, and it was shown that the injection site played a significant role in the phase inversion and drug release dynamics of ISFIs. Phase inversion occurred the most rapidly in implants formed subcutaneously, followed by those constrained in agarose, and the depots formed in PBS phase inverted the slowest (Figure 12A). Additionally, when the drug release profiles of implants formed in these three different environments were compared, the release profiles followed a similar trend with *in vivo* release occurring the most rapidly (Figure 12B). It was hypothesized that the increase in phase inversion and drug release of these implants was a result of convective removal of solvent and drug. The results of this study were further validated when the release of thymosin alpha 1(Tα1) from ISFIs fabricated from PLGA dissolved in a mixture of NMP and triacetin was evaluated both *in vivo* after subcutaneous injection in rats and *in vitro* in PBS [38]. Direct comparison of the *in vivo* and *in vitro* release of Tα1 from implants with a 1:1 mixture of NMP and triacetin showed that the implants formed *in vivo* released 100% more drug than the *in vitro* counterpart 1 day after implantation, but this deviation between *in vivo* and *in vitro* release decreased over time [38].

4.3. Effects of Injection Site on Release

While the bulk of *in vivo* studies have focused on the release of implants formed subcutaneously, one of the more promising aspects of these systems is that they can be targeted directly into the diseased tissue without the need for open surgical placement. A study by Krupka et al. evaluated the use of FPI implants for the local delivery of carboplatin, with implants percutaneously injected directly into the lesion after RF ablation [21]. The results indicated that the carboplatin-eluting implants were an effective adjunct therapy to RF ablation, showing a greater reduction in tumor diameter with time in comparison to tumors treated with RF ablation alone [21].

Despite the fact that these implants have shown to be effective both subcutaneously and as local delivery systems, the impact of different *in vivo* environments on drug release, swelling, and phase inversion has only recently been investigated. In a study by Patel et al., the behavior of FPI implants was investigated in a variety of different *in vivo* environments including subcutaneous injection, intratumoral injection in necrotic tumors, intratumoral injection in non-necrotic tumors, and intratumoral injection in ablated tumors [63]. Ultrasound was used to evaluate both the swelling and phase inversion. In all instances, the *in vivo* phase inversion was shown to be more rapid when compared with the *in vitro* system, and the ablated tumors facilitated the most rapid phase inversion, followed by non-necrotic tumors, with necrotic and subcutaneous implant polymer precipitation occurring at approximately the same rate.

The drug release from these implants was affected by both the injection site and the polymer Mw. Implants formed in ablated tumors had the highest burst release of drug followed by non-necrotic tumors, with implants formed in necrotic tumors and the subcutaneous space bursting the least. The release of drug from all environments was significantly faster than what is predicted by *in vitro* experimentation [63]. It was also shown that as Mw increased, the implants deviated less from the *in vitro* release profiles (Figure 13). An interesting note to this study is that a Pluronic tri-block co-polymer (P85) was added to the polymer solution in order to reduce burst release. *In vitro* the 18 kDa implants had the lowest release rate, with release increasing with Mw. However, the *in vivo* release increased inversely with Mw, indicating that some excipients may be less effective at modifying drug release for *in vivo* systems [63].

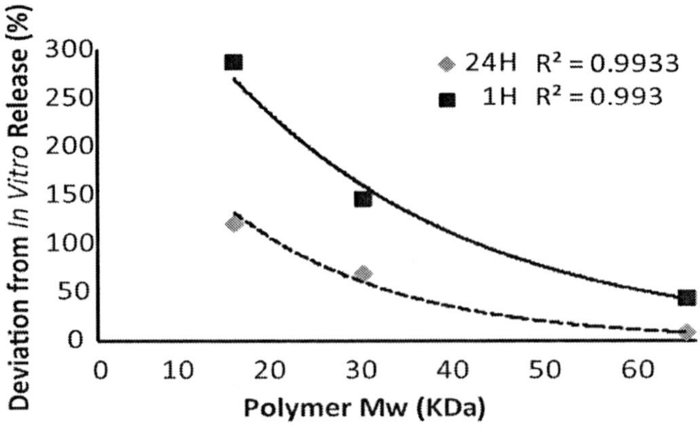

Figure 13. The effect of polymer Mw on the deviation from *in vitro* release.

Polymer swelling was also significantly different than the *in vitro* systems, with all polymer formulations becoming smaller *in vivo*. The central hypothesis for ISFIs decreasing in size with respect to time is that the solvent removal occurs at a significantly faster rate than water absorption, leading to a reduction in volume [51, 63]. This hypothesis is supported when one compares the swelling profile of implants formed in PBS to those formed in agarose and those formed subcutaneously (Figure 14). Subsequently, in an unconstrained system such as a PBS bath, implants expand without resistance from the external environment, however in agarose or tissue, the surrounding environment limits implant expansion. Our hypothesis for the difference in release due to the injection site is that the resistance to implant expansion exerted by the surrounding environment results in a reactive force on the implant that causes the solvent to be eliminated faster than in an unconstrained system.

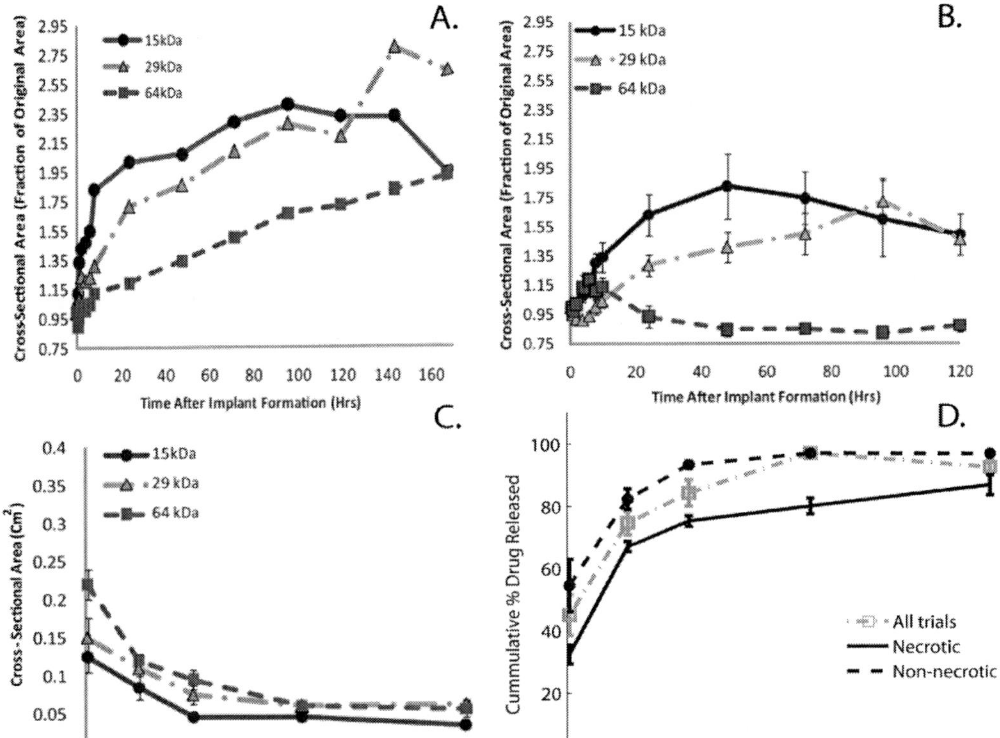

"C and D are modified from, Journal of Controlled Release, 147, R.B. Patel, L. Solorio, H. Wu, T. Krupka, A.A. Exner, Effect of injection site on in situ implant formation and drug release *in vivo*, 350–358, Copyright (2010), with permission from Elsevier."

"B modified from,Journal of Controlled Release, 143, L. Solorio, B.M. Babin, R.B. Patel, J. Mach, N. Azar, A.A. Exner, Noninvasive characterization of in situ forming implants using diagnostic ultrasound, 183–190, Copyright (2010), with permission from Elsevier."

Figure 14. Swelling behavior of nonconstrained implants formed in a PBS bath (A), swelling behavior of constrained implants formed in agarose (B), swelling behavior of subcutaneously injected implant solution (C). Release of fluorescein from implants injected in two different tumor environments, which illustrates the effect of the *in vivo* injection site on implant release behavior.

As a consequence, the effect of injection site on drug release can be attributed to two parameters: the injection site stiffness and the osmolarity of the polymer solution. The implant osmolarity plays a role in the degree of implant swelling, therefore the more the implant swells the greater the reactive forces acting on it, and consequently the greater the drug and solvent release. The second parameter affecting the implant behavior *in vivo* is the injection site stiffness. One could imagine the implant as a dry sponge loaded with dye inside a rigid container. If the sponge was allowed to absorb water but not freely expand, a bolus loss of dye would be seen. However if the walls of the container were less stiff, the reactive force caused by the expansion of the sponge would be reduced, due to energy loss to the wall, and as a consequence a lower mass of dye would be released. Additionally, if a sponge were placed in tub full of water and allowed to freely expand, even less dye would be released by the sponge [63].

This effect would play a significant role in the drug release and phase inversion dynamics of ISFIs. It could be predicted that implants injected into the subcutaneous space would release more drug than those formed in a PBS bath. Additionally, excipients and polymer formulations that increase implant osmolarity and swelling should result in a greater burst release of drug. Factors such as local edema or intratumoral pressure could also lead to enhanced release. This theory is supported not only by the studies of Patel and Solorio, but also indirectly by Liu [38, 51, 63]. When triacetin concentrations were increased in an NMP:triacetin mixture, the release of Tα1 was shown to decrease for subcutaneously formed implants, from 76.9% with a pure NMP solution to 29.3% in a 1:1 mixture of NMP:triacetin [38]. While this hypothesis nicely explains the relationship between injection site and enhanced drug release, additional factors may also play a role including local water content, bath-side composition, variations in tissue perfusion, differences in implant shape and surface area, and differences in implant diffusivity due to rapid solvent loss [17, 38, 51, 78]. Interestingly, while a difference in implant phase inversion and drug release has been reported for ISFIs *in vivo*, the rheology of the implants appears to not change significantly based on the environment, but is instead a function of the solvent used [52, 57]. While ultrasound provides a means for the noninvasive monitoring of implant shape and phase inversion, future studies utilizing MRI may provide a method of tracking solvent exchange noninvasively [57].

Conclusion

ISFIs are an exciting area of study that can be used for the continuous release of a therapeutic agent as well as local targeted delivery of drug to a diseased tissue. Since these implants are liquid solutions outside of the body, and do not precipitate to form drug-eluting depots until after exposure to an aqueous environment, they provide a means to circumvent the intrinsic limitations of pre-formed implant systems. Additionally, the simple manufacturing process results in a cost efficient system for the noninvasive administration of a variety of drugs and proteins. While the chief concern for ISFI technology has focused on solvent biocompatibility, a large number of studies have been performed to establish the safety of these implants. Advances in medical imaging have provided unique insight into the behavior of ISFIs *in vivo*, demonstrating that the environment in which the implant is injected plays a significant role in the drug release and phase inversion behavior. The insight into ISFI dynamics *in vivo* may lead to methods for the minimally invasive treatment of non-resectable tumors, or to the design of systems that can postpone or eliminate the long-term complications of diabetes. Improvements in the understanding of these implant systems may one day provide a treatment option for clinicians to treat an incalculable number of diseases.

SUMMARY

Characterization Techniques

- Dark ground images give information on diffusivity, liquid-liquid phase separation, and gel formation of a polymer phase inverting system
- Dark ground imaging tracks water diffusion and gelation fronts as they move into the polymer system
- Characterization techniques are often paired with another imaging method to provide a more complete or specific picture
- Through comparing *in vivo* and *in vitro* data, environmental effects on the system can be studied and accounted for
- EPR provides information on solvent-nonsolvent interaction, polymer degradation, and drug diffusion over time.
- Both EPR and ultrasound imaging are used *in vivo*, and these nondestructive processes allow multiple readings to be made on a single implant over a long time course as opposed to other methods which require surgical implant extraction
- Ultrasound imaging creates real time images of a polymer system and is useful for monitoring changes to an implant over long time courses

In Vitro Phase Inversion and Drug Release

- The rate of phase inversion plays a role in determining implant morphology
- FPI systems have interconnected pores that lead to burst release of drug
- SPI systems maintain a constant composition and release drug with a near zero-order release profile
- Solvent and polymer type can both be used to modify drug release and phase inversion
- Crystalline polymers can lead to burst release if used with SPI systems
- Additives can be used to alter drug release

In Vivo Phase Inversion and Drug Release

- ISFIs in vivo follow the basic principles of phase inversion and drug release with respect to the solvent strength and release
- Implants formed in vivo have significantly higher release than would be predicted through in vitro release studies
- Factors the alter implant swelling have been show to affect in vivo release and phase inversion

- The stiffness of the injection site has been shown to alter drug release in vivo
- While release and phase inversion are increased for implants formed subcutaneously, the micro-viscosity is independent of the injection site stiffness

References

[1] C. Page, M. Curtis, M. Walker, B. Hoffman, Integrated Pharmacology,3 ed,Elsevier, 2006, Philadelphila, State, 2006

[2] W.M. Saltzman, Drug Delivery: Engineering Principles for Drug Therapy,1 ed,Oxford University Press, 2001, New York, State, 2001

[3] D.Y. Arifin, K.Y. Lee, C.H. Wang, Chemotherapeutic drug transport to brain tumor, *J. Control Release* 137(3) (2009) 203-210.

[4] D.Y. Arifin, K.Y. Lee, C.H. Wang, K.A. Smith, Role of convective flow in carmustine delivery to a brain tumor, *Pharm. Res.* 26(10) (2009) 2289-2302.

[5] A.A. Exner, G.M. Saidel, Drug-eluting polymer implants in cancer therapy, *Expert Opin. Drug Deliv.* 5(7) (2008) 775-788.

[6] A.B. Fleming, W.M. Saltzman, Pharmacokinetics of the carmustine implant, *Clin. Pharmacokinet* 41(6) (2002) 403-419.

[7] S. Kunwar, S. Chang, M. Westphal, M. Vogelbaum, J. Sampson, G. Barnett, M. Shaffrey, Z. Ram, J. Piepmeier, M. Prados, D. Croteau, C. Pedain, P. Leland, S.R. Husain, B.H. Joshi, R.K. Puri, Phase III randomized trial of CED of IL13-PE38QQR vs Gliadel wafers for recurrent glioblastoma, *Neuro Oncol.* 12(8) 871-881.

[8] W.M. Saltzman, L.K. Fung, Polymeric implants for cancer chemotherapy, *Adv. Drug Deliv. Rev.* 26(2-3) (1997) 209-230.

[9] L. Solorio, R.B. Patel, H. Wu, T. Krupka, A.A. Exner, Advances In Image-Guided Intratumoral Drug Delivery Techniques, *Therapeutic Delivery* 1(2) (2010) 307-322.

[10] M. Westphal, Z. Ram, V. Riddle, D. Hilt, E. Bortey, Gliadel wafer in initial surgery for malignant glioma: long-term follow-up of a multicenter controlled trial, *Acta Neurochir.* (Wien) 148(3) (2006) 269-275; discussion 275.

[11] C.B. Packhaeuser, J. Schnieders, C.G. Oster, T. Kissel, In situ forming parenteral drug delivery systems: an overview, *European Journal of Pharmaceutics and Biopharmaceutics* 58(2) (2004) 445-455.

[12] H.B. Ravivarapu, K.L. Moyer, R.L. Dunn, Sustained activity and release of leuprolide acetate from an in situ forming polymeric implant, *AAPS PharmSciTech* 1(1) (2000) E1.

[13] H.B. Ravivarapu, K.L. Moyer, R.L. Dunn, Sustained suppression of pituitary-gonadal axis with an injectable, in situ forming implant of leuprolide acetate, *J. Pharm. Sci.* 89(6) (2000) 732-741.

[14] H.B. Ravivarapu, K.L. Moyer, R.L. Dunn, Parameters affecting the efficacy of a sustained release polymeric implant of leuprolide, *Int. J. Pharm.* 194(2) (2000) 181-191.

[15] K. Al-Tahami, M. Oak, J. Singh, Controlled delivery of basal insulin from phase-sensitive polymeric systems after subcutaneous administration: In vitro release, stability, biocompatibility, in vivo absorption, and bioactivity of insulin, *J. Pharm. Sci.*

[16] K.J. Brodbeck, J.R. DesNoyer, A.J. McHugh, Phase inversion dynamics of PLGA solutions related to drug delivery - Part II. The role of solution thermodynamics and bath-side mass transfer, *Journal of Controlled Release* 62(3) (1999) 333-344.

[17] K.J. Brodbeck, S. Pushpala, A.J. McHugh, Sustained release of human growth hormone from PLGA solution depots, *Pharmaceutical Research* 16(12) (1999) 1825-1829.

[18] S. Chen, J. Singh, Controlled delivery of testosterone from smart polymer solution based systems: in vitro evaluation, *Int. J. Pharm.* 295(1-2) (2005) 183-190.

[19] P.D. Graham, K.J. Brodbeck, A.J. McHugh, Phase inversion dynamics of PLGA solutions related to drug delivery, *Journal of Controlled Release* 58(2) (1999) 233-245.

[20] A. Hatefi, B. Amsden, Biodegradable injectable in situ forming drug delivery systems, *J. Control Release* 80(1-3) (2002) 9-28.

[21] T.M. Krupka, B.D. Weinberg, N.P. Ziats, J.R. Haaga, A.A. Exner, Injectable polymer depot combined with radiofrequency ablation for treatment of experimental carcinoma in rat, *Investigative Radiology* 41(12) (2006) 890-897.

[22] A.J. McHugh, The role of polymer membrane formation in sustained release drug delivery systems, *Journal of Controlled Release* 109(1-3) (2005) 211-221.

[23] M.A. Royals, S.M. Fujita, G.L. Yewey, J. Rodriguez, P.C. Schultheiss, R.L. Dunn, Biocompatibility of a biodegradable in situ forming implant system in rhesus monkeys, *Journal of Biomedical Materials Research* 45(3) (1999) 231-239.

[24] R.L. Dunn, J.P. English, D.R. Cowsar, D.P. Vanderbilt, Biodegradable *in situ* forming implants and methods of producing the same, *U.S. Patent* 4, 938, 763, 1990.

[25] R.L. Dunn, A.J. Tipton, Polymeric compositions useful as controlled release implants, *U.S. Patent* 5, 702, 716, 1997.

[26] R.L. Dunn, S. Garrett, The drug delivery and biomaterial attributes of the ATRIGEL technology in the treatment of periodontal disease, *Expert opinion on investigational drugs* 7(9) (1998) 1483-1491.

[27] R.M. Boom, I.M. Wienk, T. Vandenboomgaard, C.A. Smolders, Microstructures in Phase Inversion Membranes .2. The Role of a Polymeric Additive, *Journal of Membrane Science* 73(2-3) (1992) 277-292.

[28] C.A. Smolders, A.J. Reuvers, R.M. Boom, I.M. Wienk, Microstructures in Phase-Inversion Membranes .1. Formation of Macrovoids, *Journal of Membrane Science* 73(2-3) (1992) 259-275.

[29] N. Vogrin, C. Stropnik, V. Musil, M. Brumen, The wet phase separation: the effect of cast solution thickness on the appearance of macrovoids in the membrane forming ternary cellulose acetate/acetone/water system, *Journal of Membrane Science* 207(1) (2002) 139-141.

[30] A.J. McHugh, P.D. Graham, K.J. Brodbeck, Phase inversion dynamics of PLGA solutions related to drug delivery, Biomedical Materials-Drug Delivery, *Implants and Tissue Engineering* 550(1999) 41-46 376.

[31] A.J. Mchugh, D.C. Miller, The Dynamics of Diffusion and Gel Growth during Nonsolvent-Induced Phase Inversion of Polyethersulfone, *Journal of Membrane Science* 105(1-2) (1995) 121-136.

[32] A.J. Mchugh, C.S. Tsay, Dynamics of the Phase Inversion Process, *Journal of Applied Polymer Science* 46(11) (1992) 2011-2021.

[33] R. Astaneh, N. Nafissi-Varcheh, M. Erfan, Zinc-leuprolide complex: preparation, physicochemical characterization and release behaviour from in situ forming implant, *Journal of Peptide Science* 13(10) (2007) 649-654.

[34] S. Chen, D. Pederson, M. Oak, J. Singh, In vivo absorption of steroidal hormones from smart polymer based delivery systems, *J. Pharm. Sci.* 99(8) 3381-3388.

[35] S. Dhawan, R. Kapil, D.N. Kapoor, Development and Evaluation of In Situ Gel-forming System for Sustained Delivery of Insulin, *J. Biomater. Appl.*

[36] R.E. Eliaz, D. Wallach, J. Kost, Delivery of soluble tumor necrosis factor receptor from in-situ forming PLGA implants: in-vivo, *Pharm. Res.* 17(12) (2000) 1546-1550.

[37] H. Kranz, E. Yilmaz, G.A. Brazeau, R. Bodmeier, In vitro and in vivo drug release from a novel in situ forming drug delivery system, *Pharmaceutical Research* 25(6) (2008) 1347-1354.
[38] Q.F. Liu, H. Zhang, G.C. Zhou, S.B. Xie, H. Zou, Y.A. Yu, G.D. Li, D.X. Sun, G.Q. Zhang, Y. Lu, Y.Q. Zhong, In vitro and in vivo study of thymosin alpha1 biodegradable in situ forming poly(lactide-co-glycolide) implants, *International Journal of Pharmaceutics* 397(1-2) (2010) 122-129.
[39] X. Luan, R. Bodmeier, In situ forming microparticle system for controlled delivery of leuprolide acetate: influence of the formulation and processing parameters, *Eur. J. Pharm. Sci.* 27(2-3) (2006) 143-149.
[40] F.W. Okumu, N. Dao le, P.J. Fielder, N. Dybdal, D. Brooks, S. Sane, J.L. Cleland, Sustained delivery of human growth hormone from a novel gel system: SABER, *Biomaterials* 23(22) (2002) 4353-4358.
[41] O. Sartor, Eligard (R) 6: A new form of treatment for prostate cancer, *European Urology Supplements* 5(18) (2006) 905-910.
[42] S. Singh, J. Singh, Phase-sensitive polymer-based controlled delivery systems of leuprolide acetate: In vitro release, biocompatibility, and in vivo absorption in rabbits, *International Journal of Pharmaceutics* 328(1) (2007) 42-48.
[43] G. Tae, J.A. Kornfield, J.A. Hubbell, Sustained release of human growth hormone from in situ forming hydrogels using self-assembly of fluoroalkyl-ended poly(ethylene glycol), *Biomaterials* 26(25) (2005) 5259-5266.
[44] H. Ueda, M.C. Hacker, A. Haesslein, S. Jo, D.M. Ammon, R.N. Borazjani, J.F. Kunzler, J.C. Salamone, A.G. Mikos, Injectable, in situ forming poly(propylene fumarate)-based ocular drug delivery systems, *Journal of Biomedical Materials Research* Part A 83A(3) (2007) 656-666.
[45] F. Wu, T. Jin, Polymer-Based Sustained-Release Dosage Forms for Protein Drugs, Challenges, and Recent Advances, *Aaps Pharmscitech.* 9(4) (2008) 1218-1229.
[46] M. Zare, H. Mobedi, J. Barzin, H. Mivehchi, A. Jamshidi, R. Mashayekhi, Effect of additives on release profile of leuprolide acetate in an in situ forming controlled-release system: In vitro study, *Journal of Applied Polymer Science* 107(6) (2008) 3781-3787.
[47] J.R. DesNoyer, A.J. McHugh, Role of crystallization in the phase inversion dynamics and protein release kinetics of injectable drug delivery systems, *J. Control Release* 70(3) (2001) 285-294.
[48] J.R. DesNoyer, A.J. McHugh, The effect of Pluronic on the protein release kinetics of an injectable drug delivery system, *J. Control Release* 86(1) (2003) 15-24.
[49] C.S. Tsay, A.J. Mchugh, A Technique for Rapid Measurement of Diffusion-Coefficients, *Industrial and Engineering Chemistry Research* 31(1) (1992) 449-452.
[50] G.E. Gaides, A.J. Mchugh, Measurement and Analysis of Bath-Side Interfacial Concentration Gradients during Phase Inversion, *Journal of Membrane Science* 74(1-2) (1992) 83-94.
[51] L. Solorio, B.M. Babin, R.B. Patel, J. Mach, N. Azar, A.A. Exner, Noninvasive characterization of in situ forming implants using diagnostic ultrasound, *Journal of Controlled Release* 143(2) (2010) 183-190.
[52] S. Kempe, H. Metz, K. Mader, Do in situ forming PLG/NMP implants behave similar in vitro and in vivo? A non-invasive and quantitative EPR investigation on the mechanisms of the implant formation process, *J. Control Release* 130(3) (2008) 220-225.
[53] K. Mader, G. Bacic, A. Domb, O. Elmalak, R. Langer, H.M. Swartz, Noninvasive in vivo monitoring of drug release and polymer erosion from biodegradable polymers by

EPR spectroscopy and NMR imaging, *Journal of Pharmaceutical Sciences* 86(1) (1997) 126-134.

[54] D.J. Lurie, K. Mader, Monitoring drug delivery processes by EPR and related techniques - principles and applications, *Advanced Drug Delivery Reviews* 57(8) (2005) 1171-1190.

[55] L.J. Berliner, Biological Magnetic Resonance *In Vivo*Kluwer Academic, 2003, New York, State, 2003

[56] K. Mader, S. Nitschke, R. Stosser, H.H. Borchert, A. Domb, Non-destructive and localized assessment of acidic microenvironments inside biodegradable polyanhydrides by spectral spatial electron paramagnetic resonance imaging, *Polymer* 38(19) (1997) 4785-4794.

[57] S. Kempe, H. Metz, P.G.C. Pereira, K. Mader, Non-invasive in vivo evaluation of in situ forming PLGA implants by benchtop magnetic resonance imaging (BT-MRI) and EPR spectroscopy, *European Journal of Pharmaceutics and Biopharmaceutics* 74(1) (2010) 102-108.

[58] K. Mader, Pharmaceutical applications of in vivo EPR, *Physics in Medicine and Biology* 43(7) (1998) 1931-1935.

[59] K. Mader, B. Gallez, K.J. Liu, H.M. Swartz, Non-invasive in vivo characterization of release processes in biodegradable polymers by low-frequency electron paramagnetic resonance spectroscopy, *Biomaterials* 17(4) (1996) 457-461.

[60] R. Stosser, K. Mader, H.H. Borchert, W. Herrmann, G. Schneider, A. Liero, Pharmaceutical aspects of ESR investigations on drug delivery systems, tissues and living systems, *Bioradicals Detected by Esr. Spectroscopy* (1995) 301-320 337.

[61] Ng KY, T.O. Matsunaga, Ultrasound mediated drug delivery, in: B Wang, TJ Siahaan, Soltero R (Eds.), Drug Delivery: Principles and Applications, John Wiley and Sons, Inc., Hoboken, 2005, pp. 245-278.

[62] J.T. Bushberg, J.A. Seibert, E.M. Leidholdt, J.M. Boone, Ultrasound, in: Passano W (Eds.), The Essential Physics of Medical Imaging, Williams and Wilkins, Baltimore, 1994, pp. 367-416.

[63] R.B. Patel, L. Solorio, H.P. Wu, T. Krupka, A.A. Exner, Effect of injection site on in situ implant formation and drug release in vivo, *Journal of Controlled Release* 147(3) (2010) 350-358.

[64] B.F. Barton, J.L. Reeve, A.J. McHugh, Observations on the dynamics of nonsolvent-induced phase inversion, *Journal of Polymer Science Part B-Polymer Physics* 35(4) (1997) 569-585.

[65] D.F. Li, T.S. Chung, J.Z. Ren, R. Wang, Thickness dependence of macrovoid evolution in wet phase-inversion asymmetric membranes, *Industrial and Engineering Chemistry Research* 43(6) (2004) 1553-1556.

[66] C. Raman, A.J. McHugh, A model for drug release from fast phase inverting injectable solutions, *J. Control Release* 102(1) (2005) 145-157.

[67] C. Wischke, Y. Zhang, S. Mittal, S.P. Schwendeman, Development of PLGA-based injectable delivery systems for hydrophobic fenretinide, *Pharm. Res.* 27(10) 2063-2074.

[68] R.A. Jain, C.T. Rhodes, A.M. Railkar, A.W. Malick, N.H. Shah, Controlled release of drugs from injectable in situ formed biodegradable PLGA microspheres: effect of various formulation variables, *European Journal of Pharmaceutics and Biopharmaceutics* 50(2) (2000) 257-262.

[69] B.C. Hancock, P. York, R.C. Rowe, The use of solubility parameters in pharmaceutical dosage form design, *International Journal of Pharmaceutics* 148(1) (1997) 1-21.

[70] L.W. Wang, L. Kleiner, S. Venkatraman, Structure formation in injectable poly(lactide-co-glycolide) depots, *Journal of Controlled Release* 90(3) (2003) 345-354.

[71] K. Al-Tahami, A. Meyer, J. Singh, Poly lactic acid based injectable delivery systems for controlled release of a model protein, lysozyme, *Pharmaceutical Development and Technology* 11(1) (2006) 79-86.
[72] R. Astaneh, M. Erfan, J. Barzin, H. Mobedi, H. Moghimi, Effects of Ethyl Benzoate on Performance, Morphology, and Erosion of PLGA Implants Formed In Situ, Advances in Polymer Technology 27(1) (2008) 17-26.
[73] S. Dhawan, R. Kapil, D.N. Kapoor, M. Kumar, Development and evaluation of in situ gel forming system for sustained delivery of cyclosporine, *Curr. Drug Deliv.* 6(5) (2009) 495-504.
[74] F.R. Kang, J. Singh, In vitro release of insulin and biocompatibility of in situ forming gel systems, *International Journal of Pharmaceutics* 304(1-2) (2005) 83-90.
[75] S. Prabhu, L.P. Tran, G.V. Betageri, Effect of co-solvents on the controlled release of calcitonin polypeptide from in situ biodegradable polymer implants, *Drug Delivery* 12(6) (2005) 393-398.
[76] S. Singh, J. Singh, Controlled release of a model protein lysozyme from phase sensitive smart polymer systems, *International Journal of Pharmaceutics* 271(1-2) (2004) 189-196.
[77] A. Yapar, T. Baykara, Effects of Solvent Combinations on Drug Release from Injectable Phase Sensitive Liquid Implants, *Turk. J. Pharm. Sci.* 7(1) (2010) 49-56.
[78] R.B. Patel, A.N. Carlson, L. Solorio, A.A. Exner, Characterization of formulation parameters affecting low molecular weight drug release from in situ forming drug delivery systems, *Journal of Biomedical Materials Research* Part A 94A(2) (2010) 476-484.
[79] R. Bakhshi, E. Vasheghani-Farahani, H. Mobedi, A. Jamshidi, M. Khakpour, The effect of additives on naltrexone hydrochloride release and solvent removal rate from an injectable in situ forming PLGA implant, *Polymers for Advanced Technologies* 17(5) (2006) 354-359.
[80] D.C. Ma, A.J. McHugh, The interplay of membrane formation and drug release in solution-cast films of polylactide polymers, *International Journal of Pharmaceutics* 388(1-2) (2010) 1-12.
[81] S. Chhabra, V. Sachdeva, S. Singh, Influence of end groups on in vitro release and biological activity of lysozyme from a phase-sensitive smart polymer-based in situ gel forming controlled release drug delivery system, *International Journal of Pharmaceutics* 342(1-2) (2007) 72-77.
[82] H. Liu, S.S. Venkatraman, Effect of Polymer Type on the Dynamics of Phase Inversion and Drug Release in Injectable In Situ Gelling Systems, *J. Biomater. Sci. Polym. Ed.*
[83] Y. Liu, A. Kemmer, K. Keim, C. Curdy, H. Petersen, T. Kissel, Poly(ethylene carbonate) as a surface-eroding biomaterial for in situ forming parenteral drug delivery systems: A feasibility study, *European Journal of Pharmaceutics and Biopharmaceutics* 76(2) (2010) 222-229.
[84] C.B. Packhaeuser, T. Kissel, On the design of in situ forming biodegradable parenteral depot systems based on insulin loaded dialkylaminoalkyl-amine-poly(vinyl alcohol)-g-poly (lactide-co-glycolide) nanoparticles, *Journal of Controlled Release* 123(2) (2007) 131-140.
[85] R.E. Eliaz, J. Kost, Characterization of a polymeric PLGA-injectable implant delivery system for the controlled release of proteins, *J. Biomed. Mater. Res.* 50(3) (2000) 388-396.
[86] H.A. Gad, M.A. El-Nabarawi, S.S.A. El-Hady, Formulation and Evaluation of PLA and PLGA In Situ Implants Containing Secnidazole and/or Doxycycline for Treatment of Periodontitis, *Aaps Pharmscitech.* 9(3) (2008) 878-884.

[87] R. Astaneh, M. Erfan, H. Mobedi, H.R. Moghimi, Formulation of an Injectable Implant for Peptide Delivery and Studying the Effect of Polymer Molecular Weight on Its Release Behavior, *Journal of Peptide Science* 10(2004) 142-142.

[88] R. Astaneh, M. Erfan, H. Moghimi, H. Mobedi, Changes in morphology of in situ forming PLGA implant prepared by different polymer molecular weight and its effect on release behavior, *J. Pharm. Sci.* 98(1) (2009) 135-145.

[89] X. Luan, R. Bodmeier, Influence of the poly(lactide-co-glycolide) type on the leuprolide release from in situ forming microparticle systems, *J. Control Release* 110(2) (2006) 266-272.

[90] R. Astaneh, H.M. Moghimi, M. Erfan, H. Mobedi, Formulation of an injectable implant for peptide delivery, *Journal of Peptide Science* 12(2006) 241-241.

[91] Z.H. Zhang, Q.F. An, Y.L. Ji, J.W. Qian, C.J. Gao, Effect of zero shear viscosity of the casting solution on the morphology and permeability of polysulfone membrane prepared via the phase-inversion process, *Desalination* 260(1-3) (2010) 43-50.

[92] P. Radovanovic, S.W. Thiel, S.T. Hwang, Formation of Asymmetric Polysulfone Membranes by Immersion Precipitation .2. The Effects of Casting Solution and Gelation Bath Compositions on Membrane-Structure and Skin Formation, *Journal of Membrane Science* 65(3) (1992) 231-246.

[93] L. Wang, S. Venkatraman, L. Kleiner, Drug release from injectable depots: two different in vitro mechanisms, *J. Control Release* 99(2) (2004) 207-216.

[94] A.A. Exner, T.M. Krupka, K. Scherrer, J.M. Teets, Enhancement of carboplatin toxicity by Pluronic block copolymers, *J. Control Release* 106(1-2) (2005) 188-197.

[95] T.M. Krupka, D. Dremann, A.A. Exner, Time and dose dependence of pluronic bioactivity in hyperthermia-induced tumor cell death, *Exp. Biol. Med.* (Maywood) 234(1) (2009) 95-104.

[96] A. Mashak, H. Mobedi, F. Ziaee, M. Nekoomanesh, The Effect of Aliphatic Esters on the Formation and Degradation Behavior of PLGA-based In Situ Forming System, *Polymer Bulletin* DOI 10.1007/s00289-010-0386-7(2010).

[97] C.M. Deadman, I.W. Kellaway, M. Yasin, P.A. Dickinson, S. Murdan, An investigation into the influence of drug lipophilicity on the in vivo absorption profiles from subcutaneous microspheres and in situ forming depots, *J. Control Release* 122(1) (2007) 79-85.

[98] C. Matschke, U. Isele, P. van Hoogevest, A. Fahr, Sustained-release injectables formed in situ and their potential use for veterinary products, *Journal of Controlled Release* 85(1-3) (2002) 1-15.

In: Polymer Phase Behavior
Editors: T. P. Ehlers and J. K. Wilhelm, pp. 205-233
ISBN: 978-1-61324-336-7
© 2011 Nova Science Publishers, Inc.

Chapter 5

ECO-FRIENDLY (CO) POLYESTERS CONTAINING 1,4-CYCLOHEXYLENE UNITS: CORRELATIONS BETWEEN STEREOCHEMISTRY AND PHASE BEHAVIOR

Annamaria Celli,* Paola Marchese, Simone Sullalti and Corrado Berti
Department of Civil, Environmental and Materials Engineering,
University of Bologna, Bologna, Italy

ABSTRACT

Nowadays the urgency for solving plastic waste problems is inducing academic and industrial research to develop novel environmentally friendly polymers, i.e. materials produced from alternative resources, with low energy consumption, non-toxic to the environment, and biodegradable. These biopolymers should have also good physical performances. In the field of aliphatic polyesters, novel (co)polymers, containing 1,4-cyclohexylene units, appear very promising materials, which are obtainable from biomass, potentially biodegradable and characterized by good mechanical properties. Moreover, these polyesters have the interesting peculiarity that their phase behavior is strictly connected to the ratio of the two possible configurations, *cis* and *trans*, of the cyclic units. Indeed, the *trans* isomer is more rigid and symmetric than the *cis*. Highly symmetrical units tend to improve the chain packing with a consequent increment in crystallinity and crystalline perfection. On the other hand, the *cis* isomer introduces kinks into the main chain, which hinder the formation of stable crystals. Thus, at high *trans* content the polyesters are characterized by relative high degree of crystallinity, whereas at low *trans* content the polymers are amorphous. Therefore, accordingly to the final *cis/trans* ratio, the phase behavior of the homopolymers and copolymers significantly changes and the stereochemistry of the cycloaliphatic units result to be a key factor to tailor the final thermal properties of the material. In this paper the properties of some homopolymers and copolymers, containing the 1,4-cyclohexylene units with different

* Via Terracini 28, 40131 Bologna, Italy. e-mail: annamaria.celli@unibo.it.

cis/trans ratio, are discussed just in terms of the correlations between stereochemistry and phase behavior.

1. Introduction

In recent years there has been a great interest on environmentally and safety problems, which become more and more urgent in front of the amounts of not biodegradable waste and widespread pollution. As to the plastics, great difficulties are connected to the substitution of traditional polymers, which are generally prepared from non-renewable feedstock and are resistant to biodegradation. Biobased chemicals and sustainable materials, instead, are now a priority.

In this context all-aliphatic polyesters are excellent candidates to have a significant role in the next generation of plastics. Indeed, they often derive from renewable resources and potentially have some good properties, such as low toxicity and biodegradability. Moreover, with respect to the widely appreciated industrial thermoplastics, such as poly(butylene terephthalate) (PBT) and poly(ethylene terephthalate) (PET), aliphatic polyesters show improved capabilities, in terms of UV stability, as they do not contain any functional groups that absorb UV light, resistance to weather, heat and humidity, which make them ideal polymers in outdoor applications.

It is known, however, that the mechanical properties of aliphatic polyesters sometimes fail with respect to the properties of aliphatic-aromatic polyesters. Poly(R)-3-hydroxybutyrate (PHB), for example, is a brittle material and poly(alkylene dicarboxylate)s are characterized by very low melting peak temperatures (about ~ 40-80°C) [1,2].

Introduction of alicyclic units to the main chain of the polymer can be a way to increase the rigidity of the macromolecular chains. Recently, for example, Liu and Turner [3] describe the preparation of a systematic series of random copolyesters using different cycloaliphatic diesters, which are, for the Authors, the most suitable monomers to achieve high T_g values, up to 155°C. Moreover, conformational transitions of cyclohexylene rings in the backbone originate secondary relaxations in dynamical mechanical spectrum, which contribute to improve the performances of the materials.

1,4-cyclohexane dicarboxylic acid (CHDA), dimethyl-1,4-cyclohexane dicarboxylate (DMCD), and 1,4-cyclohexane dimethanol (CHDM) (see Scheme 1) are the monomers commercially available to introduce the 1,4-cyclohexylene units in polyesters. The synthesis and properties of polyesters and copolyesters containing these alicyclic moieties were studied at the beginning of the eighties by Eastman Chemical Company, interested to develop materials with excellent tensile strength, stiffness and impact properties as well as materials to be used as improved hot melt adhesives. After the first patents [4,5], now there is a new interest for the polyesters deriving from cyclic diacid/diesters and diols or containing aliphatic C_6 rings, especially after the recent improvements in the preparation procedure [6-9].

In our opinion, today the use of the cyclic monomers of Scheme 1 to prepare all-aliphatic polyesters presents further advantages. Although CHDA, DMCD, and CHDM are now obtained from petroleum resources, however, they can be prepared from bio-based terephthalic acid, starting from limonene and other terpenes [10]. Therefore, polymers derived, for example, from DMCD and a diol obtainable from biomass (as 1,4-butanediol, which can be prepared, for example, from succinic acid [11]) can be considered fully sustainable materials. Moreover, we have observed that the presence of the 1,4-cyclohexylene units along a macromolecule does not hinder the attack of microorganisms in some homopolymers and copolymers [12]. Therefore, the polyesters containing the 1,4-cyclohexylene units can be considered biodegradable materials and are very promising, environmentally friendly polyesters.

Again, it is notably that the 1,4-cyclohexylene unit shows another remarkable peculiarity: it can have two possible configurations, *cis* and *trans*, as described in Scheme 2 for the monomeric unit of the polyester prepared from DMCD and 1,4-butanediol, called PBCHD. From the literature, which studied the incorporation of cycloaliphatic monomers in polyesters, polyamides, polyurethanes [13-18], and from some of our researches [19-21], it results that the isomeric ratio of the cycloaliphatic residues along the chains is the key factor which determines the phase behavior of the materials. This is a crucial point, which in our opinion has not been explored fully in the literature. Indeed, although there are numerous reports about the properties of various cyclic ester containing polyesters, there are few papers of all-aliphatic (co)polyesters based on a systematic variation of the stereochemistry of the cyclic monomers.

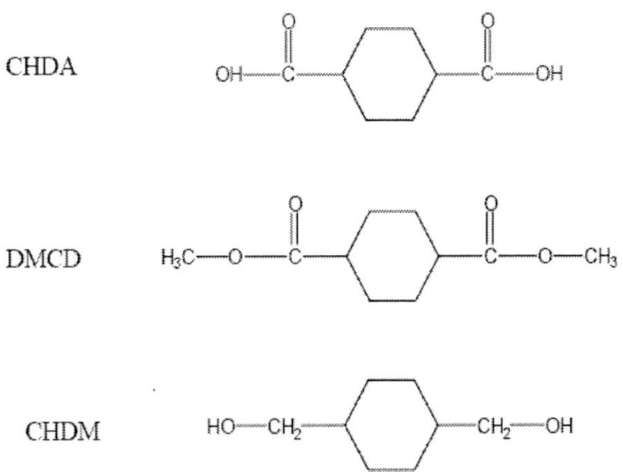

Scheme 1. Molecular structures of cyclic monomers.

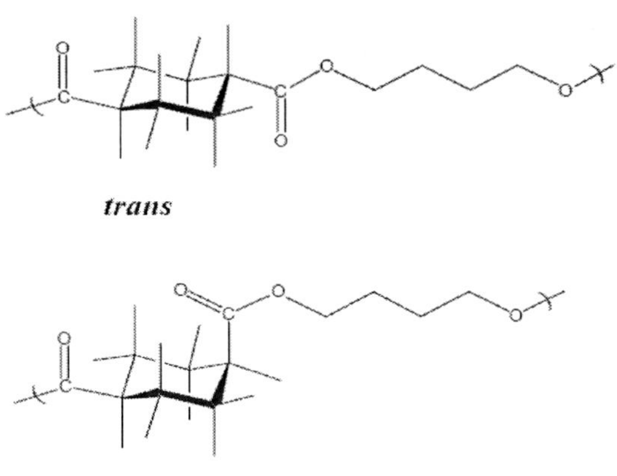

Scheme 2. *Trans* and *cis* configurations of the monomeric unit in PBCHD.

For example, for CHDA in ref. 22 only a *cis/trans* ratio of about 40/60 has been investigated, whereas in ref. 23 and 24 only *cis/trans* ratios of 80/20 and 70/30, respectively, have been used. Instead, we believe it should be possible to obtain completely different properties of the copolymers using a different *cis/trans* ratio of the starting materials.

Therefore, this paper focus on the preparation and characterization of polyesters and copolyesters from a cyclic diester, DMCD, with a systematic variable *cis/trans* ratio, from about 80/20 to 0/100 mol%. The goal is a wide discussion on the correlations existing between the stereochemistry of the 1,4-cyclohexylene ring and the final phase behavior of the novel materials.

2. Experimental Part

2.1. Materials

Two commercial samples of dimethyl-1,4-cyclohexane dicarboxylate (DMCD), with 22% and 100% of *trans* isomer content, 1,4-butanediol (BD), 1,4-cyclohexane dimethanol (CHDM) with 66 mol% of *trans* isomer, 1,3-propanediol, dimethyl terephthalate (DMT), dimethyl adipate (DMA), titanium tetrabutoxide (TBT) (all from Aldrich chemicals) were high purity products, used as received. Scheme 1 describes the monomers containing an alicyclic unit.

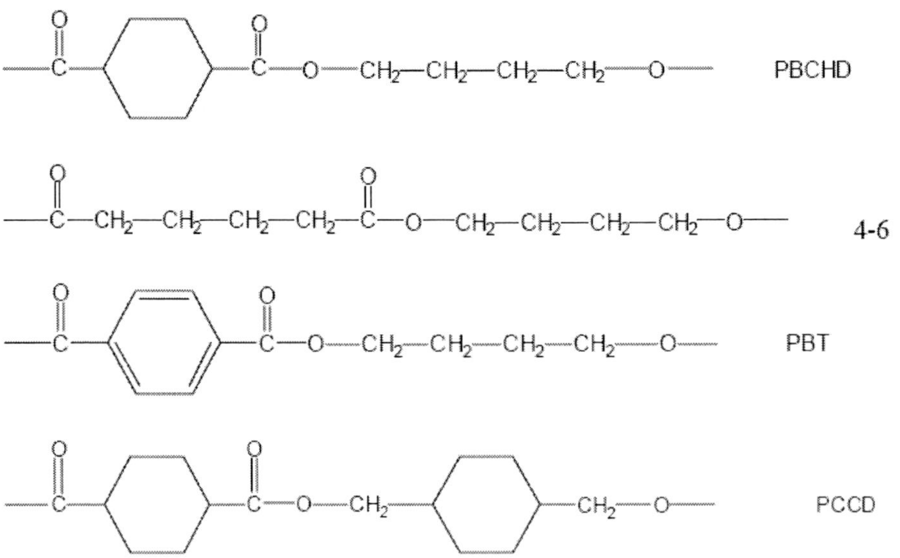

Scheme 3. Molecular structure of the repeating units of poly(butylene-1,4-cyclohexanedicarboxylate) (PBCHD), Poly(butylene adipate) (4-6), poly(butylene terephthalate) (PBT), and poly(1,4-cyclohexylenedimethylene 1,4-cyclohexanedicarboxylate) (PCCD).

Poly(buthylene terephthalate) (PBT) was a commercial product supplied by General Electric and was used as a reference material. Poly(propylene terephthalate) (PPT) was synthesized in our laboratory [25]. Ecoflex was a commercial product by BASF.

In the text the polyesters prepared from DMCD and BD are named with the code PBCHD-xx, (see Scheme 3) where xx indicates the percentage of the aliphatic rings, derived from DMCD, in *trans* configuration.

The polyesters derived from DMCD and CHDM are indicated with the code PCCD-Dxx-Eyy (see Scheme 3), where xx indicates the percentage of C_6 rings deriving from CHDM (D) in *trans* configuration and yy the percentage of C_6 rings deriving from DMCD (E) in *trans* configuration. The PCCD samples here considered are all prepared from CHDM containing 66 mol % of *trans* isomer.

The copolymers derived from BD, DMA, and DMCD are named (4-6)-*co*-$PBCHD_{xx}$-a/b, where xx indicates the percentage of the aliphatic rings, derived from DMCD, in *trans* configuration and a/b is the feed molar ratio of the DMA/DMCD.

2.2. Sample Preparation

All the syntheses are two stage polycondensations carried out in the presence of TBT as catalyst. The syntheses of PBCHD and PCCD samples are described in ref 19 and 20, respectively. For 4-6 the procedure of synthesis is similar to those discussed in ref. 2 for analogous poly(alkylene dicarboxylate)s. As example, the syntheses of a PBCHD specimen and of a (4-6)-*co*-PBCHD copolymer are here reported.

2.2.1. Synthesis of Poly(Butylene 1,4-Cyclohexanedicarboxylate) (PBCHD)

DMCD (32.24 g, 0.171 mol), BD (20.5 g, 0.228 mol), TBT (0.02 g, 0.059 mmol) were placed into a round-bottomed wide-neck glass reactor (250 ml capacity). The reactor was closed with a three-neck flat flange lid equipped with a mechanical stirrer and a torque meter which gives an indication of the viscosity of the reaction melt. The reactor was immersed into a silicone oil bath preheated to 200°C. The first stage was conducted at atmospheric pressure under nitrogen atmosphere and the mixture was allowed to react for 120 min under stirring with continuous removal of water. The second stage was started by gradually reducing the pressure to 0.02 mbar while the temperature was raised to the final value of 220°C. These conditions were reached within 60 min, using a linear gradient of temperature and pressure, and maintained for 180 min.

2.2.2. Synthesis of (4-6)-co-$PBCHD_{90}$-50/50 Copolyester

DMA (16.20 g, 0.093 mol), DMCD 100% *trans* (16.22 g, 0.081 mol), DMCD 22% *trans* (2.40 g, 0.012 mol), BD (20.12 g, 0.223 mol), TBT (0.02 g, 0.047 mmol) were placed into a round bottom wide-neck glass reactor (250 ml capacity). The reactor was closed with a three-necked flat flange lid equipped with a mechanical stirrer and a torque meter which gives an indication of the viscosity of the reaction melt. The reactor was immersed into a silicone oil bath preheated to 200°C. The first stage was conducted at atmospheric pressure under nitrogen atmosphere and the mixture was allowed to react for 90 min under stirring with continuous removal of water. The second stage was started by gradually reducing the pressure to 0.2 mbar while the temperature was raised to the final value of 220°C. These conditions were reached within 90 min, using a linear gradient of temperature and pressure, and maintained for 120 min.

The feed DMA/DMCD molar ratios used for the syntheses of the (4-6)-*co*-PBCHD copolymers are 30/70, 50/50, and 70/30 in order to obtain copolyesters with different compositions.

2.3. Characterization

The ^1H NMR spectra were recorded at room temperature on samples dissolved in $CDCl_3$ using a Varian Mercury 400 spectrometer, the proton frequency being 400 MHz.

Molecular weights (expressed in equivalent polystyrene) were determined by gel permeation chromatography (GPC), using a Hewlett Packard Series 1100 liquid chromatography instrument equipped with a PL gel 5μ Mixed-C column. Chloroform was used as eluent and a calibration plot was constructed with polystyrene standards.

The thermogravimetric analysis (TGA) was performed using a Perkin-Elmer TGA7 thermobalance under nitrogen atmosphere (gas flow 40 ml/min) at 10°C·min^{-1} heating rate from 50°C to 900°C.

The calorimetric analysis was carried out by means of a Perkin-Elmer DSC6, calibrated with high purity standards. The measurements were performed under nitrogen flow. In order to cancel the previous thermal history, the samples (ca. 10 mg) were initially heated to different temperatures, varying from 160 to 260°C according to the sample characteristics, kept at high temperature for 1 min, and then cooled to a temperature range varying from -70 to 20°C at 10°C·min^{-1}. After this thermal treatment, the samples were analyzed by heating to 160-260°C at 10°C·min^{-1} (2nd scan). During the cooling scan the crystallization temperature (T_{CC}) and the enthalpy of crystallization (ΔH_{CC}) were measured. During the 2nd scan the glass transition temperature (T_g), the cold crystallization temperature (T_{ch}) and enthalpy (ΔH_{ch}), the melting temperature (T_m) and the enthalpy of fusion (ΔH_m) were measured. T_g was taken as the midpoint of the heat capacity increment associated with the glass-to-rubber transition.

Specimens for dynamic mechanical measurements were obtained by injection molding in a Mini Max Molder (Custom Scientific Instruments) equipped with a rectangular mold (30 x 8 x 1.6 mm^3). The molded samples were rapidly cooled in water and then dried in a oven at 50°C under vacuum overnight. Dynamic mechanical measurements were performed with a dynamic mechanical thermal analyzer (Rheometrics Scientific, DMTA IV), operated in the dual cantilever bending mode, at a frequency of 3 Hz and a heating rate of 3 °C·min^{-1}, over a temperature range from -150 to a final temperature varying from 100 to 150°C, according to the sample characteristics.

3. Results and Discussion

3.1. Preparation and Characterization of Homopolymers

3.1.1. Synthesis

The synthesis of samples with different *cis/trans* ratio of the 1,4-cyclohexylene rings, derived from DMCD, is possible since the final *cis/trans* ratio depends on the isomeric content of the diester. Two DMCD monomers, characterized by 100 and 22 mol% of *trans* isomer respectively, are available on the market. An adequate physical mixture of these two components enables us to obtain the desired stereochemistry in the final polymer.

Isomerization reactions, which can change the initial *cis/trans* content towards the thermodynamically stable 34/66 mol%, could take place. However, they are favored when the synthesis or thermal treatments are carried out at temperatures higher than 260°C and for longer than 1 h, and in the presence of acid groups. Indeed, the use of 1,4-cyclohexane dicarboxylic acid (CHDA) as monomer gives rise to a 5-7% degree of isomerization, indicating a catalytic effect of the carboxylic acid towards isomerization [26]. These are the reasons why all the syntheses were performed by starting from DMCD (and not CHDA), and the temperatures did not exceed 220-240°C. Therefore, during the syntheses from DMCD, isomerisation is virtually absent. Moreover, for the syntheses of PCCD, we have verified that CHDM is unable to isomerise in the experimental conditions used.

The *cis/trans* isomeric ratio in polymers can be evaluated by ^{1}H NMR analysis. An example of an ^{1}H-NMR spectrum of PCCD is reported in Figure 1. The ratio of the areas of the signals centred at 2.3 (trans isomer) and 2.5 (cis isomer) ppm has been used to calculate the trans percentage [22, 27].

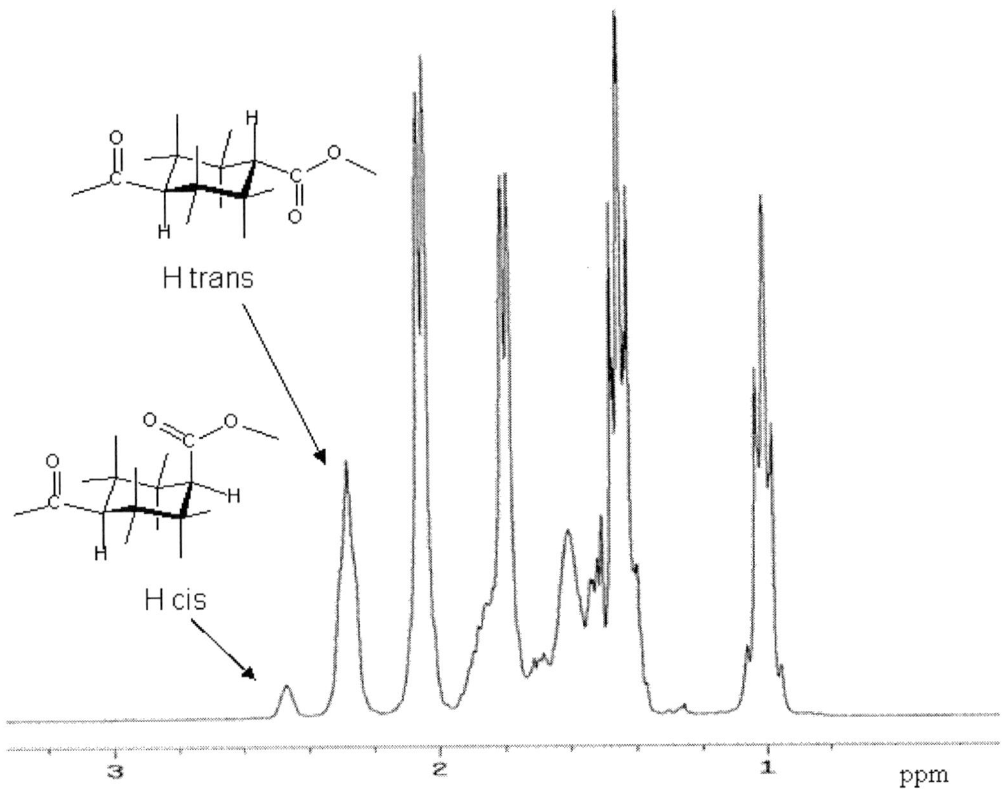

Figure 1. ^{1}H-NMR spectrum of PCCD-D66-E90 sample, with the indication of the signal used to calculate the *cis/trans* ratio.

Table 1. Molecular characteristics and thermal properties of PBCHD samples, with different *cis/trans* ratio, compared with those of a PBT specimen

Sample	*trans* %[a] in DMCD units of the polymer	$M_w \cdot 10^{-3}$ [b]	M_w/M_n [b]	T_{CC} [c] °C	ΔH_{CC} [c] $J \cdot g^{-1}$	T_g [d] °C	T_m [d] °C	ΔH_m [d] $J \cdot g^{-1}$
PBCHD-20	24	57.0	2.2	-	-	-12	-	-
PBCHD-50	52	88.6	2.8	-	-	-7	-	-
PBCHD-70	72	77.6	2.3	79	34	-2	122	22
PBCHD-80	80	78.4	2.3	106	37	1	132-141	27
PBCHD-90	91	54.9	2.3	130	45	6	150-158	37
PBCHD-100	100	73.4	2.5	149	48	10	165-171	47
4-6	-	90.0	2.5	32	67	-58	52-57	70
PBT	-	47.7	2.4	189	48	42[e]	224	43

[a] Calculated by ^1H NMR.
[b] Measured by GPC in $CHCl_3$.
[c] Measured in DSC during the cooling scan at 10°C·min^{-1}.
[d] Measured in DSC during the 2nd heating scan at 10°C·min^{-1}.
[e] From ref. 28.

Table 2. Molecular characteristics and thermal properties of PCCD samples, with different *cis/trans* ratio

Sample	*trans* %[a] *in DMCD* units of the polymer	$M_w \cdot 10^{-3}$ [b]	M_w/M_n [b]	T_{CC} [c] °C	ΔH_{CC} [c] $J \cdot g^{-1}$	T_g [d] °C	T_{ch} [d] °C	ΔH_{ch} [d] $J \cdot g^{-1}$	T_m [d] °C	ΔH_m [d] $J \cdot g^{-1}$
PCCD-D66-E20	24	43.3	2.2	-	-	38	-	-	-	-
PCCD-D66-E50	52	83.7	2.7	-	-	50	-	-	-	-
PCCD-D66-E70	66	71.7	2.2	-	-	55	142	3	181	4
PCCD-D66-E80	81	75.2	2.2	154	23	60	-	-	200-205	23
PCCD-D66-E90	91	56.9	2.3	188	29	65	-	-	218-226	30
PCCD-D66-E100	97	62.0	2.1	204	33	-	-	-	230	34

[a] Calculated by ^1H NMR.
[b] Measured by GPC in $CHCl_3$.
[c] Measured in DSC during the cooling scan at 10°C·min^{-1}.
[d] Measured in DSC during the 2nd heating scan at 10°C·min^{-1}.

The final *trans* content varies from 24 to 100 mol% for both the series of PBCHD and PCCD samples, as reported in Table 1 and 2.

The molecular weight data, calculated by GPC, show that all the samples have significantly high and similar molecular weights, that is they are suitable for comparison in terms of thermal behaviour.

3.1.2. Effect of the Molecular Structure on Thermal Properties

The effect of the introduction of the 1,4-cyclohexylene moieties to an aliphatic macromolecular chain can be highlighted by comparing the thermal properties of the poly(butylene adipate), which is named 4-6, with those of PBCHD. Figure 2, for example, shows the DSC traces of 4-6 and PBCHD-100 and Table 1 reports the calorimetric data. It is noteworthy that both polymers are semicrystalline, with a high degree of crystallinity and a significant capability of crystallizing. Indeed, crystallization exotherms from the melt are very sharp and intense peaks. However, crystallization and melting temperatures of PBCHD-100 are shifted to about 100°C higher than those of 4-6. The T_g values are -58°C for 4-6 and vary from -12 to 10°C for all the PBCHDs, i.e. differ of about 50-70°C. All these data confirm that the macromolecular chains are characterized by a notably high rigidity in the presence of the alicyclic units [3]. Moreover, as it will be discussed in the following section, the chair (preferred) or boat conformations of the 1,4-cyclohexylene units are suitable for good chain packing with the formation of very stable and perfect crystals. However, this conclusion is valid only for the PBCHD samples at high *trans* percentage (see section 3.1.3).

In any case, it is evident that the introduction of the 1,4-cyclohexylene units to the aliphatic macromolecules causes very interesting properties, such as relatively high melting temperatures and good chain rigidity.

In order to have a complete overview of the properties of alicyclic polyesters, it should be interesting to compare their thermal behavior with that of the traditional aliphatic-aromatic polyesters. Indeed, PBCHD is the all-aliphatic counterpart of the aliphatic-aromatic PBT (see Scheme 3). The difference in chemical structures is connected to the substitution of the 1,4-cyclohexylene group with the terephthalate unit of PBT. This difference causes lower T_g, T_C and T_m values in PBCHD than in PBT, as shown in Table 1 [28] . Indeed, in PBT, and in general in polyesters based on terephthalic acid, the coplanarity between the carbonyl and phenyl groups restricts the rotational angles about C_{phenyl}-CO to 0 and 180°, even if rotations about the terephthaloyl residue virtual bond, resulting in nonplanar conformations, are also probable, with a barrier which increases for rotation angles increasing from 0 to 90° [29,30]. The planar conformation favors the molecular packing of the chains in the crystal, and enhances the attractive intermolecular interactions between the ester groups of neighboring chains. As a result, the aromatic polyesters exhibit high melting points [31]. The low flexibility of the chains, moreover, induces a high T_g value. On the other hand, in PBCHD the presence of the aliphatic ring, in chair or boat conformations, excludes the planarity of the CO-C_6 system; moreover the cyclohexyl groups are conformationally more mobile than the rigid phenyl ring. For these reasons the melting temperatures and T_g values are considerably lower [24].

A similar behavior can be observed in Table 3 where two polyesters derived from 1,3-propanediol and DMCD (named PPCHD) are compared with a sample of poly(propylene terephthalate) (PPT) [25]. Also in this case, the molecular structures differ for aliphatic and aromatic rings. It is worth noting that the aliphatic rings inside the chains induce higher flexibility and, thus, notably lower T_g values. At the same time, the capability of crystallizing in PPCHD is significantly reduced with respect to PPT and the melting peak is more than 80°C lower.

Analogously, in ref. 22 PET and the homologous, aliphatic poly(ethylene 1,4-cyclohexanedicarboxylate) (PECHD) are compared and some data collected in Table 3. It is confirmed that the presence of terephthalate units makes the chains more rigid and improve the crystallizability of the material.

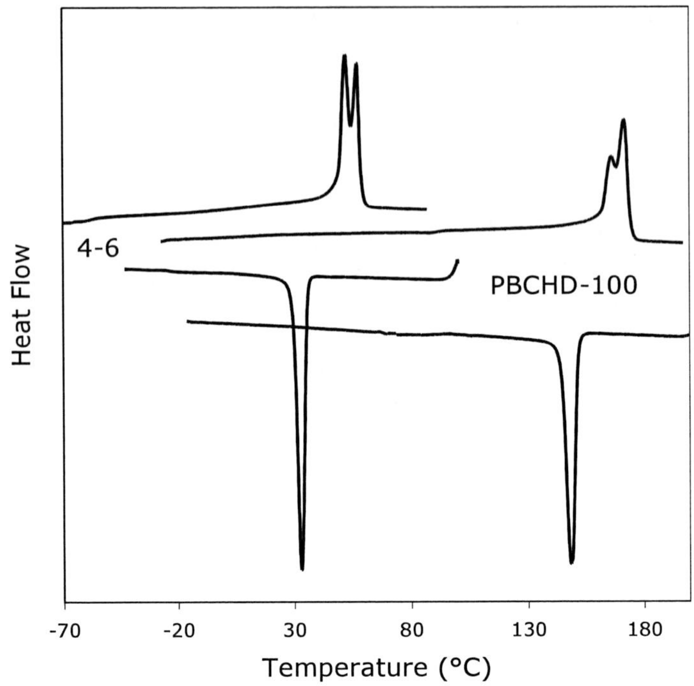

Figure 2. DSC thermograms for 4-6 and PBCHD-100 homopolymers.

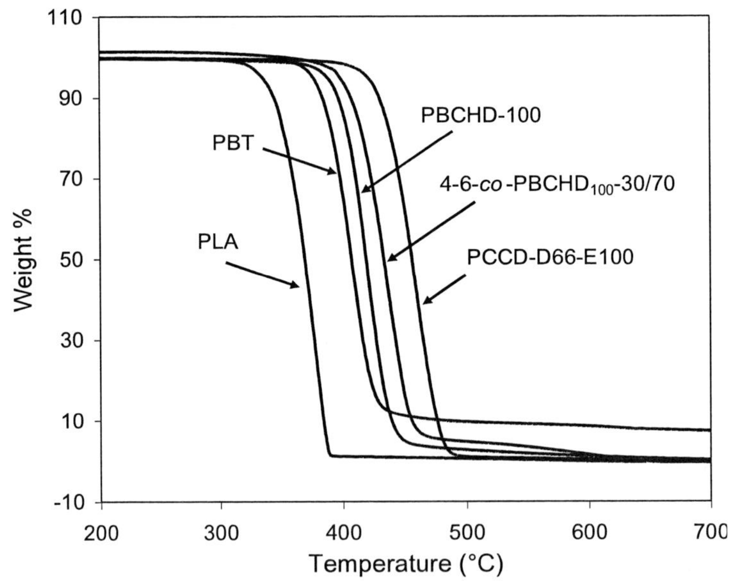

Figure 3. Thermogravimetric curves, obtained in nitrogen at 10°C·min^{-1} for different aliphatic and aromatic polyesters and copolyesters.

Table 3. Molecular characteristics and thermal properties of PPCHDs, with different *cis/trans* ratio, and PECHD, compared with those of PPT and PET specimens

Sample	*trans* %[a] in DMCD units of the polymer	$M_w \cdot 10^{-3}$ [b]	M_w/M_n [b]	T_{Ch} [c] °C	ΔH_{Ch} [c] $J \cdot g^{-1}$	T_g [c] °C	T_m [c] °C	ΔH_m [c] $J \cdot g^{-1}$
PPCHD-50	53	62.4	2.4	-	-	-1	-	-
PPCHD-90	87	65.6	2.5	80	28	7	140	29
PPT [d]	-	65.0	2.3	-	-	49	223	68
PECHD [e]	60	55.3	2.3	-	-	14	-	-
PET [e]	-	34.3	2.4	-	-	82	252	37

[a] Calculated by ^1H NMR
[b] Measured by GPC.
[c] Measured in DSC during the 2nd heating scan.
[d] From ref. 25.
[e] From ref. 22.

Finally, it must be remarked that alicyclic polyesters and copolyesters, from DMCD, are also characterized by a high thermal stability. Figure 3 shows some example of thermogravimetric curves for a sample of purified polylactic acid (PLA) (from Natureworks), a PBT and some polyesters and copolyesters containing the 1,4-cyclohexylene units. It is evident that the most stable materials are exactly these latter. This result indicates that the substitution of the aromatic ring with an aliphatic ring improves the thermal stability of the materials. The same observations have been made, for example, by Wang et al. [23] for copolymers based on PET: the introduction of the 1,4-cyclohexylene rings causes an increment of the thermal stability of PET of 5-10 degrees, as a function of the aliphatic ring content. Moreover, also the substitution of the aliphatic sequence, in PBCHD, with another aliphatic ring, in PCCD, contribute to further stabilize the material. In a similar way, the thermal stability of some aliphatic-aromatic random copolyesters increases as the cyclic units, obtained from CHDM, increase. [32].

3.1.3. Analysis of the Phase Behaviour for PBCHD and PCCD Homopolymers

The two isomers of the 1,4-cyclohexylene units, *cis* and *trans* (see Scheme 1), strongly influence the solid-state behaviour and produce in a polymers a wide variety of properties. Figure 4 and 5 show the thermal behavior of PBCHD samples with different *cis/trans* ratio and Table 1 reports the calorimetric data.

It must be remarked that only certain PBCHD samples crystallize during the cooling scan from the melt at $10°C \cdot min^{-1}$, in particular only PBCHD-70, -80, -90, and -100, i.e. only the samples with a *trans* content $\geq$ 70%. For these samples, the exothermal peaks become more and more intense and narrow as the *trans* content increases. Accordingly, T_{CC} and ΔH_{CC} have a notable increment. In particular, from PBCHD-70 to PBCHD-100 T_{CC} increases by about 70°C and the enthalpy doubles. These differences appear very significant and indicate that the *trans* configuration of the C_6 ring along the chain improves the capacity of the samples to crystallize. On the other hand, it is significant that the PBCHD-50 and PBCHD-20, with the lowest percentages of *trans* stereoisomer, are not able to rearrange towards an ordered state at all.

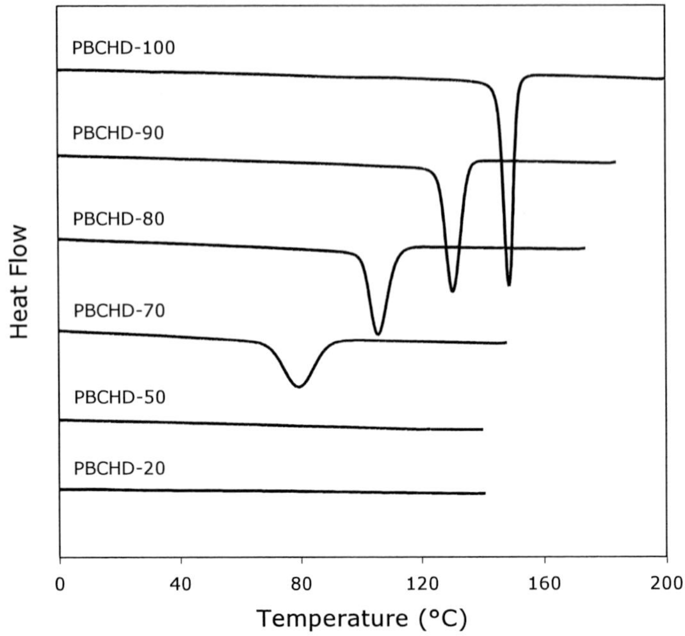

Figure 4. DSC thermograms of PBCHD samples during the cooling scan from the melt at 10°C·min^{-1}.

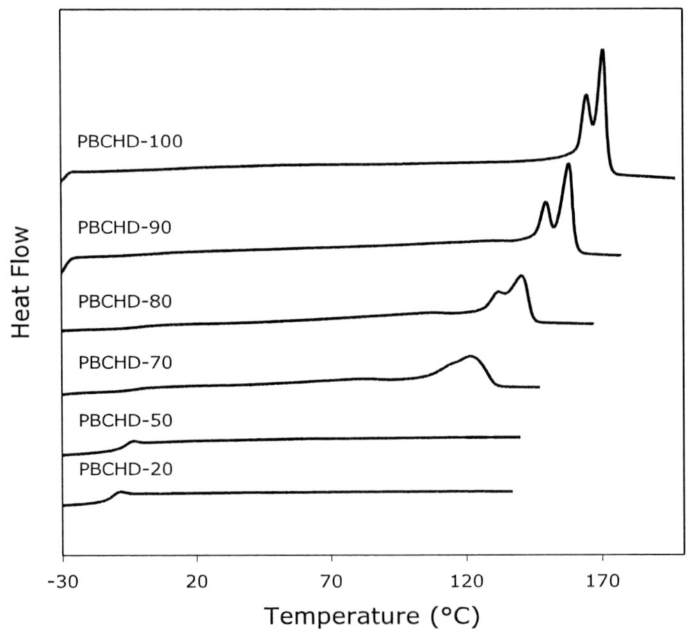

Figure 5. DSC thermograms of PBCHD samples during the 2nd heating scan at 10°C·min^{-1}.

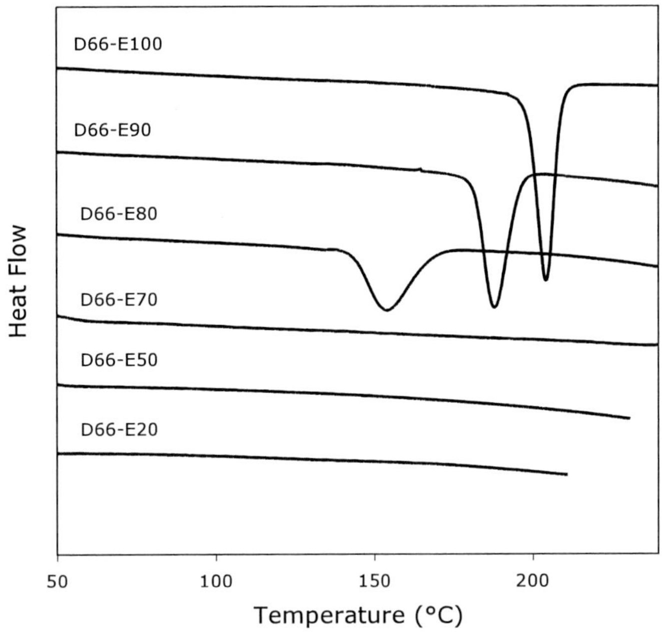

Figure 6. DSC thermograms of PCCD samples during the cooling scan from the melt at 10°C·min^{-1}.

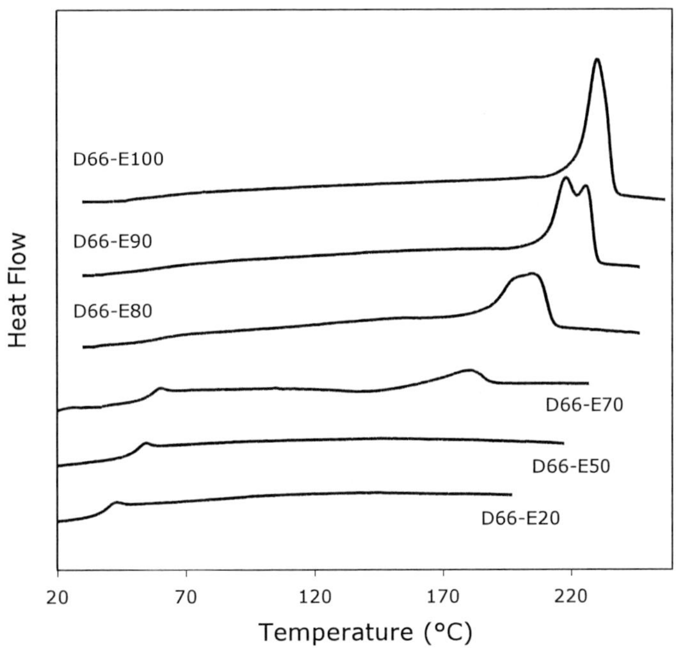

Figure 7. DSC thermograms of PCCD samples during the 2nd heating scan at 10°C·min^{-1}.

The 2nd heating scans (see Figure 5) confirm that some PBCHD samples (-20 and -50) have the characteristics of completely amorphous materials. Indeed, they do not show any evidence of cold crystallization and following melting process during the heating scan in the experimental conditions used. On the other hand, the samples with a *trans* content higher than or equal to 70% are semicrystalline, with melting temperatures which increase considerably with the increment of the *trans* content, from 122°C for PBCHD-70 up to 165-170°C for PBCHD-100. Correspondently, ΔH_m also reaches high values. For the melting process too, these variations are very significant and correspond to drastic changes in the final properties of the materials.

The tendency of the aliphatic rings in *trans* configuration to favour crystallizability can be explained by considering that in this configuration the polymeric chain assumes a "stretched" form and a high symmetry. These are conditions favourable to the chain packing. On the other hand, the *cis* isomer introduces kinks into the chain, which hinder the formation of stable crystals. Literature widely supports these observations. For example, in liquid crystalline polyesters only the *trans* isomer is compatible with melt anisotropy [33]. The *cis* units represent "kinks" which disrupt the rigid-rod main–chain structure, leading to non-ordered polymers [34,35]. Moreover, in partially cycloaliphatic polyamides, characterized by the insertion of 1,4-cyclohexylene rings along the chains, from solid-state NMR studies and WAXD analyses the *cis* isomer is found only in the amorphous regions, whereas the *trans* isomer is located both in the crystalline and in the amorphous phases [16,17]. This means that the crystals are formed only by the *trans* isomer, whereas the *cis* isomer is fully excluded from the crystals. Again, Kricheldorf and Schwarz [13], who studied the thermodynamically controlled *cis/trans* ratio of polyesters deriving from CHDA, observed that the favoured *trans* content of the crystalline material is 100%. Moreover, the formation of stable crystals of *trans* isomer seems to be the driving force for the observed *cis*→*trans* isomerization in various polyesters [13,36].

Therefore, it is confirmed that the stereochemistry of the 1,4-cyclohexylene ring plays a fundamental role in determining the characteristics of the phase behaviour of PBCHD.

A similar behaviour has been found in PCCD samples. From the data of Table 2 and from Figure 6 it is noteworthy that only certain samples crystallize during the cooling scan from the melt at 10°C·min^{-1}. In particular only D66-E80, D66-E90, and D66-E100 samples – in other words, only those samples with the highest *trans* content – crystallize. As the *trans* content increases, the temperature at which crystallisation from the melt occurs (T_{CC}) and enthalpy (ΔH_{CC}) tend also to increase. Therefore, the *trans* conformation of the C_6 rings along the chain improve the capacity of the samples to crystallize, in terms of the crystallization rate, crystallinity and crystal perfection. It is significant that the other samples, with a lower percentage of *trans* stereoisomer, are totally unable to rearrange towards an ordered state.

Figure 7 shows the 2nd heating scans of the PCCD samples. It is confirmed that some PCCDs (D66-E20 and D66-E50) have the characteristics of completely amorphous materials. D66-E70 sample, instead, presents a certain capacity to reach a small amount of order (the enthalpies of crystallization and melting are about 4 J·g^{-1}). On the other hand, for *trans* content equal to and higher than about 80% the PCCD is semicrystalline, with crystallization and melting temperatures (T_{CC} and T_m) increasing notably, up to 204°C and 230°C respectively, for practically all the *trans* material. Correspondingly, ΔH_m also reaches a high value (34 J/g). Therefore, the stereochemistry of the 1,4-cyclohexylene rings plays a fundamental role in determining the crystallinity and crystallizability of PCCD too.

As regards the glass transition temperature, Table 1 and 2 show the experimental data for PBCHDs and PCCDs, respectively. It is necessary to emphasize that for the samples with the highest level of crystallinity (PBCHD-100, PCCD-D66-E100) the determination of T_g is affected by a too high degree of uncertainty as the percentage of amorphous phase is limited.

In any case, it is evident that the T_g values increase with the increment of the *trans* percentage. For example, from PBCHD-20 to PBCHD-100, T_g increases by about 30°C (from -12 to 10°C). For PCCDs T_g increases by about 25°C (from 38 to 65°C) with the increment of the *trans* content from 20 to 90%. Similarly, in polyamides obtained from CHDA and various diamines, T_g increases up to about 17°C when the *trans* content of the 1,4-cyclohexylene moieties changes from 23 to 100% [37].

In this case, to explain the effect of the stereochemistry of the cycloaliphatic ring on the glass transition, it is necessary to recall that the most important factors influencing the T_g values are chain flexibility, symmetry, and steric hindrance and bulkiness of the side groups attached to the backbone chain. In amorphous polymers it has been observed that an increment of symmetry can induce an increment of T_g, as seen for example in 1,4 ring substituted nylon 66 copolyamides which have higher T_g values with respect to 1,2 and 1,3 ring substituted copolyamides [38]. Indeed, in 1,4 ring polymers a better molecular fit is achieved in the polymer backbone, resulting in a better chain packing and improved orientation, which would restrict the movement of the chains upon heating. The same consideration can be made for PBCHD and PCCD, where the *trans* isomer is more symmetrical than the *cis*. Moreover, in the presence of high *trans* % the high T_g values observed can be attributed also to the notable level of crystallinity which creates numerous impediments to the chain mobility in the amorphous state.

3.1.4. Relationship between Molecular Structure and Phase Behavior

3.1.4.1. PBCHD

In order to emphasize the effect of the *trans* isomer content on the thermal properties of our aliphatic polyesters, a statistical approach has been used.

The chemical structure of PBCHD has been analyzed by considering that the 1,4-cyclohexanedicarbonyl moiety is the unit whose characteristics (in terms of *cis* (*c*) or *trans* (*t*) isomer) have a great influence on the thermal transitions typical of the crystalline state. By focusing the attention on the sequence of three monomeric units, i.e. the sequence of three C_6 rings, eight different triads are possible, based on the configuration of the rings: *ttt*, *ctt*, *tct*, *ttc*, *cct*, *tcc*, *ctc*, *ccc*. As the polymer chain is very long, it can be considered as formed by a high number of triads and each triad can assume one of the eight configurations described above, in a random distribution. Therefore, the probability (P) of finding a particular triad, for example the probability P_{tct} of finding the *tct* triad, within the chain, is calculated from the product of the probability (f_c or f_t) of finding the *cis* or *trans* configuration for each ring:

$$P_{tct} = (f_t \cdot f_c \cdot f_t) \cdot 100 \quad (1)$$

f_t and f_c are calculated by the *trans* and *cis* content of the DMCD unit in the polymer, reported in Table 1, divided by 100. Table 4 shows the probabilities of finding the eight possible triads within the chains in the PBCHD samples which show the capacity to crystallize.

Table 4. Probability of finding the *ttt, ctt, tct, ttc, cct, tcc, ctc, and ccc* triads in a sequence of three cycloaliphatic rings along the PBCHD chain for the semicrystalline samples

Sample	P_{ttt} %	P_{ctt} %	P_{tct} %	P_{ttc} %	P_{cct} %	P_{tcc} %	P_{ctc} %	P_{ccc} %
PBCHD-70	37.3	14.5	14.5	14.5	5.6	5.6	5.6	2.2
PBCHD-80	51.2	12.8	12.8	12.8	3.2	3.2	3.2	0.8
PBCHD-90	75.4	7.5	7.5	7.5	0.7	0.7	0.7	0.1
PBCHD-100	100	0.0	0.0	0.0	0.0	0.0	0.0	0.0

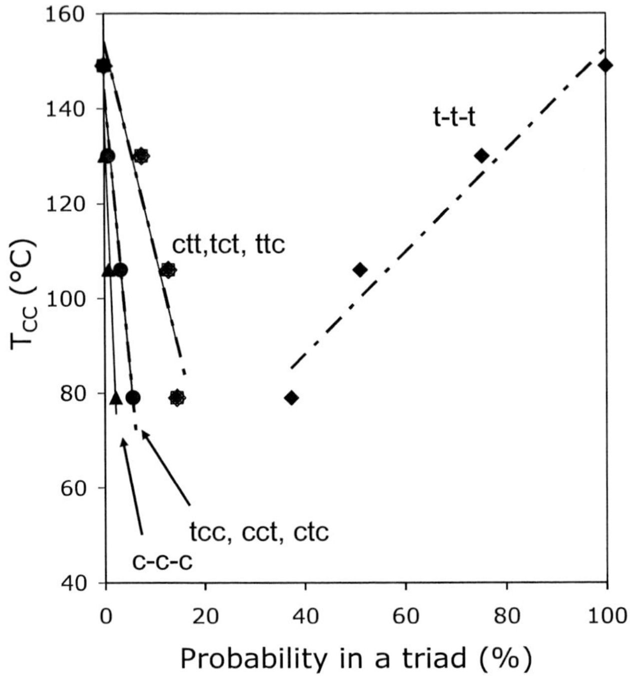

Figure 8. Trends of T_{CC} vs. the probabilities of finding the *ttt, ctt, tct, ttc, cct, tcc, ctc, and ccc* triads in the sequence of three 1,4-cyclohexylene rings along the PBCHD chain.

The correlations between the T_{CC} and T_m values for the semicrystalline PBCHD samples and the probabilities of finding the above-mentioned triads are shown in Figure 8 and 9, respectively. For PBCHD-80, -90 and -100, which show a double melting peak, only the temperature of the first melting peak is reported, due to the fact that it corresponds to the melting of the crystals actually formed at T_{CC} during the cooling scan. Indeed, the second melting peak was attributed to melting-recrystallization and remelting processes occurring during the calorimetric scan.

Firstly, it is notable that linear trends, with good correlation coefficients, are always obtained. Moreover, it is notable that both T_{CC} and T_m increase considerably with the increment of the percentage of *ttt* triads. For example, T_{CC} varies from 79 to 149°C and T_m from 122 to 165°C when the percentage of the *ttt* triads changes from about 37 to 100%. Therefore, a high percentage of three *trans* sequences significantly improves the crystalline perfection and crystallizability of PBCHD.

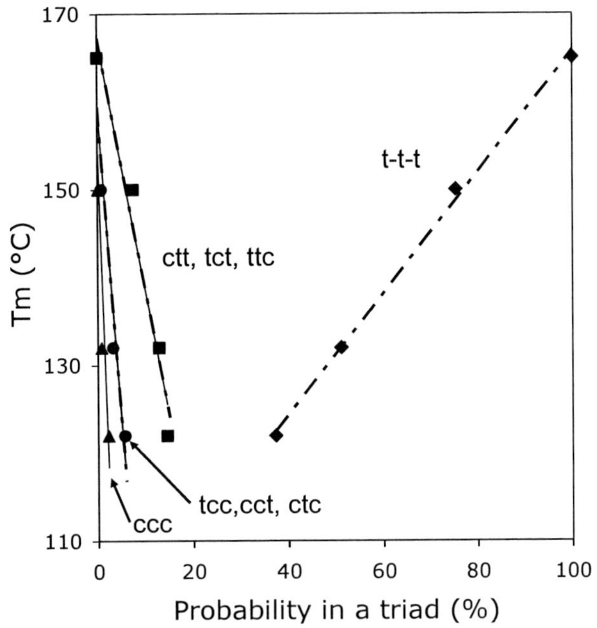

Figure 9. Trends of T_m vs. the probabilities of finding the *ttt, ctt, tct, ttc, cct, tcc, ctc, and ccc* triads in the sequence of three 1,4-cyclohexylene rings along the PBCHD chain.

Table 5. For PCCD probability of finding the *ttt, ctt, tct, ttc, cct, tcc, ctc, and ccc* triads in the D-E-D sequence

Sample	P_{ttt} %	P_{ctt} %	P_{tct} %	P_{ttc} %	P_{cct} %	P_{tcc} %	P_{ctc} %	P_{ccc} %
D66-E20	10.5	5.4	33.1	5.4	17.1	17.1	2.8	8.8
D66-E50	22.7	11.7	20.9	11.7	10.8	10.8	6.0	5.5
D66-E70	28.7	14.8	14.8	14.8	7.6	7.6	7.6	3.9
D66-E80	35.3	18.2	8.3	18.2	4.3	4.3	9.4	2.2
D66-E90	39.2	20.2	4.4	20.2	2.2	2.2	10.4	1.2
D66-E100	42.3	21.8	1.3	21.8	0.7	0.7	11.2	0.3

On the other hand, all the other triads (*ctt, tct, ttc, cct, tcc, ctc, ccc*) cause a sharp and significant decrement of both T_{CC} and T_m: the decrement is even greater when two or, particularly, three rings in *cis* configuration are present. This result is further evidence that the *cis* configuration acts as a disturbance in the crystallization process of PBCHD: the presence of "kinks" along the macromolecular chains prevents a high crystalline perfection and high melting peak. The "disturbance" of the *cis* form is so great that the presence of two rings in *trans* configuration in the *ctt, tct,* and *ttc* triads are not sufficient to favor the formation of stable crystals.

3.1.4.2. PCCD

For PCCD the description of the molecular structure is more complicated than for PBCHD: the two units characteristic of the PCCD are the 1,4-cyclohexanedicarbonyl unit (called E), deriving from DMCD, and the 1,4-cyclohexanedimethyleneoxy unit (called D), deriving from CHDM. Attention has been focused on a sequence of three 1,4-cyclohexylene rings: D-E-D and E-D-E, as represented in Scheme 4. \

Firstly, the D-E-D sequence was analysed. Table 5 shows the probabilities of finding the eight possible triads within the chains in the analyzed samples, following eq. (1).

A first correlation between molecular structure and thermal properties is described in Figure 10, where the T_m values (Table 2) are reported as a function of the probability of finding the *ttt, ctt, tct, ttc, cct, tcc, ctc, and ccc* triads (Table 5). Also in this case, for the samples presenting a double melting peak, only the first peak is reported, due to the fact that, for the possible melting-recrystallization process, it corresponds to the melting of the crystals actually formed at T_{CC} during the cooling scan.

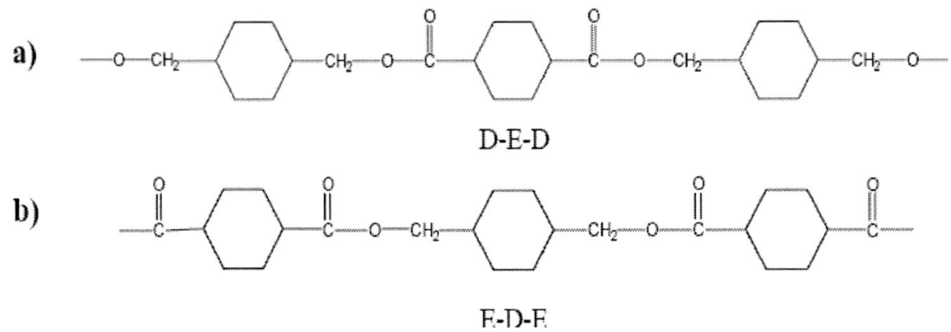

Scheme 4. Representation for PCCD of the different sequences of three 1,4-cyclohexylene rings considered for the determination of the probability of finding the triads reported in Tables 5 and 6.

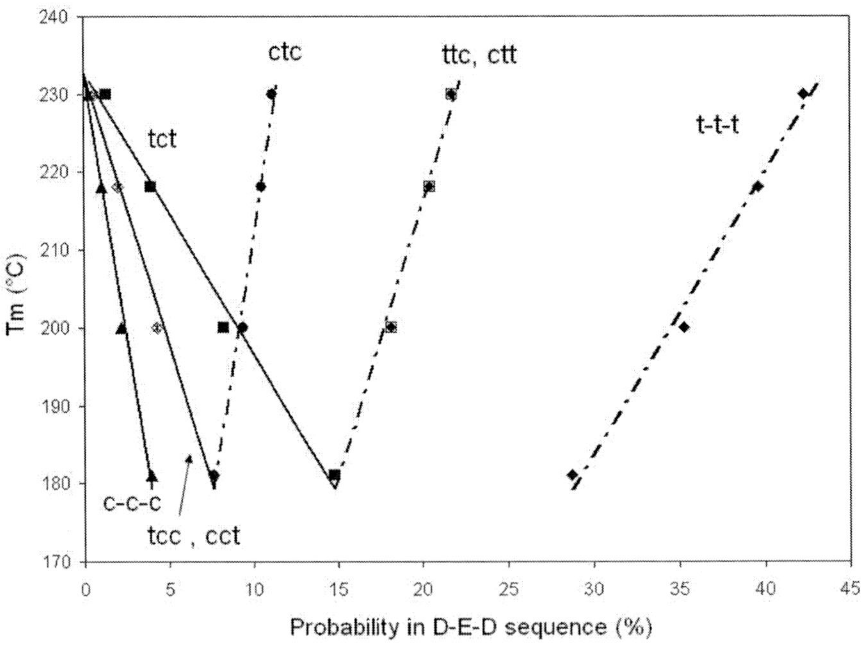

Figure 10. Trends of T_m vs. the probabilities of finding the *ttt, ctt, tct, ttc, cct, tcc, ctc, and ccc* triads in the D-E-D sequence of PCCD.

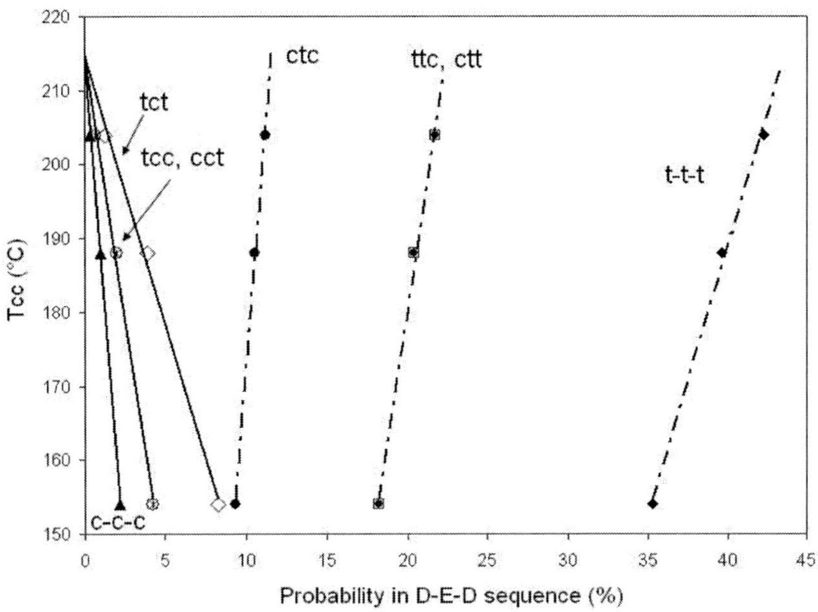

Figure 11. Trends of T_{CC} vs. the probabilities of finding the *ttt, ctt, tct, ttc, cct, tcc, ctc, and ccc* triads in the D-E-D sequence of PCCD.

Table 6. Probability of finding the *ttt, ctt, tct, ttc, cct, tcc, ctc, and ccc* triads in the E-D-E sequence

Sample	P_{ttt} %	P_{ctt} %	P_{tct} %	P_{ttc} %	P_{cct} %	P_{tcc} %	P_{ctc} %	P_{ccc} %
D66-E20	3.8	12.0	2.0	12.0	6.2	6.2	38.1	19.6
D66-E50	17.8	16.5	9.1	16.5	8.4	8.4	15.2	7.8
D66-E70	28.7	14.8	14.8	14.8	7.6	7.6	7.6	3.9
D66-E80	43.3	10.2	22.3	10.2	5.2	5.2	2.4	1.2
D66-E90	53.5	5.9	27.5	5.9	3.1	3.1	0.7	0.3
D66-E100	62.1	1.9	32.0	1.9	1.0	1.0	0.1	0.03

Firstly, it is noteworthy that linear trends, with good correlation coefficients, are obtained. Moreover, it is notable that T_m increases with the increment of the percentage of four triads (*ttt, ctt, ttc, and ctc*). For example, T_m varies from 181 to 230°C when the percentage of the *ttt* triads changes from about 29 to 42%. The same trend is observed for the *ctt* and *ttc* triads, i.e. for triads containing two consecutive rings in *trans* configuration. This result suggests that a high percentage of two or three *trans* sequences has a significant role in improving the crystalline perfection of PCCD. The observation is in perfect agreement with the previous discussion and with the literature results.

Moreover, by observing Figure 10, it may also be seen that the *ctc* triads, i.e. the triads containing only the unit E in *trans* configuration, improve T_m significantly, independently of the fact that the other two rings are in *cis* configuration. On the other hand, a small increment of the number of the the *tct* triads, where two D units are in *trans* configuration, causes a rapid decrement of T_m. This result is important because it is a first indication of a different effect of the stereochemistry of the D and E units on the PCCD thermal properties. Indeed, it is sufficient that the E unit is in *cis* configuration that crystal perfection decreases. Therefore, it seems that the configuration of the E unit has a very important effect in influencing the trend of T_m.

Analysis of the remaining curves shows that T_m decreases as the percentages of the *ccc, tcc,* and *cct* triads increase. Also in this case, for the sequences of two or three rings in *cis* configuration, the result is expected: the presence of "kinks" along the macromolecular chains prevents a high crystal perfection and high melting peak.

Given these results, the effect of the percentage of triads on the crystallization temperature T_{CC}, measured in DSC during the cooling scan from the melt (Table 2), is also significant. In Figure 11 it is evident that the presence of the *ttt, ctt, ttc, and ctc* triads causes an increment of T_{CC}. The trends are the same with respect to those observed in Figure 10, indicating that a high crystallization capacity and crystal perfection are improved by the presence of two and three sequences of rings in *trans* and, more significantly, by the presence of an isolated ring, belonging to E unit, in *trans* configuration. On the contrary, two and three sequences of rings in *cis* and, more significantly, an isolated ring belonging to E unit, in *cis* configuration inhibit the crystallization process.

Therefore, in order to emphasise the different effect of the stereochemistry of the C_6 ring, depending on whether it belongs to E or D unit, we have analyzed the molecular structure of the PCCD samples focusing on the E -D- E sequences (Scheme 4, b). In this case two E units and only one D unit have been considered. The probability of finding the eight possible triads (*ttt*, *ctt*, *tct*, *ttc*, *cct*, *tcc*, *ctc*, *ccc*) has been calculated and the results reported in Table 6.

Figure 12 describes the trend of T_m as a function of the probability of finding the different triads in the E-D-E system. Now an important difference may be seen with respect to the trends reported in Figure 10 and 11: the increment of T_m occurs only for two kinds of sequences, i.e. *ttt* and *tct*. All the other sequences (*ccc, cct, ctc, tcc, ttc,* and *ctt)* induce a decrement of T_m. It is notable that for T_{CC} the trend is exactly the same. Therefore, the *ccc, cct, ctc, tcc, ttc, and ctt* triads are responsible of lower crystal perfection and lower tendency to crystallise. Then, in this case, the presence of two consecutive rings in *trans* configuration (triads *ttc* and *ctt*) is not determinant in improving crystallizability. Indeed, the configuration of the ring of E unit appears to be the most important factor influencing crystallisation capacity: its *trans* configuration favours crystallization, whereas its *cis* configuration is unfavourable.

As regards the glass transition, Figure 13 describes the trend of T_g as a function of the probability of finding the different triads in the D-E-D sequences. It is notable that the same results discussed for T_m and T_{CC} (Figures 10 and 11) are obtained. Analogously, for the E-D-E sequences the trend of T_g vs the probability of finding the different triads is exactly the same of that shown in Figure 12 for T_m and that obtained for T_{CC}. As a conclusion, T_g increases as the % of *ttt* triads increases and when the 1,4-cyclohexylene ring in E unit is in *trans* configuration.

The results obtained can be discussed in terms of chain mobility. It is noteworthy that in unit E the 1,4-cyclohexylene moiety is linked to two ester groups, where the carbon atoms are characterized by sp^2 hybridisation, with planar geometry and bond angles of 120°. This structure has a certain degree of rigidity. On the other hand, in unit D the rings are connected to two $-CH_2-$ groups, which are characterised by carbon atoms with sp^3 hybridisation, tetrahedral geometry and bond angles of about 109°. In this case, each carbon atom possesses a higher mobility and can assume different but equivalent orientations in space. Therefore, the most rigid structure, the E unit, is found to have the highest influence in determining the thermal properties.

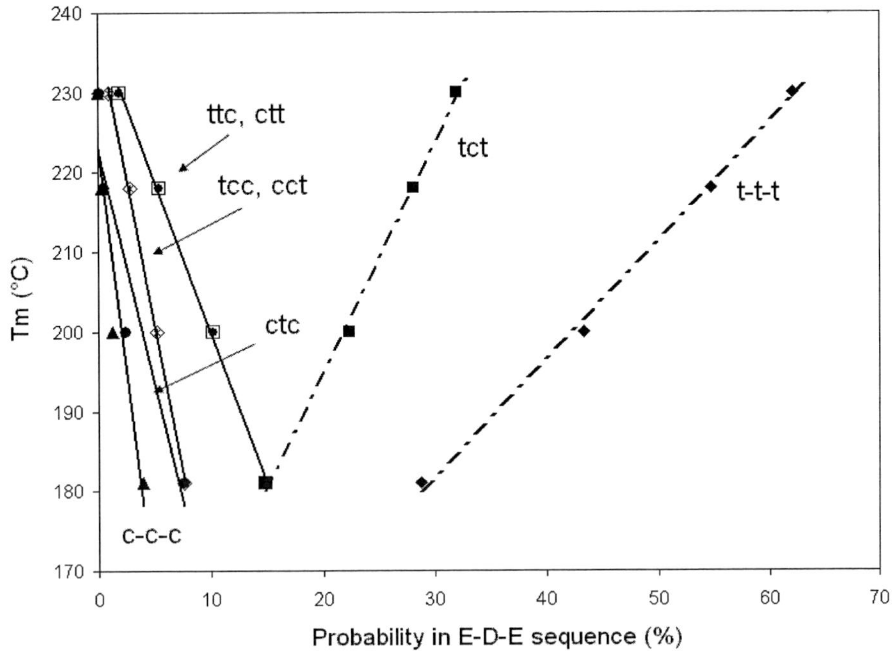

Figure 12. Trends of T_m vs. the probabilities of finding the *ttt, ctt, tct, ttc, cct, tcc, ctc, and ccc* triads in the E-D-E sequence of PCCD.

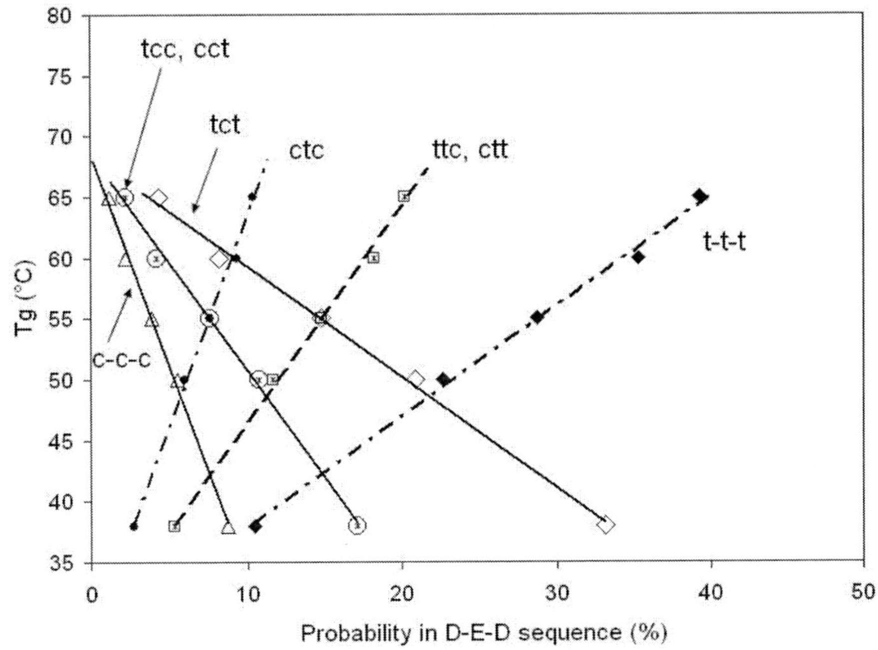

Figure 13. Trends of T_g vs. the probabilities of finding the *ttt, ctt, tct, ttc, cct, tcc, ctc, and ccc* triads in the D-E-D sequence.

a)

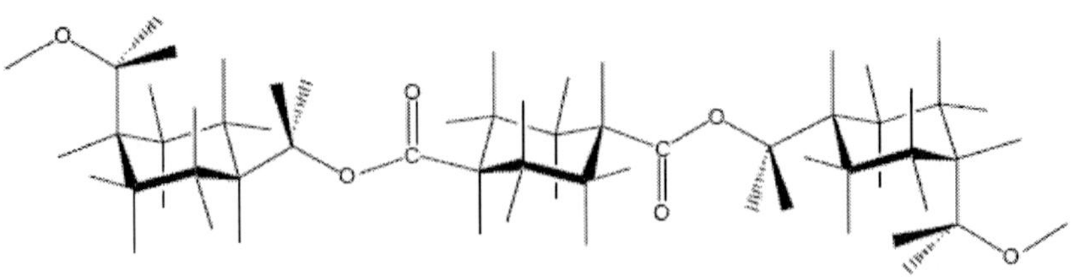

cis D – *trans* E – *cis* D

b)

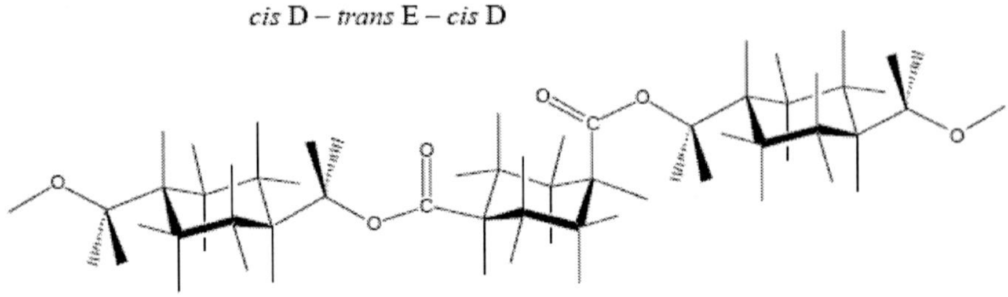

trans D – *cis* E – *trans* D

Scheme 5. Two configurations of D-E-D sequence.

Moreover, the configurations of the two sequences D-E-D are shown in Scheme 5: *ctc* and *tct.* The influence of these two sequences on the thermal transitions, previously discussed, is significant because the first sequence (a) causes an increment of T_g, T_{CC} and T_m , whereas the second (b) induces a decrement of the T_g, T_{CC} and T_m. This means that when the aliphatic ring in E unit is in *trans* configuration an improvement of the thermal properties occurs, whereas when the C_6 ring is in *cis* configuration the thermal properties are lower.

In particular, the rigidity of E unit in *trans* configuration (a) gives a high symmetry to *ctc* triad, improving chain packing, crystalline order and also T_g. On the other hand, the *cis* configuration of E unit (b) is characterized by a high steric hindrance of the two substituent groups. This leads to less probability of chain folding and crystallizability for *tct* than *ctc* triad.

3.2. Preparation and Characterization of Copolymers

3.2.1. Molecular Characterization

Some novel aliphatic, random copolyesters were synthesized from DMCD, BD, and DMA by following the procedure described in the Experimental Part. They are called (4-6)-*co*-PBCHD$_{xx}$-a/b, where xx indicates the percentage of the aliphatic rings, derived from DMCD, in *trans* configuration and a/b is the feed molar ratio of the DMA/DMCD.

The molecular structure is reported in Scheme 6, while Table 7 describes all the samples prepared.

The chemical structure was analyzed by ^{1}H NMR spectroscopy. For all the samples the spectra were found to be consistent with the expected structures. The composition was calculated, but it was not possible to evaluate the degree of randomness and the average length of the sequences. However, since we observed no differences in the reactivity of BD with respect to the two diesters during the synthesis of the two homopolymers (4-6 and

PBCHD), we hypothesize that the copolyesters have a random distribution of the sequences along the chain, as demonstrated for aliphatic-aromatic (4-12)-*co*-PBT samples [39]. From the ratio of the contributions of the protons of the cycloaliphatic ring in *trans* and in *cis* configurations, the trans percentage of the cycloaliphatic units inside the copolymers has been also determined.

As a result, the copolyesters differentiate for molar composition and for *trans* mol % of the PBCHD units (see Table 7). In particular, the *trans* content varies from 50 to 100 mol %, as in the case of the corresponding PBCHD homopolymers previously described. It is also evident that the molecular weights of all the samples are very high. In this way, the samples can conveniently be compared in order to study their thermal behavior.

3.2.2. Analysis of the Phase Behaviour of Copolymers

One of reasons of the great interest towards (4-6)-*co*-PBCHD copolyesters is due to the fact that they are the corresponding aliphatic materials of the poly(butylene terephthalate-*co*-butylene adipate) (PBTA) (see Scheme 6). This is an aliphatic-aromatic copolymer, commercialized with the trade name of Ecoflex. PBTA is characterized by the presence of terephthalate groups and, thus, for the reasons already discussed, it has relatively high transition temperatures: T_g is located at about -30°C and melting temperature at 120°C. Therefore, the introduction of aromatic rings to the aliphatic chains of 4-6 induces a notably improvement in thermal characteristics. Moreover, PBTA has good mechanical properties and excellent thermal stability [40]. As regards its biodegradation, Witt et al. concluded that there is no indication of environmental risk when it is involved in the composting process [41], in spite of the presence of not-biodegradable units. Therefore, Ecoflex is described as biodegradable plastic ideal for trash bags and disposable packaging.

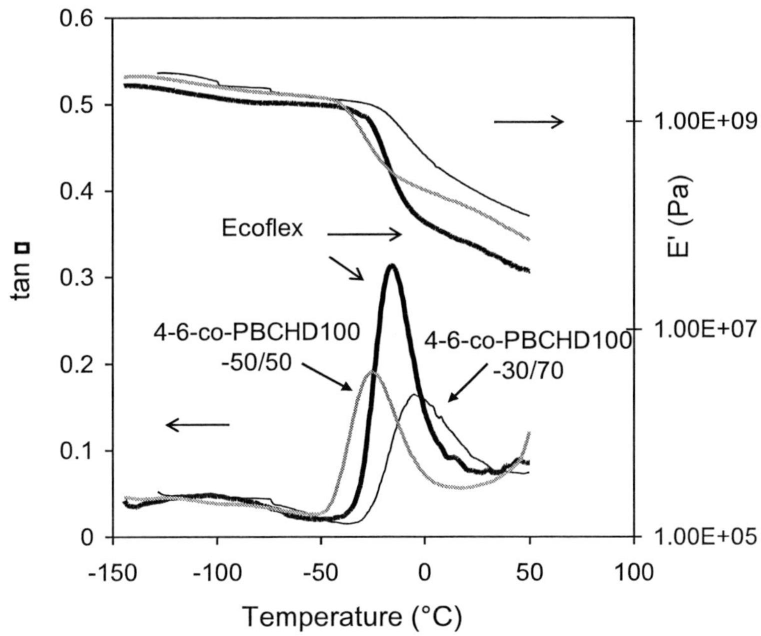

Figure 14. DMTA spectra of Ecoflex, (4-6)-*co*-PBCHD$_{100}$-70/30, and (4-6)-*co*-PBCHD$_{100}$-50/50 copolymers.

Indeed, the novel (4-6)-*co*-PBCHD copolyesters are characterized by mechanical properties not too different with respect to those of Ecoflex, as shown in Figure 14, and, also the thermal properties are not poor, as discussed later (see Table 8). Moreover, since our copolymers are all-aliphatic, their biodegradability is potentially really good [12]. As a consequence, the new (4-6)-*co*-PBCHD copolyesters are excellent candidates to cover important roles in the future applications of polymers.

As to the relationships between stereochemistry and phase behavior, Figure 15 and 16 reports some examples of DSC curves (cooling and heating scans), obtained on (4-6)-*co*-PBCHD-30/70 samples, with *trans* percentage varying from 100 to 50 mol %. In this case, the copolymers at high percentage of PBCHD units are taken in account, because the effect of the stereochemistry of the cycloaliphatic ring is more evident.

As already discussed for PBCHD homopolymers, the capacity of the samples to crystallize is found to be to strongly modified. For example, the copolymer containing PBCHD units with 50 mol% of *trans* isomer is a fully amorphous material, which is not able to crystallize either during the cooling scan from the melt or during the subsequent heating scan. By increasing the *trans* content to 70 mol %, the copolymer gains the capacity to organize itself into a more ordered structure, partially during the cooling scan and partially during the heating scan. Samples at high *trans* contents (90 and 100 mol %) crystallize with narrow peaks, reaching level of crystallinity similar to that of the homopolymer, although T_{CC} values are significantly lower (T_{CC} =97°C for (4-6)-*co*-PBCHD$_{100}$-30/70 and 149°C for PBCHD$_{100}$). Correspondently, the T_m values also increase with the *trans* content, as evident from an analysis of all the data of Table 8.

Different observations can be made for the T_g values, reported in Table 1 and 8. Indeed, if the T_g data of PBCHD samples are compared as a function of the *trans* content, it is evident that T_g decreases with the decrement of the *trans* content, due to the low flexibility of the 1,4-cyclohexylene ring in trans configuration. From PBCHD$_{100}$ to PBCHD$_{50}$ T_g varies from 10 to -7°C. For the (4-6)-*co*-PBCHD copolymers, instead, this trend is not observed. For example, all the (4-6)-*co*-PBCHD-70/30 samples have experimental T_g values of about -50°C, independently of the fact that the *trans* content varies from 100 to 50 mol %. All the (4-6)-*co*-PBCHD-50/50 copolymers have T_g values of about -40°C and all the (4-6)-*co*-PBCHD-30/70 copolymers have T_g values of about -25°C. In this case, the effect of the stereoregularity of the aliphatic ring on the chain flexibility seems to be absent.

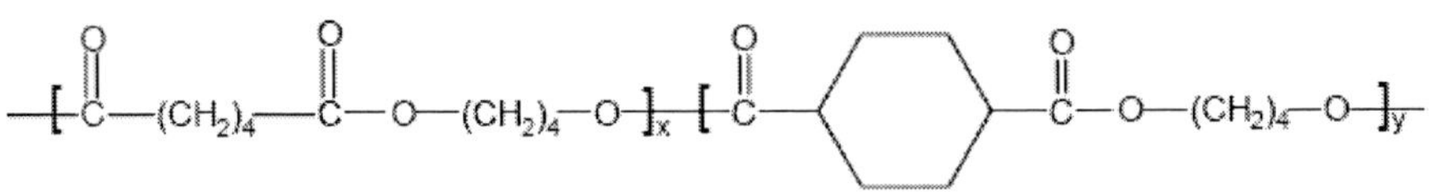

(4-6)-*co*-PBCHD

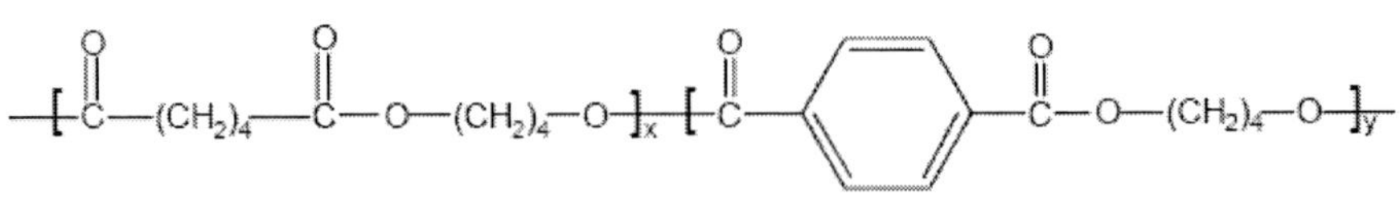

PBTA – trade name: Ecoflex

Scheme 6. Molecular structures of the novel (4-6)-*co*-PBCHD copolyesters and Ecoflex.

Table 7. Molecular characteristics of the copolyesters.

Sample	(4-y)/PBCHD molar ratio[a]	*trans* % of the ring in the polymer[a]	M_w x 10^{-3} [b]	M_w/M_n [b]
4-6	-	-	90.0	2.5
Ecoflex	-	-	121.0	2.6
(4-6)-*co*-PBCHD$_{100}$-70/30	64/36	100	95.4	2.1
(4-6)-*co*-PBCHD$_{100}$-50/50	47/53	100	73.6	2.3
(4-6 –*co*-PBCHD$_{100}$-30/70	24/76	100	97.4	2.2
(4-6)-*co*-PBCHD$_{90}$-70/30	65/35	88	84.9	2.0
(4-6)-*co*-PBCHD$_{90}$-50/50	47/53	87	87.8	2.2
(4-6)-*co*-PBCHD$_{90}$-30/70	24/76	90	99.4	2.4
(4-6)-*co*-PBCHD$_{70}$-70/30	65/35	70	86.3	2.0
(4-6)-*co*-PBCHD$_{70}$-50/50	47/53	68	88.5	2.3
(4-6)-*co*-PBCHD$_{70}$-30/70	24/76	72	95.3	2.4
(4-6)-*co*-PBCHD$_{50}$-70/30	65/35	53	76.4	2.1
(4-6)-*co*-PBCHD$_{50}$-50/50	46/54	56	65.1	2.1
(4-6)-*co*-PBCHD$_{50}$-30/70	24/76	56	128.3	2.2

[a] Calculated by ^{1}H NMR.
[b] Measured by GPC in $CHCl_3$.

Table 8. Thermal data of the copolyesters

Sample	T_{CC} [a] °C	ΔH_{CC} [a] J·g^{-1}	T_g [b] °C	T_{ch} [b] °C	ΔH_{ch} [b] J·g^{-1}	T_m [b] °C	ΔH_m [b] J·g^{-1}
4-6	32	67	-58	-	-	52 and 57	70
Ecoflex	80	21	-31	-	-	124	11
(4-6)-*co*-PBCHD$_{100}$-70/30	14	28	-49	-	-	53	24
(4-6)-*co*-PBCHD$_{100}$-50/50	66	33	-43	-	-	94	32
(4-6)-*co*-PBCHD$_{100}$-30/70	97	41	-23	-	-	125 and 134	40
(4-6)-*co*-PBCHD$_{90}$-70/30	2	30	-49	-	-	43	27
(4-6)-*co*-PBCHD$_{90}$-50/50	41	25	-40	-	-	80	26
(4-6)-*co*-PBCHD$_{90}$-30/70	74	29	-25	-	-	119	27
(4-6)-*co*-PBCHD$_{70}$-70/30	-20	4	-50	-9	19	30	23
(4-6)-*co*-PBCHD$_{70}$-50/50	-	-	-40	19	14	56	16
(4-6)-*co*-PBCHD$_{70}$-30/70	27	16	-25	21	5	86	22
(4-6)-*co*-PBCHD$_{50}$-70/30	-	-	-52	-2	5	18	6
(4-6)-*co*-PBCHD$_{50}$-50/50	-	-	-41	-	-	-	-
(4-6)-*co*-PBCHD$_{50}$-30/70	-	-	-26	-	-	-	-

[a] Measured by DSC (cooling scan at 10°C/min).
[b] Measured by DSC (2nd heating scan at 10°C/min).

Indeed, it is reasonable that an effect of the *cis/trans* ratio of the 1,4-cyclohexylene ring should be present mainly in the copolymers rich in PBCHD, i.e. in 30/70 copolymers. However, the presence of the PBCHD crystalline phase [21] creates a more rigid matrix and originates an amorphous phase whose composition is not perfectly correspondent to the theoretical one, but richer in 4-6 units. The 4-6 units can lead to a greater flexibility of the chain and, thus, to lower T_g data. Therefore, opposite effects are present: the *trans* isomer and the crystalline phase increase the rigidity of the system and cause an increment of T_g, whereas the 4-6 units, which are preferentially in the amorphous state, induces a decrement of T_g. As a result, the final T_g value could be a constant.

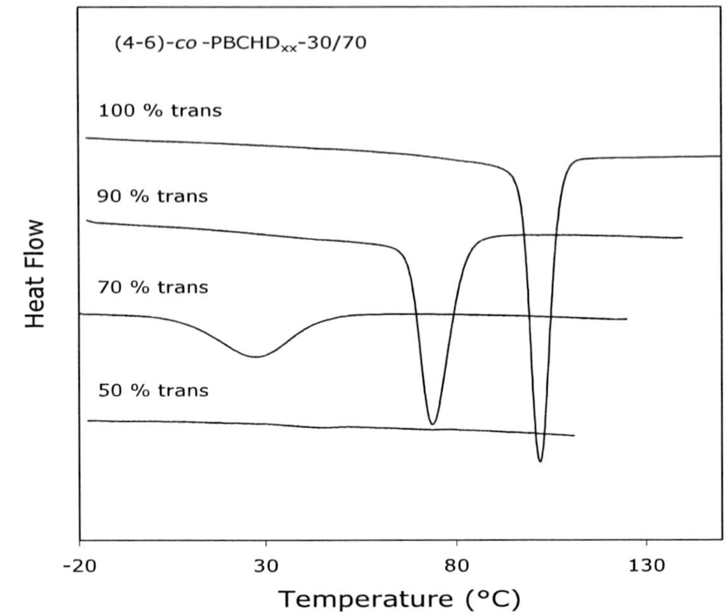

Figure 15. DSC traces of the cooling scans at 10°/min for the 4-6-*co*-PBCHD copolymers.

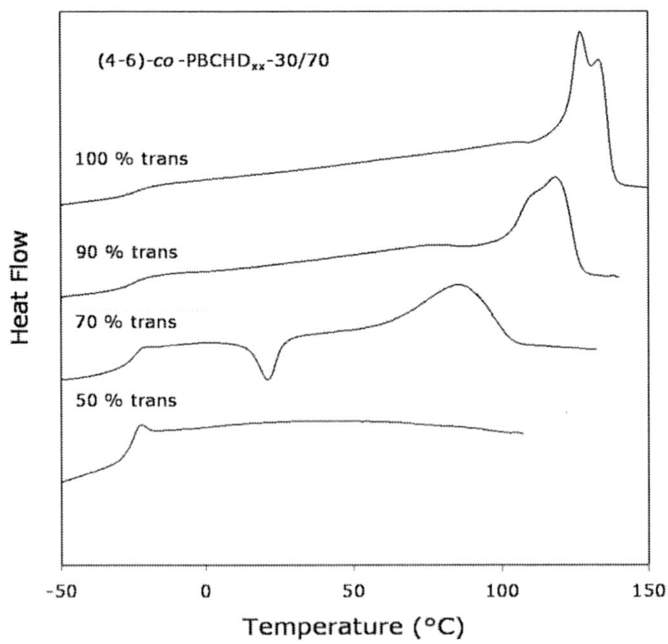

Figure 16. DSC traces of the 2nd heating scans at 10°/min for the 4-6-*co*-PBCHD copolymers.

CONCLUSIONS

The thermal behaviour of polyesters and copolyesters, containing the 1,4-cyclohexylene units, have been analyzed in order to highlight the correlations existing between stereochemistry of the aliphatic ring and phase behaviour. It is noteworthy that significant relationships have been found for all the materials here analyzed (PBCHD, PCCD, (4-6)-*co*-PBCHD).

In particular, with the increment of the *trans* content the polymers change from completely amorphous to semicrystalline material. Correspondingly, T_g increases of about 20-25°C in PBCHD and PCCD with the increment of the *trans* content from 20 to 100 mol%. At the same time, T_m increases of about 40-50°C with the increment of the *trans* amount from 70 to 100 mol%. Moreover, the polymers in all *trans* conformation are able to reach a very high level of crystallinity and crystal perfection. This behaviour has been attributed to the better chain packing which occurs in the presence of the more "stretched" *trans* conformation. Indeed, the rings in *cis* conformation cause “kinks” and a lower degree of symmetry.

An analysis of the molecular structure of PCCDs shows that the stereochemistry of 1,4-cyclohexylene ring deriving from DMCD is the main element responsible for the thermal properties. This result is due to the higher rigidity of the 1,4-cyclohexanedicarbonyl unit with respect to 1,4-cyclohexanedimethyleneoxy unit, deriving from CHDM. Therefore, when the most rigid structure is in *trans* conformation, T_g, T_m and crystallization rate tend to increase, whereas the *cis* conformation of the 1,4-cyclohexanedicarbonyl unit tends to induce a decrement of T_g and crystallizability.

Significant results have been found in copolymers too: in this case, it is possible to easily modulate the final properties of the materials not only by changing the molar composition, but also varying the *cis/trans* ratio of the alicyclic units inside the chains.

As a conclusion, PBCHD, PCCD, and (4-6)-*co*-PBCHD copolymers present the great advantages of having cycloaliphatic units which, according to their stereochemistry, allow the material to show modulated thermal properties and good physical performance, and a fully aliphatic main chain, which could provide the favourable biodegradability and resistance to weather, light, heat and water, typical of the aliphatic polyesters.

REFERENCES

[1] El-Hadi, A.; Schnabel, R.; Straube, E.; Müller, G.; Henning, S. Polymer Testing 2002, 21, 665-674.

[2] Berti, C.; Celli, A.; Marchese, P.; Marianucci, E.; Barbiroli, G.; Di Credico, F. *e-Polymers* 2007, 057, 1-17.

[3] Liu, Y.; Turner, R. *J. Polym. Sci, Part A: Polym. Chem.* 2010, 48, 2162-2169.

[4] Jackson Jr., W.J.; Darnell, W.R. *US Patent* 4327206, 1982.

[5] Jackson Jr., W.J.; Darnell, W.R. *US Patent* 4342862, 1982.

[6] Brunelle, D. J.; Jang, T. *U.S. Patent* 6,084,055, 2000.

[7] Brunelle, D.J.; Berti, C.; Celli, A.; Colonna, M.; Fiorini, M.; Marianucci, E.; Messori, M.; Pilati, F.; Sisti, L.; Marchese, P. *U.S. Patent* 6,828,410, 2003.

[8] Brunelle, D.J.; Jang, T. *Polymer* 2006, 47, 4094-4104.

[9] Turner, S.R.; Seymour, R.W.; Dombroski J.R.. In: Scheirs, J.; Long, T.E. Editors. Modern polyesters. Chichester, Wiley; 2003, 267-292.

[10] Berti, C.; Binassi, E.; Colonna, M.; Fiorini, M.; Kannan, G.; Kanaram, S.; Mazzacurati, M. US 20100168461.

[11] Bechthold, I.; Bretz, K.; Kabasci, S.; Kopitzky, R.; Springer, A. *Chem. Eng. Technol.* 2008, 31, 647-654.

[12] Berti, C.; Celli, A.; Marchese, P.; Barbiroli, G.; Di Credico, F.; Verney, V.; Commereuc, S. *Europ. Polym. J.* 2009, 45, 2402-2412.

[13] Kricheldorf, H.R.;Schwarz, G. *Makromol. Chem.* 1987,188,1281-1294.

[14] Vanhaecht, B.; Teerenstra, M.N.; Suwier, D.R.; Willem, R.; Biesemans, M.; Koning, C.E. *J. Polym. Sci. Part A: Polym. Chem.* 2001, 39, 833-840.

[15] Vanhaecht, B.; Rimez, B.; Willem, R.; Biesemans, M.; Koning, C.E. *J. Polym. Sci. Part A: Polym. Chem.* 2002, 40, 1962-1971.

[16] Koning, C.; Vanhaecht, B.; Willem, R.; Biesemans, M.; Goderis, B.; Rimez, B. *Macromol. Symp.* 2003, 199, 431-442.

[17] Vanhaecht, B.; Willem, R.; Biesemans, M.; Goderis, B.; Basiura, M.; Magusin, P.C.M.M.; Dolbnya, I.; Koning, C.E. *Macromolecules* 2004, 37, 421-428.

[18] Joseph, M.D.; Savina, M.R.; Harris, R.F. *J. Appl. Polym. Sci.* 1982, 44, 1125-1134.

[19] Berti, C.; Celli, A.; Marchese, P.; Marianucci, E.; Barbiroli, G.; Di Credico, F. *Macromol. Chem. Phys.* 2008, 209, 1333-1344.

[20] Berti, C.; Binassi, E.; Celli, A.; Colonna, M.; Fiorini, M.; Marchese, P.; Marianucci, E.; Gazzano, M.; Di Credico, F.; Brunelle, D.J. *J. Polym. Sci. Part B: Polym. Phys.* 2008, 46, 619-630.

[21] Berti, C.; Celli, A.; Marchese, P.; Marianucci, E.; Sullalti, S.; Barbiroli, G. *Macromol. Chem. Phys.* 2010, 211, 1559-1571.

[22] Sánchez-Arrieta, N.; Martínez di Ilarduya, A.; Alla, A.; Muñoz-Guerra, S. *Europ. Polym. J.* 2005, 41,1493-1501.

[23] Wang, L.; Xie, Z.; Bi, X.; Wang, X.; Zhang, A.; Chen, Z., Zhou, J.; Feng, Z. *Polym. Degrad. Stab.* 2006, 91, 2220-2228.

[24] Sandhya, T.E.; Ramesh, C.; Sivaram, S. *Macromolecules* 2007, 40, 6906-6915.

[25] Berti, C.; Bonora, V.; Colonna, M.; Lotti, N.; Sisti, L. *Europ. Polym. J.* 2003, 39,1595-1601.

[26] Colonna, M.; Berti, C.; Binassi, E.; Celli, A.; Fiorini, M.; Marchese, P.; Messori, M.; Brunelle, D.J. *Polym Int* 2011, accepted for publication.

[27] Matsuda, H.; Nagasaka, B.; Asakura, T. *Polymer 2003*, 44, 4681-4687.

[28] Marchese, P.; Celli, A.; Fiorini, M. *J. Polym. Sci. Part B: Polym. Phys.* 2004, 42, 2821-2832.

[29] Tonelli, A.E. *J. Polym. Sci, Polym. Lett. Ed* 1973, 11, 441-447.

[30] Tonelli, A.E. *J. Polym. Sci, Polym. Phys.* 2002, 40, 1254-1260.

[31] Gonzalez, C.C.; Riande, E.; Bello, A.; Pereña, J.M.. *Macromolecules* 1988, 21, 3230-3234.

[32] Ki, H.C.; Ok Park, O. *Polymer* 2001, 42,1849-1861.

[33] Kwolek, S.L. ; Luise, R.R. *Macromolecules* 1986, 19, 1789-1796.

[34] Osman, M.A. *Macromolecules*1986, 19, 1824-1827.

[35] Reck B.; Ringsdorf, H.; Gardner, K.; Starkweather Jr., H. *Makromol. Chem.* 1989, 190, 2511-2526.

[36] Lenz, R.W.; Go S. *J. Polym. Sci: Polym. Chem. Ed* 1973, 11, 2927-2946.

[37] Srinivasan, R.; McGrath, J.E. *Polym. Prepr.* 1992, 33, 503-504.

[38] Ridgway, J.S. *J. Polym. Sci, Polym. Chem.* 1970, 8, 3089-3111.

[39] Berti, C.; Celli, A.; Marchese, P.; Barbiroli, G.; Di Credico, F.; Verney, V.; Commereuc, S. *Europ. Polym. J.* 2008, 44, 3650-3661.

[40] Herrera, R.; Franco, L.; Rodríguez-Galán, A.; Puiggalí, J. *J. Polym. Sci. Part A: Polym. Chem.* 2002, 40, 4141-4157.

[41] Witt, U.; Einig, T.; Yamamoto, M.; Kleeberg, I.; Deckwer, W. -D.; Müller, R. -J. *Chemosphere* 2001, 44, 289-299.

In: Polymer Phase Behavior
Editors: T. P. Ehlers and J. K. Wilhelm, pp. 235-264
ISBN: 978-1-61324-336-7
© 2011 Nova Science Publishers, Inc.

Chapter 6

The Features of Partitioning Behavior of Recombinant Amino Acid Dehydrogenases in Aqueous Two-phase Systems

Hamid Shahbaz Mohammadi*[*] and Eskandar Omidinia
Department of Biochemistry,
Pasteur Institute of Iran, Tehran, Iran

Abstract

Partitioning in aqueous two-phase systems (ATPS) is a proved technology for separating and purifying of enzymes. The goal of this study was to evaluate the applicability of polymer-salt ATPS based on polyethylene glycol (PEG)/K_2HPO_4-KH_2PO_4 as a putative method to isolate and recovery of recombinant amino acid dehydrogenases (AADHs). The partition behaviors of three models of AADHs namely phenylalanine dehydrogenase (PheDH), proline dehydrogenase (ProDH) and Leucine dehydrogenase (LeuDH) in two-phase partitioning systems prepared by PEG-4000/K_2HPO_4-KH_2PO_4 were investigated. The influence of different process parameters such as polymer molecular weight, type and concentration of salt, pH, phase volume ratio (V_R), tie-line length (TLL), type and concentration of inorganic salts, temperature, and cell extract loading on system phase behavior and extraction behavior were evaluated. Furthermore, the efficiency of partition behaviors was analyzed by SDS-PAGE method. The best optimal system for model AADHs with regard the partition coefficient (K_E), recovery (R%) and yield (Y%) was: 9.0% (w/w) PEG-4000, 18.0% (w/w) K_2HPO_4-KH_2PO_4, 8% (w/w) NaCl and a TIL of 52.3% (w/w). The partition parameters were as follows; PheDH (K_E=51.4, R=84.7%, Y=92.5) LeuDH (K_E=81.8, R=94.5%, Y=95.34) and ProDH (K_E=73.4, R=91.6%, Y=94.83). Three target enzymes showed to be partitioned in favor of the PEG-4000 rich top-phase. PEG-4000 proved to have a stabilizing effect on the enzymes of interest. K_2HPO_4-KH_2PO_4 was selected as the phase forming salt because

[*] Corresponding author: Hamid Shahbaz Mohammadi. Postal address: Department of Biochemistry, Pasteur Institute of Iran, Pasteur st., 12th Farvardin st., Tehran, 13164, Iran. Telefax: +98 21 66402770.
E-mail: Sh_hamid_biochem@yahoo.com and h_Shahbazmohamadi@pasteur.ac.ir.

of its ability to enhance the hydrophobic difference between the phases. It was found that the partitioning was not affected by V_R, while PEG-4000 concentration and K_2HPO_4-KH_2PO_4 concentration had significant effects on separation behavior. Longer TLL and higher pH resulted in better partitioning into the top phase. Addition of sodium chloride to the ATPS proved to be suitable to increase the recovery of target enzymes. Collectively, the observed partition behaviors of the model AADHs showed that developed ATPS can be a promising system for partitioning and potential recovery of recombinant AADHs.

Keywords: Amino acid dehydrogenase (AADH); polyethylene glycol 4000 (PEG-4000); K_2HPO_4-KH_2PO; Partition behavior

1. Introduction

When two incompatible polymers such as polyethylene glycol (PEG) and dextran or one polymer and one salt such as potassium phosphate are dissolved at certain concentrations in an aqueous solution, the solution separates into two distinct liquid phases, one rich in one polymer and the other rich in the other polymer or salt. The critical phase separating concentration is called the binodial line on the phase diagram [1]. Separation techniques based on these two-phase mixtures have come to be known as aqueous two-phase systems (ATPS) [2, 3]. The use of ATPS in bioprocessing research has been well documented. ATPS has been successfully applied for the separation of various biological products (e.g. protein, DNA and RNA), metal ions, dyes, drug molecules and small organic species. Such partitioning systems have many desirable characteristics [4]. The basis of separation in these aqueous systems is the selective distribution of substances between two phases. There are many factors influencing the partitioning of bioactive materials in ATPS. The partition coefficient is affected by the characteristics of the target molecules, i.e., charge, molecular weight, conformation, surface hydrophobicity, concentration and environmental conditions, such as phase forming polymers or salts, pH, temperature, type of buffer, temperature and ionic strength, speed of phase separation as well as the mode of separation [5, 6] (Figure 1). Thus, different factors can be manipulated to achieve the desired partition conditions. The contributions of these factors to the partition coefficient (K) can be summed up in logarithmic terms, according to Albertsson [7], as follows:

$$\text{In } K = \text{In } K^0 + \text{In } K_{elec} + \text{In } K_{hydrop} + \text{In } K_{biosp} + \text{In } K_{size} + \text{In } K_{conf} + \ldots$$

Wherein elec, hydrop, biosp, size and conf stand for electrostatic, hydrophobic, bio-specific, size and conformational effects and K^0 represents other factors. Two-phase systems have been studied extensively to describe the mechanisms responsible for phase separation. Although, the current thermodynamic models such as lattice theory, Flory-Huggins theory and Mc-Millan-Mayer solution theory help in understanding and predicting the factors influencing partition behavior, the driving force is not yet mentioned in literatures. Hence, due to the difficulties in predicting the behavior of molecules within two-phase systems, experimentation is necessary to define an optimal system [8-10]. At present, partitioning in ATPS has proven to be a valuable tool for separating and purifying biological material and especially proteins with regard to scale of operation and low processing time [11, 12]. ATPSs were introduced as a separation method for biomolecules in the mid-1950's by Albertsson. Albertsson (1956) originally showed that microorganisms, cell walls, chloroplasts and other biological molecules can be selectively partitioned between the phases of an ATPS [1]. ATPS is ideal technology where clarification, concentration, and partial purification can be integrated in one step. This technique can be made highly selective and can be suited for continues operation in large scale, thus allowing wider biotechnological applications [2, 13]. Selective separation in two phases is the key point of this technique. Therefore, it is necessary to know the partition behavior of the desired biomolecules in these systems [14]. Among many types of ATPS, polymer-polymer and polymer-salt ATPS have been most commonly used for the partition of biological macromolecules. Since the bulk of both phases consists mainly of water (80-90%) and most polymers have a stabilizing effect on the protein tertiary structure [15, 16]. However, the polymer-salt ATPS has more advantages than the polymer-polymer systems due to the large differences in density, greater selectivity, lower viscosity, lower cost and the larger relative size of the drops. In fact, the high viscosity and high cost of the polymer-polymer systems restrict their uses in the industry scale. In addition, due to the size differences between the smaller molecules and the polymer, the thermodynamic behavior of polymer-salt based systems is also more complicated [17]. In these ATPS, top phase contains a polymer, and the lower phase contains inorganic salt. In such two-phase systems, partitioning of biomaterials and different components occurs between the phases, and therefore separation can be achieved. Separation conditions are created in which the desired molecule partitions preferentially to the top polymer rich-phase and unwanted material to the bottom phase. The cell mass (cell debris) is also collected at the interface between the phases [1] (Figure 2). In view of these, polymer-salt two-phase partitioning is a promising technique for downstream processing of biomolecules due to the viability of its industrial applications [18]. Most of the reported research works on polymer-salt based two-phase systems were made using polyethylene glycol with salts. As described in pervious works from our laboratory, polyethylene glycol 6000 (PEG-6000)/ammonium sulfate [19] and PEG-6000/K_2HPO_4/KCl [20] systems were applied to partition of recombinant phenylalanine dehydrogenase (PheDH). From these studies, this conclusion can be safely drawn that ATPS partitioning is very effective for the separation and purification of amino acid dehydrogenases (AADHs). However, based on our understanding of the potential of ATPS, there is still a need for more knowledge on design and scale up ATPS for AADHs. In the present study, we used PEG-4000 to form aqueous systems with salt to enhance the selectivity of partitioning of recombinant AADHs. Since, PEG-4000 is a water-soluble polymer that is biodegradable and non-toxic, and can be used for the separation of biomaterials. PEG-4000/salt based systems can be designed to increase the partitioning of soluble proteins to the polymer phase, which results in an enrichment of target protein [2].

AADHs (EC 1.4.1.X) are one of the important industrial enzymes that catalyze the reversible oxidative deamination of amino acids to form the corresponding keto acids, using NAD^+ or $NADP^+$ as cofactors. These biocatalysts constitute a valuable and interesting group of coenzyme-dependent enzymes found in both prokaryotes and eukaryotes [21, 22]. The AADHs have a great industrial potential because of their broad applications in industry and medicine. The stereospecific reaction catalyzed by the AADHs makes them ideal biocatalysts for the synthesis of chiral amino acids for dietary or pharmaceutical applications. In addition, members of this family have shown interesting potential as analytical reagents for quantitative determination of blood amino acids, ammonia or urea in clinical applications [23, 24]. In view of the increasing commercial applications of AADHs, development of efficient and scalable alternative purification methods for the recovery of these enzymes is of great interest. Definitely, this will extend the use of this family of enzymes for a wider range of biotechnological processes. It is well known that the ready access to target enzyme facilities its practical biotechnological uses.

The main objective of this communication was to investigate the partition behavior of recombinant AADHs in PEG-4000/salt based systems to recovery these enzymes. In order to determine appropriate partition conditions for this process, the distribution behavior and the relation between protein partitioning and ATPS parameters need to be known. For this reason, to obtain a comprehensive knowledge of the phase behavior of the AADHs as well as to develop an efficient separation system, three recombinants AADH including phenylalanine, leucine and proline dehydrogenases were chosen as model enzymes of this family. On the other hand, applicability of the PEG-4000/salt system for recovery and differential partitioning of AADHs was tested using experimental data of three representative model enzymes. To improve the selectivity of enzyme partitioning the effects of various parameters such as the phase composition, system pH, tie-line length, phase volume ratio, type and concentration of neutral salts (NaCl, KCl) and sample loading will be discussed in this paper. The large variety of biotechnological applications of AADHs was the main reason for investigating of PEG-4000-salt systems in this work. The experimental results of this study will help us further to utilization of this particular ATPS in targeted extracting of AADHs.

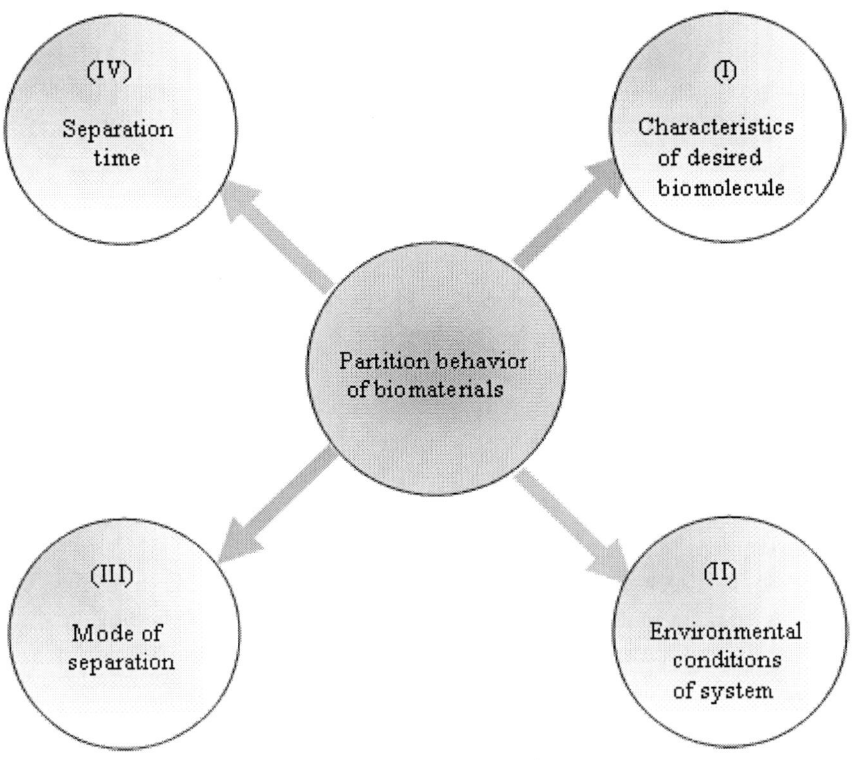

Figure 1. Parameters involved in partition behavior of biomaterials.

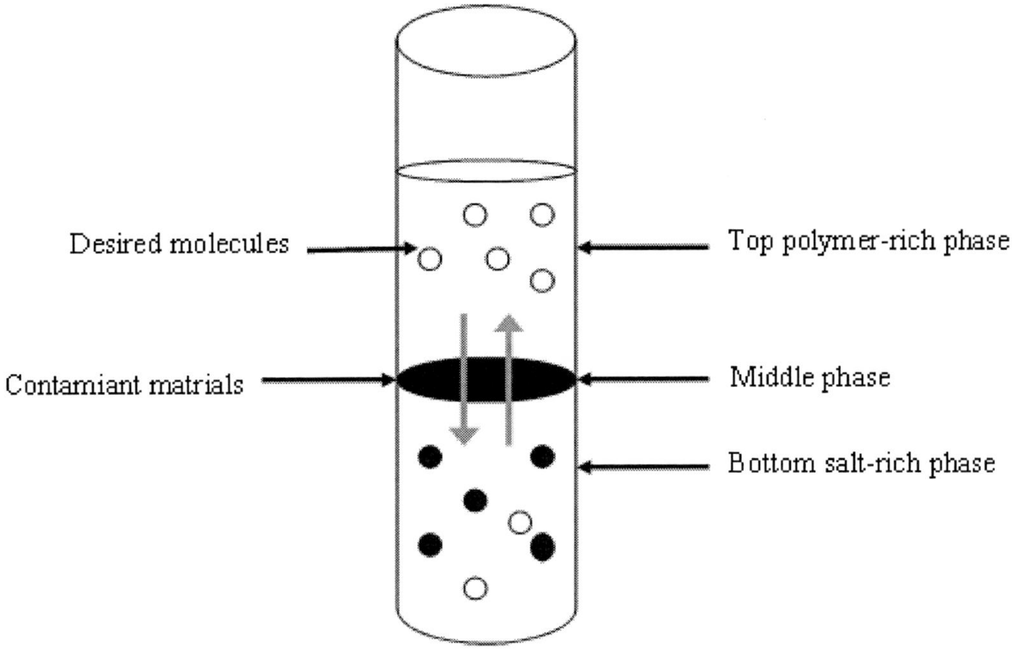

Figure 2. A schematic picture of the aqueous two-phase extraction.

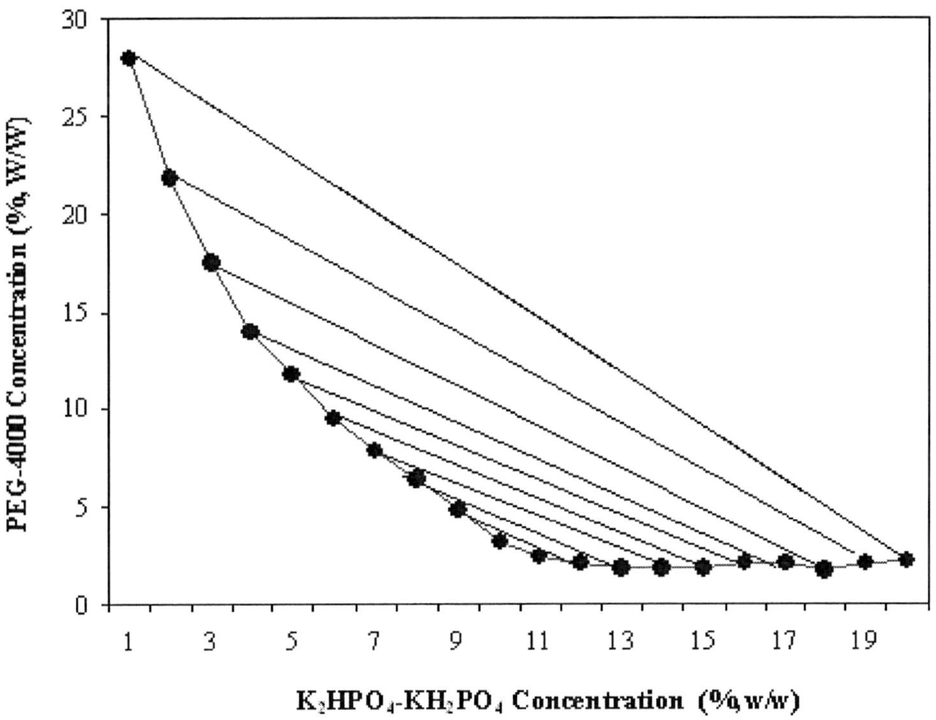

Figure 3. Phase diagram for the ATPS composed of PEG-4000/K_2HPO_4-KH_2PO_4 at 25 ^{0}C and pH 8.0.

2. EXPEIMNTAL PROCEDURES

2.1. Materials

Vectors pET16b and pET23a (Novagen, Inc. Madison, USA) was used for expression. Polyethylene glycol (PEG) with average molecular weight (MW) of 1000, 2000, 4000, 6000, 8000, 10000 and 20000 were purchased from Merck (Darmstadt, Germany). The salts used to prepare ATPS were obtained from Merck (Darmstadt, Germany) and Sigma-Aldrich (St. Louis, MO, USA). NAD^+ and NADH were from Sigma-Aldrich (St. Louis, MO, USA) and L-phenylalanine, L-proline and L-leucine were obtained from Merck (Darmstadt, Germany). INT (2-(p-iodophenyl)-3-(p-nitrophenyl)-5-phenyltetrazolium chloride) and PMS (phenazine methosulfate) were purchased from Sigma-Aldrich Corp (USA). The all other chemicals and biochemical reagents were of laboratory grade, and double-distilled water was used throughout the experiments.

2.2. Production of Recombinant AADHs

2.2.1. Production of Recombinant Bacillus badius PheDH

E. coli BL21 (DE3) cells harboring plasmid pET16b with recombinant *B. badius* PheDH activity were cultivated in shake flasks containing Luria-Bertani (LB) medium supplemented with 0.1 mg /ml of ampicillin at 37 °C and 140 rpm. After 16 h, 10 ml culture was transferred as a seed into 1 L of LB medium in baffled culture flasks and shaken at 37 °C until an OD_{600}=0.8 was reached, then cooled to approximately 23 °C by stirring the flasks in an ice-water bath for 4 min. The T_7 promoter was induced by addition of 0.5 mM sterile isopropyl-β-D-thiogalactopyranoside (IPTG) and shaking at 23 °C for 6 h. After fermentation, cells were harvested by centrifugation at 3500 rpm for 20 min, washed with 0.9% NaCl solution, centrifuged and suspended in 0.1 M potassium phosphate buffer (pH 7.0) containing 0.1 mM EDTA and 2-mercaptoethanol. Cell rupture was achieved by sonication for 20 min total at 9-KHz. The resulting homogenate was dialyzed against the same buffer and centrifuged at 3800 rpm at 4 °C for 60 min to clarity. The supernatant was used as a crude enzyme solution in the partition experiments [20].

2.2.2. Production of Recombinant Pseudomonas putida POS-F84 ProDH

E. coli BL21 (DE3) plysS cells bearing pET23aPDHPF5 construct were cultivated overnight in LB medium containing 100 mg /ml of ampicillin at 37 °C and 150 rpm. 100 ml preculture broth was transferred into 1 L of LB medium in culture flasks and incubated at 37 °C and 150 rpm. When cell density reached an OD_{600} of 0.6-0.8, ProDH enzyme was expressed by the addition of 0.5 mM sterile IPTG. After 6 h induction at 23 °C, cells were harvested, washed twice with 0.9% NaCl solution and stored at -20 °C for further uses. Bacterial pellet was suspended in the lysis buffer (50 mM Tris-HCl, 50 mM NaCl, 10 mM EDTA, pH 8.0), mechanically disrupted by sonication in pulse sequence of 15s on and 10 s off and clarified by centrifugation at 4000 rpm for 1 h. The precipitate (inclusion bodies) containing recombinant ProDH enzyme was washed twice with wash buffer (50 mM Tris-HCl, 50 mM NaCl, 10 mM EDTA, pH 8.0, 1% Triton X-100). The washed pellet was resuspended in 50 mM Tris-HCl (pH 8.0), 100 mM NaCl, 10 mM EDTA, 10% glycerol and 0.1 mM DTT (buffer A) containing 8 M urea and incubated in 4 °C with continuous stirring for 24 h to solubilize the inclusion bodies. Any insoluble material was removed by centrifugation at 4000 rpm at 4 °C for 1 h. Refolding was performed by stepwise dialysis against descending concentrations of urea. The unfolded recombinant ProDH was first dialyzed against buffer A supplemented with 4 M, 2 M and then without urea. The buffer was changed every 24 h. For reconstitution, the renaturated enzyme was dialyzed overnight at 4 °C in the buffer A containing 0.15 mM FAD. The dialysate was centrifuged at 10000 rpm at 4 °C for 1 h. The supernatant solution containing renaturated proteins were used for further purification [25].

2.2.3. Production of Recombinant Bacillus cereus LeuDH

Recombinant strains *E. coli* TG1 transformed with plasmid pUC 18 were grown at 37 °C for 10–12 h with shaking (200 rpm) in LB's broth supplemented with sodium ampicillin (0.1 mg/ml). The cells were harvested in the late exponential phase (OD_{600}=1.0) by centrifugation and the biomass was washed 0.85% NaCl and stored at -20 °C. The washed cells were suspended in 50 mM Tris-HCl buffer (pH 7.5) containing 1 mM EDTA and 0.01% 2-mercaptoethanol and disrupted by ultrasonication. The cell lysate was centrifuged at 3500 rpm for 1 h at 4 °C, and the supernatant was used for partition experiments [26].

2.3. Partitioning Studies of Model Recombinant AADHs in Polymer-Salt ATPS

2.3.1. Phase Diagram Determination for PEG-4000/K_2HPO_4-KH_2PO_4

Phase diagram for the PEG-4000/K_2HPO_4-KH_2PO_4 systems was determined by the cloud-point method [1]. A salt solution of known concentration was added drop-wise to a known amount of a concentrated stock of polymer until turbidity and a two-phase system was formed. The criterion for deciding the presence of an ATPS is the appearance of a clear interface between the two phases. When the interface is perpendicular to the tube wall, the ATPS has reached equilibrium. The composition of this mixture was noted and provided a point on the binodal curve. Then, 0.1 M potassium phosphate buffer (pH 8.0) was added drop-wise to the tube to get a clear one-phase system and more salt solution was added again to afford a biphasic system. The composition of this mixture was noted and so on [20].

2.3.2. Partition Experiment

In all experiments performed, phase systems were prepared by adding defined amounts of solid components (PEG and inorganic salt), 0.1 M potassium phosphate buffer (pH 8.0) and cell free extract to make up the final mass of 10 g. The total system compositions were selected according to the binodal diagrams obtained in our laboratory (Figure 3). Systems were gently shaken for 1 h and then centrifuged at 3500 rpm for 40 min at 25 ^{0}C to complete phase separation. After this treatment, the two-phases became clear and transparent and the interface was well defined. Samples of the top and bottom phases were carefully extracted and analyzed for enzyme and protein concentration [20]. The individual effects of polymer concentration, salt addition, pH, TLL on the enzyme activity were also tested. The extraction experiments were done in triplicates and the average values were reported. The strategy used to obtain the optimal parameters of ATPS partitioning are shown in Figure 4.

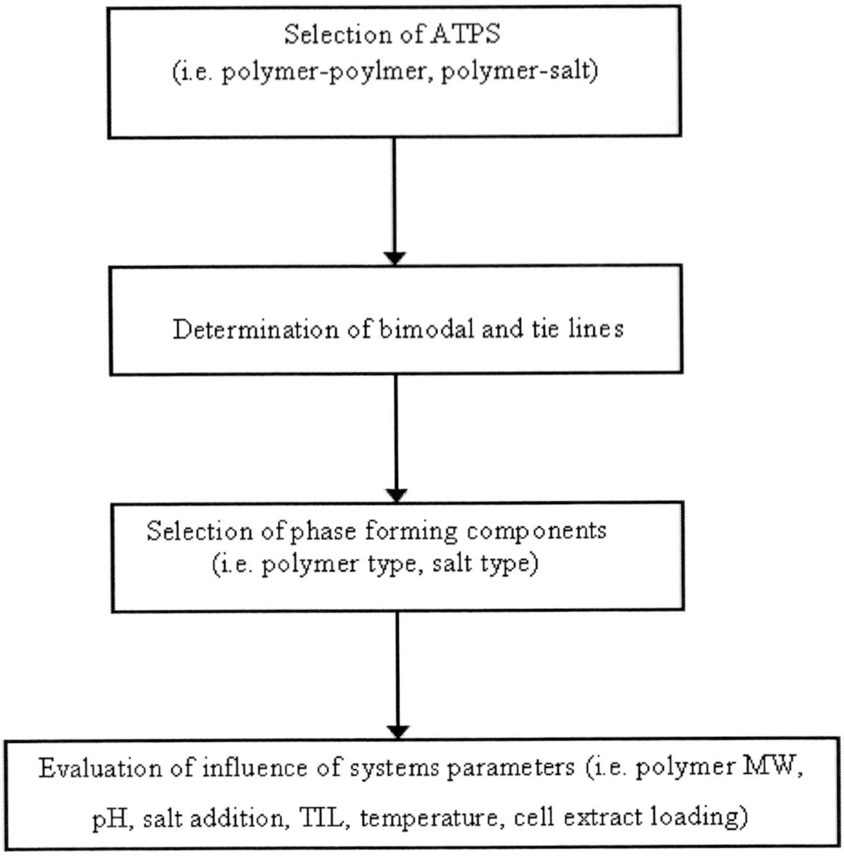

Figure 4. Systemic strategy for evaluation of partition features of desired AADHs in ATPS.

2.3.3. Parameters Affecting Partition Behavior of Recombinant Enzymes

The separation performance of PEG-4000/salt ATPS was evaluated by considering the following parameters [1, 2];

Partition coefficient (K_E or K_P) is defined as the ratio of the enzyme activity or protein concentration in the top phase divided by the correspondent value in the bottom phase.

$$K_E = A_t/A_b \qquad (1)$$

$$K_P = C_t/C_b \qquad (2)$$

where A_t and A_b denote the AADH activity in the top and bottom phase, respectively; C_t and C_b represent the concentrations of AADH in the top and bottom phase, respectively.

Recovery (R%): is the ratio of the enzyme activity in the top phase (A_t) to the initial activity added to the system (A_{ori}). Note that the sum of the activity in the top and the bottom phase is not necessarily equal to the activity in the original cell lysate.

$$R\% = A_t/A_{ori} \qquad (3)$$

Yield (Y%): is calculated as:

$$Y(\%) = 100V_tK_E/(V_t K_E + V_b) \quad (4)$$

where V_t and V_b are the volumes of the top and bottom phase, respectively.

Specific activity (*SA*): is determined as the enzyme activity (A, U/ml) in the phase sample divided by the total protein concentration (C, mg/ml) and is expressed in U/mg of protein.

$$SA = A/C \quad (5)$$

Purification factor (PF): is calculated as the specific activity in the top phase (SA_t) divided by the initial specific activity in the initial extract (SA_{ori}).

$$PF = SA_t/SA_{or} \quad (6)$$

Extraction efficiency (*E*%): is estimated using the equation:

$$E = V_tC_t/(V_bC_b + V_tC_t) \quad (7)$$

Volume ratio (V_R): V_R is the relative volume of the two phases:

$$V_R = V_t/V_b \quad (8)$$

Tie-line length (TLL): is described as the straight lines that connect the composition values of the top and bottom phase in equilibrium. They will be on different sides of the binodial curve and never cross each other. It was calculated according to the following formula:

$$TLL(\%, W/W) = \sqrt{(P_T - P_B)^2} + \sqrt{(S_T - S_B)^2} \quad (9)$$

Selectivity (*S*): is defined as the ratio of K_E to K_P:

$$S = K_E/K_P \quad (10)$$

where P_T and P_B represent the polymer concentrations in the top and bottom phase, respectively. S_T and S_B are the salt concentrations in the top and bottom phase, respectively.

Notice that the partitioning experiments were carried out in triplicate and the average values were given in our research. The partition coefficient, yield and recovery were key parameters for the selection of the optimum condition for enzyme separation.

2.4. Activity Assay of AADHs

2.4.1. Determination of PheDH Activity

Enzyme activity was assayed spectrophotometrically (double-beam UV-visible spectrophotometer, Shimadzu 1601 PC, Japan) by monitoring the reduction of NAD^+ at 340 nm. Mixture assay contained 10 mM L-phenylalanine, 100 mM glycine-KCl–KOH buffer (pH 10.4), 2.5 mM NAD^+ and the enzyme solution in a total volume of 1 ml. One unit of PheDH activity (U) is expressed as the amount of enzyme catalyzing the formation of 1 μmol NADH per minute under the assay conditions [27].

2.4.2. Determination of ProDH Activity

ProDH activity was measured using the proline: INT (2-(p-iodophenyl)-3-(p-nitrophenyl)-5-phenyltetrazolium chloride) oxidoreductase assay which was performed by INT as a terminal electron acceptor and Phenazine methosulfate (PMS) as a mediator electron carrier. The standard reaction mixture was composed of 100 mM Tris-HCl, 10 mM $MgCl_2$, 10% glycerol, pH 8.5, 200 mM L-proline, 0.2 mM FAD, 0.4 mM INT, 0.08 mM PMS and the enzyme in a total volume of 1ml. The increase in absorbance at 490 nm was estimated and corrected for blank values lacking proline. Furthermore, all values were corrected for the low rate of enzyme-independent proline oxidation observed in assay mixtures containing all components except enzyme. One unit (U) of ProDH activity was defined as the quantity of enzyme, which transfers electrons from 1 μmol of proline to INT per minute at 25 °C [25].

2.4.3. Determination of LeuDH Activity

The standard reaction mixture for the oxidative deamination contained 200 mM of L-leucine, 0.3 mM of NAD^+, 100 mM of glycine-KCl-NaOH buffer (pH 10.9) and enzyme in a final volume of 1.0 ml. The substrate was replaced by water in a blank. Incubation was performed at 30 °C in a cuvette with a 1 cm light path. The reaction was stated by addition of NAD^+ and monitored by measuring the initial absorbance at 340 nm with a Shimadzu 1601 PC double-beam UV-visible spectrophotometer at 30 °C. One unit of enzyme activity was defined as the amount of enzyme that catalyzed the formation of 1 μmol of NADH per minute in the oxidative deamination with a molar absorption coefficient of 6220 M^{-1} cm^{-1} [26]

2.5. Protein Assay

The concentration of total protein was measured by the method of Bradford using bovine serum albumin (BSA) as a standard [28]. The appropriate blank system phase was employed to correct for minor interference from polymer and salt.

2.6. Purity Analysis of the Separated Enzymes from ATPS

The purified samples were analyzed by sodium dodecyl sulphate polyacrylamide gel electrophoresis (SDS-PAGE). The samples were bodied for 5 min with the presence of 1% SDS, 80 mM 2-mercatoethonal, 100 mM Tris-HCl buffer (pH 6.8) and 15 % glycerol and loaded onto a 12% SDS ployacrylamide gel. Electrophoresis was proceeded at 10 mA until the marker reached the bottom of the separation gel. The migrated proteins were stained with a staining solution composed of 0.25% (w/v) Coomassie Brilliant Blue, 45% (v/v) methanol and 10% (v/v) acetic acid. The gel was destained by washing in a mixture of 7.5% (v/v) acetic acid and 12.5% (v/v) methanol [29]. The silver staining method was also used for some gels [29].

3. RESULTS AND DISCUSSION

To investigate the effectiveness of PEG-4000/salt ATPS for partitioning and recovery of recombinant AADHs, we analyzed the partitioning behavior of three model enzymes: PheDH, LeuDH and ProDH in PEG-4000/K_2HPO_4-KH_2PO_4 two-phase systems. The conventional ATPS processes for the extraction of intracellular proteins contain several discrete steps, including cell disruption by mechanical means, disruptate clarification by centrifugation, and following ATPS step. Figure 5 depicts the process flow diagram of partitioning experiments in PEG-4000-salt systems.

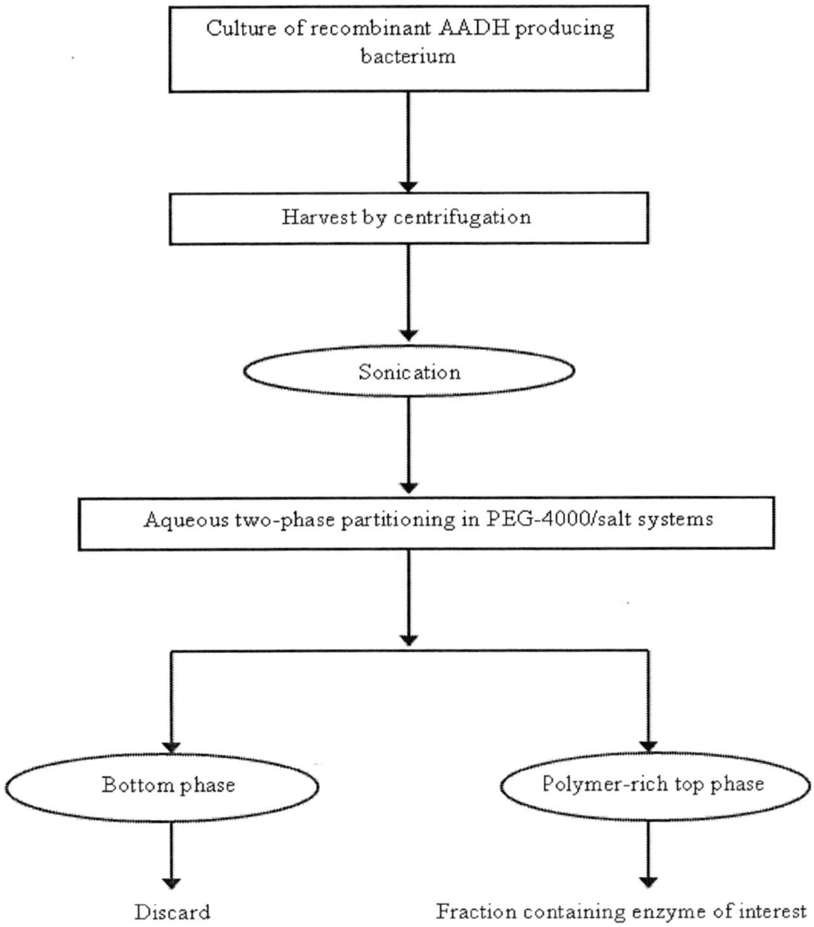

Figure 5. Flow chart for the separation and recovery of recombinant AADHs in ATPS.

Table 1. Effect of phase-forming concentrations in PEG-4000/K_2HPO_4-KH_2PO_4 ATPS on partitioning behavior of model recombinant AADHs

No	Total composition (%, w/w)		K_E			R %		
	PEG-4000	K_2HPO_4-KH_2PO_4	PheDH	LeuDH	ProDH	PheDH	LeuDH	ProDH
1	5.2	9.30	10.2	11.6	8.7	39.7	41.2	38.5
2	6.0	10.35	12.1	10.6	13.2	43.4	40.2	42.7
3	7.2	11.44	15.0	14.3	17.8	50.0	48.5	53.6
4	7.5	12.0	25.4	26.0	29.3	61.2	59.6	62.4
5	7.7	14.0	30.5	31.2	40.6	65.8	67.3	70.2
6	7.8	14.5	36.4	35.0	50.7	70.4	73.2	76.8
7	8.0	15.0	33.4	38.4	60.4	68.3	75.2	78.3
8	8.3	16.0	29.6	40.2	66.3	55.4	78.7	80.2
9	8.5	16.5	37.7	50.4	72.4	76.0	85.5	87.0
10	8.7	17.0	44.2	57.0	70.3	90.2	93.0	88.3
11	8.8	17.4	40.6	53.0	62.0	86.4	89.5	83.5
12	9.0	18.0	51.4	81.8	73.4	98.5	120.6	110.6
13	10.3	18.4	50.5	70.1	51.0	95.4	108.3	99.0
14	11.0	19.0	49.1	73.2	60.2	93.0	105.7	102.5
15	11.5	19.4	45.0	75.2	62.3	90.2	106.5	106.0
16	11.7	19.8	46.2	78.4	70.0	88.2	95.3	90.3
17	12.0	20.0	48.3	69.0	71.2	80.4	87.6	80.8
18	12.5	20.4	40.2	61.8	67.7	75.4	79.5	71.4

[a] All systems were prepared with 0.1 M potassium phosphate buffer at pH 8.0.
[b] K_E and R represent the partition coefficient and yield of recombinant AADHs, respectively.
[c] Each data is the average of three independent experiments.

It should be noted that the reason for the analogous behaviors of the mentioned model enzymes in different experiments is the fact that the AADHs have nearly similar physico-chemical features. Therefore, an analogous reasoning was used to explain the partition behaviors of the enzymes of interest.

3.1. Phase Behaviors of the PEG-4000/Salt ATPS

3.1.1. Phase Diagrams

Partitioning between the two phases of an ATPS is a complex function guided by several factors such as interactions of the partitioned substance and the phase components (e.g. hydrophobic and electrostatic interactions), steric and conformational effects and also the phase composition. As a matter of fact, unfortunately there is not much information about the exact mechanism of partitioning and can not easily be predicted. Generally, in order to define the adequate system the phase composition is chosen according to the phase diagram and partitioning of the interested molecule is investigated [30]. In this study, the selection of the desired system was preformed as described above. To prepare phase diagrams of PEG-4000/salt ATPS, K_2HPO_4-KH_2PO_4 was selected as phase forming salt, since pervious results have shown that AADH enzymes like PheDH partition predominantly to the top phase when potassium phosphate salt is used. Furthermore, the main criteria for selecting PEG as the most suitable polymer were: low cost and fast approach to equilibrium. Concentrations of PEG-4000 and K_2HPO_4-KH_2PO_4 systems were chosen based on the phase diagram (shown in Figure 1).

3.1.2. Effect of Concentrations of Phase-Forming Components on Partitioning Behavior

The influence of different concentrations of PEG-4000 and K_2HPO_4-KH_2PO_4 on partitioning of PheDH, LeuDH and ProDH were evaluated. In order to select the ATPS with the best capability of recovery and separating recombinant AADHs, we calculated the K_E and R parameters after one extraction process. Table 1 presents the partition coefficients and yields for PheDH, LeuDH and ProDH, as a function of PEG-4000 and salt concentrations in ATPS. As can be seen, the best partition coefficients and recoveries for PheDH, LeuDH and ProDH were measured in the system containing 9.0% (w/w) PEG-4000 and 18.0% (w/w) K2HPO4-KH2PO4. The high partition coefficients were obtained for model enzymes in PEG-4000/K_2HPO_4-KH_2PO_4, thus showing the feasibility of partitioning and recovery of AADHs in these systems. The optimized system favored the distribution of AADH into the polymer-rich phase, which is free and better for protein stability. Another important requirement of ATPS for an effective extraction is that contaminant proteins should not be extracted in the same phase as the target enzyme. Therefore, along with a high K_E value, a high R value should be desirable. Based on these criteria (K_E, R), system 12, which gave the highest extraction efficiency was chosen for the following experiment. Furthermore, the partition coefficient of LeuDH was higher than PheDH and ProDH. This behavior could be assigned to the hydrophobic character of LeuDH. Generally, the partition behavior of a protein in polymer-salt ATPS depends on the hydrophobicity of the protein. Proteins with hydrophobic characteristics will show a higher affinity for the polymer-rich top phase, which is hydrophobic than the salt rich phase [31]. Similar behavior was observed for proteinase in PEG-1000/$MgSO_4$ [32] and model proteins (ovalbumin and myoglobin) in PEG-4000/poly(acrylic acid) [13].

3.1.3. Effect of Phase-Forming Polymer Molecular Weight on Partition Behavior

Using the previous findings, the ATPS composed of 9.0% PEG-4000 and 18.0% K_2HPO_4-KH_2PO_4 was selected to investigate the effect of molecular mass of PEG on partition behavior. The partitioning and extraction of biomaterials are strongly depended on the PEG molecular mass [33]. According to the Albertsson's comments, the selection of the best molecular weight (MW) of polymer is generally the first step in the ATPS experiments [7]. It has been demonstrated that the MW of PEG affects the composition of phases and the number of polymer-protein interactions [34]. This influence is usually attributed to hydrophobic interactions between the chains of PEG and the hydrophobic area of the bio-molecule. Therefore, it can be said that besides PEG exclusion effect, the partitioning in also governed by the chemical structure of protein. A high hydrophobical surface of a protein exposed to solvent is a factor that favors the partition equilibrium displacement to the PEG–rich phase. Hydrophobicity is also changed by the type and MW of polymer. Meanwhile, the PEG-salt interactions must be considered, for example, in the PEG/phosphate systems due to the repulsion between the PEG and the phosphate molecules; the PEG phase has higher self energy than the phosphate phase [35].

At high MW, the preferential interaction between the PEG and the protein domain is decreased due to the exclusion of polymer from the protein, and this leads to a decrease in K_E. Moreover, when the PEG phase becomes too ordered (at higher polymer size) due to the reduced available water content, the steric exclusion effect increases. On the other hand, polymer acquires a more compact conformation with intramolecular hydrophobic bonds in solution and precludes the partition of biomaterial into the top phase. This would be resulted in significant movement of target protein to the salt rich phase, high viscosity and bad reproducibility. Low molecular mass is also unsuitable for adequate partitioning because that the exclusion effect decreases and as a result the polymer can attract all proteins (contaminant and desired proteins) to the upper phase. However with consideration of all these descriptions, there is no general rule about the mechanism governing partition and even in some studies these parameters show opposite results [36, 37]. At MW, the preferential interactions between the PEG and the protein domain decrease and this lead to reduce in K_E. One reason could be the increase in polymer MW increases the exclusion effect, thus the polymer acquires a more compact conformation with intramolecular hydrophobic bonds and hinders the partition of biomaterial into the top phase. This would be resulted to significant transfer of target protein to the salt rich phase, high viscosity and bad reproducibility. Low molecular mass is also unsuitable for adequate partitioning because that the exclusion effect decreases and as a result the polymer can attract all proteins (contaminant and desired proteins) to the upper phase. Therefore, it can be said that the intermediate molecular mass of PEG is the best choice for ATPS experiments. However, with consideration all these descriptions, there is no general rule about the mechanism governing partition and in some studies, these parameters show opposite results [14]. Figure 6 illustrates the effects of different PEGs on the K_E, *Y*, and *R* of the model AADHs. As anticipated, low MW of PEG (1000) and the much higher PEG MW (20000) were unsuitable for adequate partitioning. All the enzymes showed the same behaviors. The reason for the analogous behaviors of the mentioned model enzymes is the fact that the AADHs have nearly similar physico-chemical features. Therefore, an analogous reasoning was used to explain the partition behaviors of the enzymes of interest. Increasing PEG MW from 1000 to 4000 in the biphasic systems resulted in higher separation of target enzymes into the top phase, obviously due to the increase in solubility of desired enzymes in the upper phase. Conversely, when the PEG molecular weight increased from 4000 to 20000, the partition efficiency of the enzymes of interest decreased. This behavior is an agreement with an exclusion effect owing to the diminution of the free volume in the upper phase. The highest partition parameters (*R*, *Y*, K_E) were obtained by PEG-4000. These results also suggest that the AADHs have a great hydrophobic surface which enhances enzyme-polymer interactions. On the other hand, the high presence of hydrophobic regions at the molecular surface of enzyme tended to increase its preference for separation into the polymer phase. Our findings were supported by other literatures [38]. Similar behaviors were observed for ß-glucosidase in PEG-4000/phosphate [39], C-phycocyanin in PEG4000/phosphate [40], and glycomacropeptide in PEG-4000/sodium citrate [41]. In conclusion, based on the obtained results in this sub-section, the PEG MW should be kept at 4000 for the next experiments.

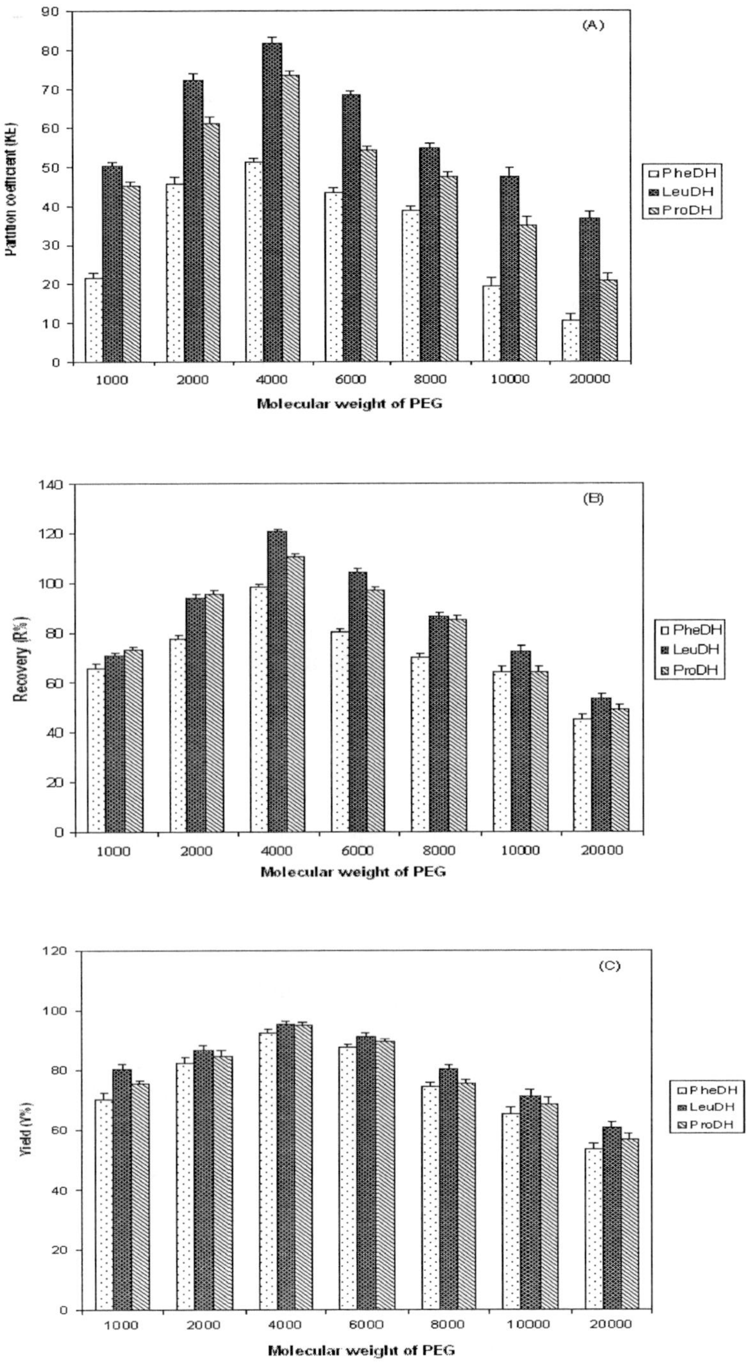

Figure 6. Effects of PEG molecular weight on partition coefficient (A), recovery (B) and yield (C) of PheDH, LeuDH and ProDH in systems containing PEG 9.0% (w/w) and 18.0% (w/w) K_2HPO_4-KH_2PO_4 (pH 8.0). The partition experiments were carried out in triplicate to estimate experimental errors.

Table 2 Influence of salt type on the partitioning behavior of recombinant model AADHs in PEG-4000/salt ATPS

No.	salt (w/w, %)	K_E			R (%)			Y (%)		
		PheDH	LeuDH	ProDH	PheDH	LeuDH	ProDH	PheDH	LeuDH	ProDH
1	NH_4NH_2COO	30.0	23.4	21.3	72.8	68.5	59.4	81.0	69.6	60.1
2	NaH_2PO_4	28.65	35.6	28.6	76.4	83.4	71.2	73.0	65.3	67.5
3	$(NH_4)C_2O_4$	29.3	33.6	39.4	81.0	88.5	77.1	75.4	77.5	79.4
4	Na_2HPO_4-NaH_2PO_4	34.5	47.5	26.8	83.5	92.3	65.4	68.4	61.2	59.6
5	K_2HPO_4-KH_2PO_4	73.4	81.8	51.4	110.6	120.6	98.5	92.5	95.34	94.83
6	Na_2SO_4	9.5	19.4	26.6	52.3	64.3	48.5	52.0	48.4	49.5
7	$C_4H_{47}Na_2O_4$	35.6	26.6	31.2	82.3	69.5	75.4	75.6	79.4	76.5
8	K_2HPO_4	52.3	64.5	46.7	96.0	105.6	89.4	85.3	91.2	92.5
9	$(NH_4)H_2PO_4$	36.0	47.5	38.5	64.7	75.4	70.6	87.0	74.0	81.2
10	Na_2HPO_4	42.5	56.4	34.5	66.5	79.5	51.4	81.3	69.5	84.0
11	$(NH_4)_2HPO_4$	50.0	73.4	39.5	96.4	114.5	90.4	85.1	85.7	90.3
12	Na_2CO_3	22.0	36.8	16.4	74.6	84.5	63.4	71	66.7	69.0
13	Na_3PO_4	14.16	18.5	11.4	45.3	49.7	38.5	41.86	37.5	45.3
14	$(NH_4)_2SO_4$	54.4	77.5	43.5	100.3	116.4	84.5	86.0	81.4	86.7

3.1.4. Effect of Salt Type

The influences of the salt type on partition behavior of the model AADHs were investigated. As seen in Table 2, the salt type had a significant impact on the partitioning efficiency of PheDH, LeuDH and ProDH. $(NH_4)_2HPO_4$, $(NH_4)_2SO_4$, K_2HPO_4 and K_2HPO_4-KH_2PO_4 exhibited better partition parameters for three model enzymes in terms of partition coefficient, recovery and yield ($K_E > 50$, $R > 95\%$ and $Y\% > 90\%$). The first reason was related to the higher salting-out power of K_2HPO_4-KH_2PO_4 that promoted the portioning of model enzymes to the top phase, and provided that it is not subjected to the excluded volume effects of PEG. On the other hand, the stabilizing effect of salts depends on the nature of anion and cation. Generally, anions with a higher valence such as phosphate are better salting-out agents than anions with a lower valence because the higher valence hydrates more water and thus decreases the amount of water available to hydrate polymer. In contrast, weakly hydrated cations such as potassium have the most stabilizing effect as well [42]. The second possible reason was that compared with other phase-forming salts, K_2HPO_4-KH_2PO_4 resulted in an appropriate medium pH for the target enzyme. Therefore, K_2HPO_4-KH_2PO_4 was chosen as the salt phase for further experiment.

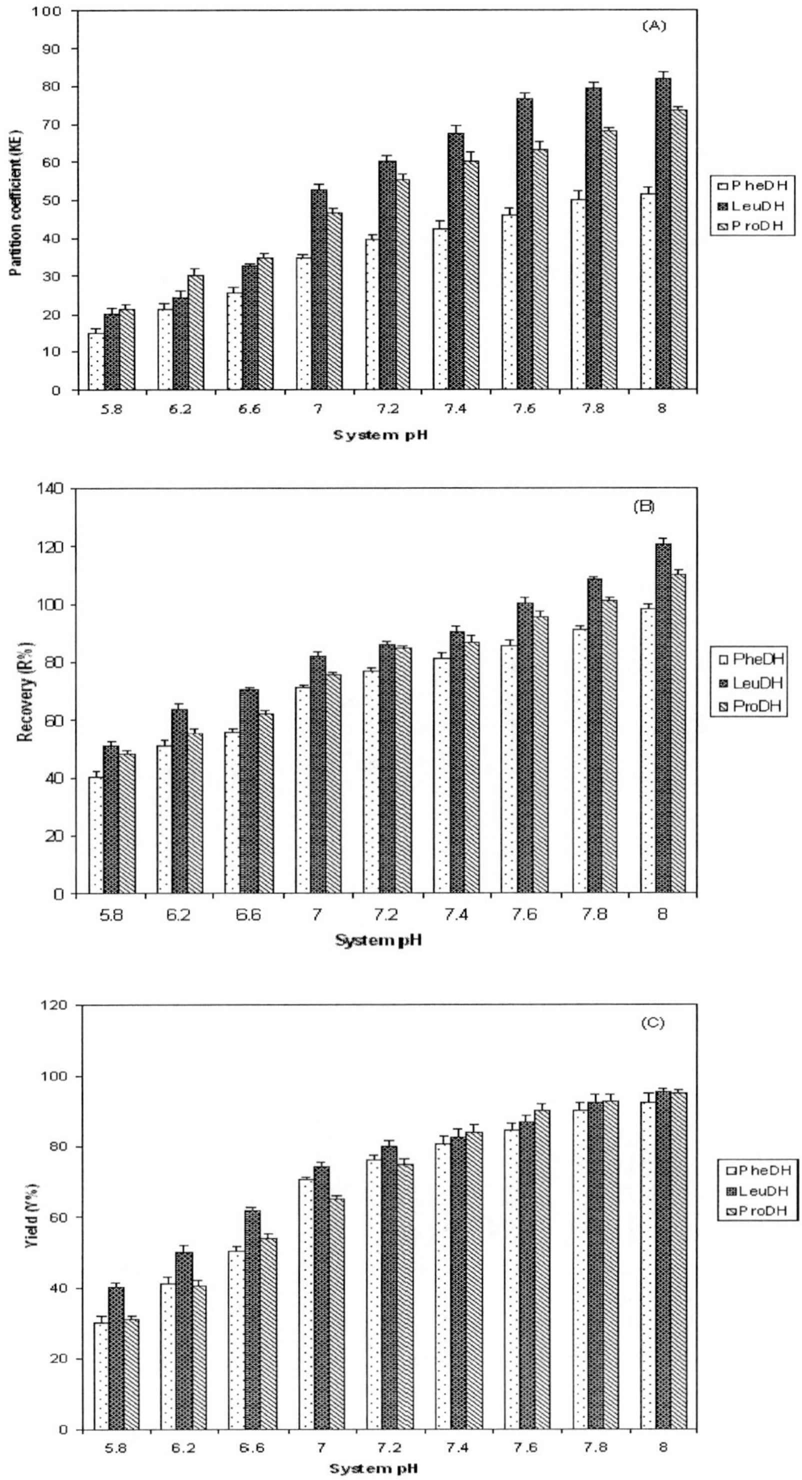

Figure 7. Effects of system pH on partition coefficient (A), recovery (B) and yield (C) of PheDH, LeuDH and ProDH in systems containing PEG-4000 9.0% (w/w) and 18.0% (w/w) K_2HPO_4-KH_2PO_4 (pH 8.0). The partition experiments were carried out in triplicate to estimate experimental errors.

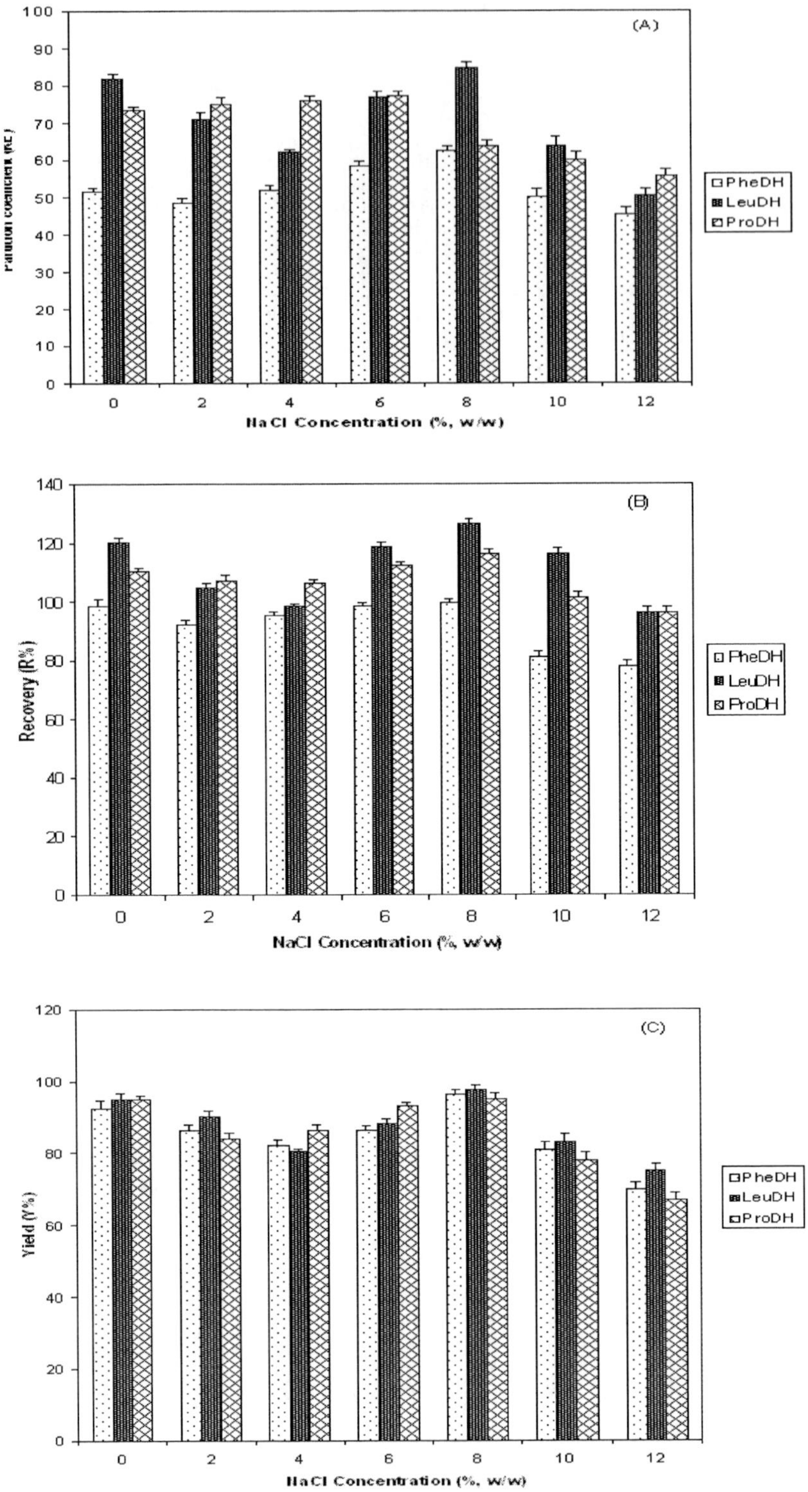

Figure 8. Effects of NaCl concentration on partition coefficient (A), recovery (B) and yield (C) of PheDH, LeuDH and ProDH in PEG-4000 9.0% (w/w) and 18.0% (w/w) K_2HPO_4-KH_2PO_4 (pH 8.0) systems. The partition experiments were carried out in triplicate to estimate experimental errors.

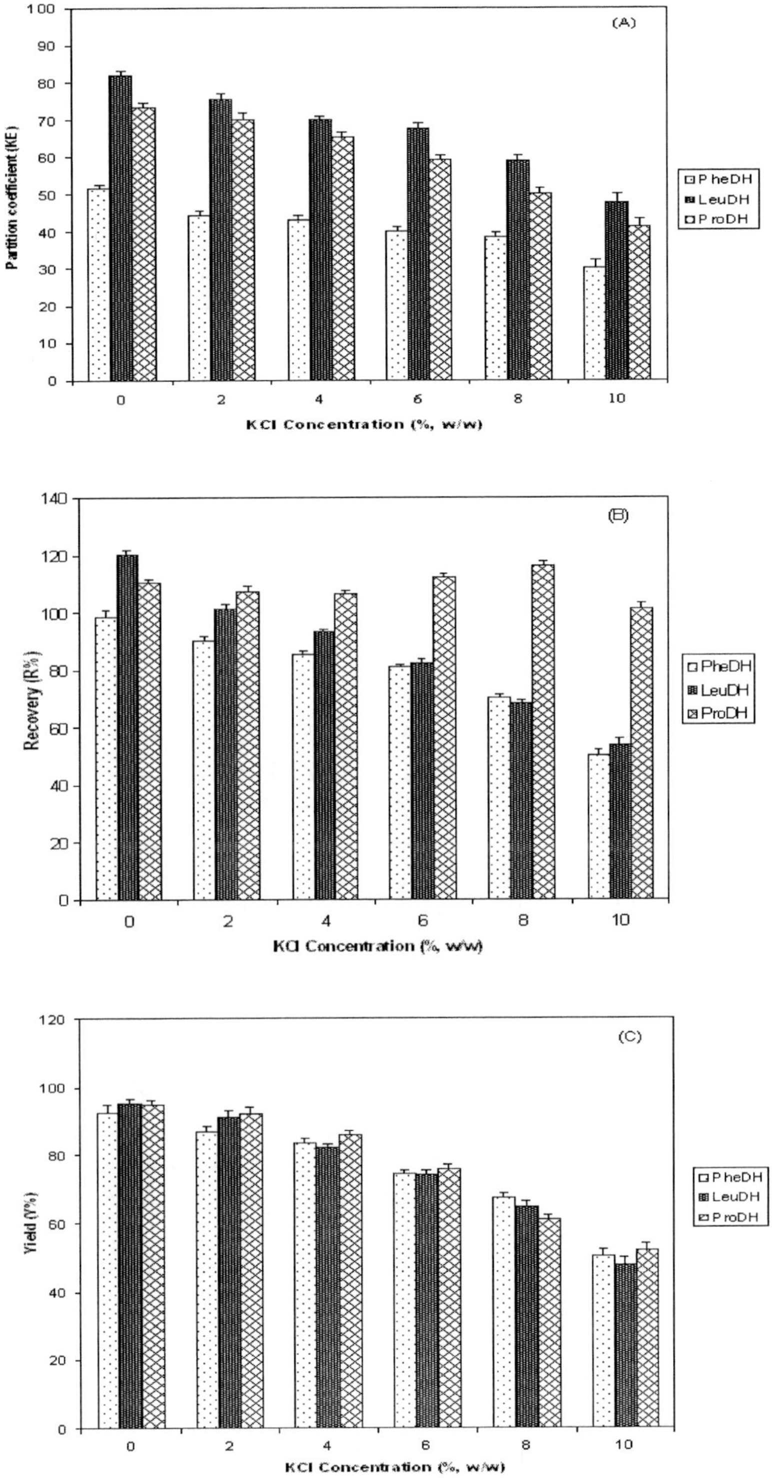

Figure 9. Effects of KCl concentration on partition coefficient (A), recovery (B) and yield (C) of PheDH, LeuDH and ProDH in PEG-4000 9.0% (w/w) and 18.0% (w/w) K_2HPO_4-KH_2PO_4 (pH 8.0) systems. The partition experiments were carried out in triplicate to estimate experimental errors.

3.1.5. Effect of System pH on Partition Behavior

The influence of pH of the phosphate buffer on partition behavior of model enzymes was also analyzed. Another important factor which influences the partitioning of biomolecules is the system pH [1-3]. The partition is determined by the kind of existence ions and the ratio among them. The pH influences the ionizable surface group of proteins and therefore, affects their seprartion behavior. Figure 7 shows the effect of pH on the partitioning parameters of model AADHs. The optimal values for the partition coefficient, recovery and yield of model enzymes were obtained at pH 8.0. This could be explained by the fact that the p*I* of AADHs is acidic [21]. In general, negatively charged proteins prefer the upper phase in PEG-salt systems, while positively charged proteins normally partition selectively to the bottom phase. Therefore, our results were in agreement with this rule since the isoelectric point (pI) of AADHs around 5 (PI_{PheDH}= 5.1, PI_{LeuDH}= 5.0, PI_{ProDH}=5.49) [21, 42] and were more negatively charged at this pH. This behavior also could be explained assuming Albertsson's equation, which takes into account electrostatic and non- electrostatic (van der waals) molecular interactions [7]:

$$\ln K_p = \ln K^0_p + \left(\frac{Z_P F}{RT}\right) \Delta\Psi \tag{11}$$

where K_p and K^0_p are the partition coefficient at a given of pH, and the isoelectric point (p*I*) and $\Delta\Psi$ interfacial potential between the top and the bottom phases $(\Psi_{top} - \Psi_{bottom})$ which influences the part target bio-molecule. The Z_P, F, R and T the target the net protein charge, Faraday constant, universal gas constant and absolute temperature, respectively. Above the P*I* of AADHs, these enzymes were negatively charged and PEG behaved as a positively charged molecule and thereby polyanions of the target enzyme were attracted by the PEG–rich phase. As result, target enzymes gave partitioned the result, op phase. This phenomenon was observed by others as well [43], confirming the generality of our findings.

3.1.6. Effect of Added Salts on Partition Behavior

The partition behaviors of recombinant AADHs in ATPS containing 9% (w/w) PEG-4000 and 18% (w/w) K_2HPO_4-KH_2PO_4 (pH 8.0) with different inorganic salts such as NaCl, and KCl was studied. The aim of this stage was to take advantage of their stronger partitioning into one of the phases and their ability to exhibit specific interactions such as electrostatic interactions and hydrophobic forces with the target biomolecule. The salts are often used to improve and direct selectivity partitioning between the phases [45]. When salts added to the system, there is a decrease in the total mass of water in order to keep constant the final composition. On the other hand, as the concentrations of salt increases, the volume of the top phase decreases and this leads the PEG rich phase more concentrated. Generally, the addition of neutral salts changes the phase diagram, alters the properties of the partitioning solute, shortens the phase separation time and also induces the displacement of equilibrium partitioning to the upper phase [1, 7]. Although adding salts have been reported to be beneficial to target products, negative results have been obtained as well [46]. The effects of NaCl on the partition coefficient, recovery and yield of PheDH, LeuDH and ProDH are shown in Figure 8. High values of the partition parameters were obtained when 8% of NaCl was added. As can be seen, there was no regular relation between the different concentrations of NaCl and changing the partition patterns of PheDH, LeuDH and ProDH. These behaviors could be caused by the unequal distribution of cation (Na^+) and anion (Cl^-) ions between top and bottom phase, which resulted in a potential difference between both phases. On the other hand, the uneven distribution of neutral salt ions with different positions in the Hofmeister or lyotropic series in phases of PEG-phosphate system was a consequence of different affinity of ions for the bottom and top phases which created two distinct ionic medium in both phases [2]. The influences of KCl addition on the partition behavior of three desired enzymes are depicted in Figure 9. As illustrated in figures 9 (A, B and C), inclusion of KCl concentration decreased the extraction efficiency of PheDH, LeuDH and ProDH. Our obtained data suggested that KCl could not improve effectively the partitioning of recombinant AADHs. Also, the volume ratio in the current work was remained practically constant for the all neutral salt at different concentration (unpublished data). The best partitioning behaviors for three recombinant AADHs (PheDH, LeuDH and ProDH) were obtained in the ATPS of 9% (w/w) PEG-4000, 18% (w/w) K_2HPO_4-KH_2PO_4 and 8% (w/w) NaCl at pH 8.0.

Table 3. Tie-line length (TLL) and phase volume ratio (V_R) of ATPS containing PEG-4000)/K_2HPO_4- KH_2PO_4 (pH 8.0)

No.	Total composition (%, w/w)		Tie-line length (%,w/w)	Phase volume ratio (V_R)
	PEG-4000	K_2HPO_4-KH_2PO_4		
1	5.2	9.30	15.1	0.25
2	6.0	10.35	19.2	0.25
3	7.2	11.44	26.1	0.25
4	7.5	12.0	29.2	0.25
5	7.7	14.0	35.3	0.25
6	7.8	14.5	36.6	0.25
7	8.0	15.0	40.2	0.25
8	8.3	16.0	42.5	0.25
9	8.5	16.5	44.2	0.25
10	8.7	17.0	46.4	0.25
11	8.8	17.4	48.0	0.25
12	9.0	18.0	52.3	0.25
13	10.3	18.4	54.0	0.25
14	11.0	19.0	56.4	0.25
15	11.5	19.4	58.2	0.25
16	11.7	19.8	60.5	0.25
17	12.0	20.0	64.3	0.25
18	12.5	20.4	67.6	0.25

Partitioning parameters of these systems were calculated in Table 1.

3.1.7. Effect of TIL on Partition Behavior

The influence of Tie-line on the partitioning behaviors of recombinant target AADHs in PEG-4000/ K_2HPO_4-KH_2PO_4 ATPS was also investigated. The effect of polymer and salt concentrations on K_E of proteins is determined as a function of TIL. Various TLL describes the different selectivity or partition pattern for different protein components [47]. Generally, a short TIL is composed of low polymer-salt concentrations, while a long TIL is composed of high polymer-salt concentrations. Different tie lines exhibit different partitioning behavior for different protein components. It is obvious that the investigation of different selectivity conditions is crucial. Table 3 shows the calculated TILs for all investigated ATPSs. The increasing of TLL (Table 3) caused to better partitioning of the model AADHs into the top phase and subsequent increase of the partition coefficient, recovery and yield values (Table 1). This can be attributed to the salting out effect of potassium phosphate, which caused molecular exclusion of the enzymes of interest to move to the upper phase. In contrast, when salt concentration increased from 9% (w/w) to 20.4% (w/w), model enzymes were partitioned at the bottom phase. Therefore, it was decided to select the ATPS with a TLL of 52.3% (w/w) for partitioning and recovery of recombinant AADHs.

3.1.8. Effect of System Temperature on Partition Behavior

Under an identical conditions of phase composition and optimized experimental conditions, the influence of temperature on the extraction efficiency of the recombinant model AADHs was evaluated. As shown in Figure 10, as the temperature increased from 15 to 25 °C, the extraction efficiency of all model enzymes raised. This behavior could be explained based on temperature effect on the PEG hydrophobicity [48].

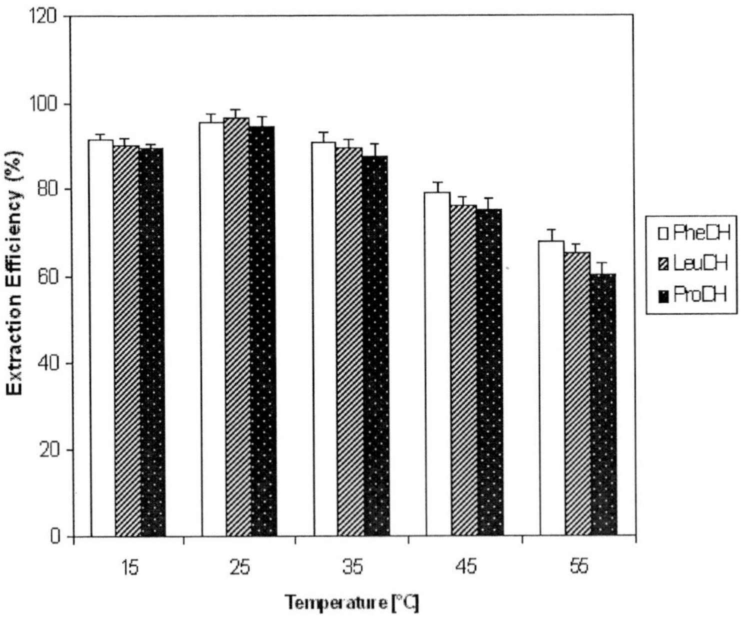

Figure 10. Temperature dependence of the extraction efficiency of recombinant PheDH, LeuDH and ProDH in system composed of 9.0% (w/w) PEG, 18.0% (w/w) K_2HPO_4-KH_2PO_4 and 8% (w/w) NaCl at pH 8.0.

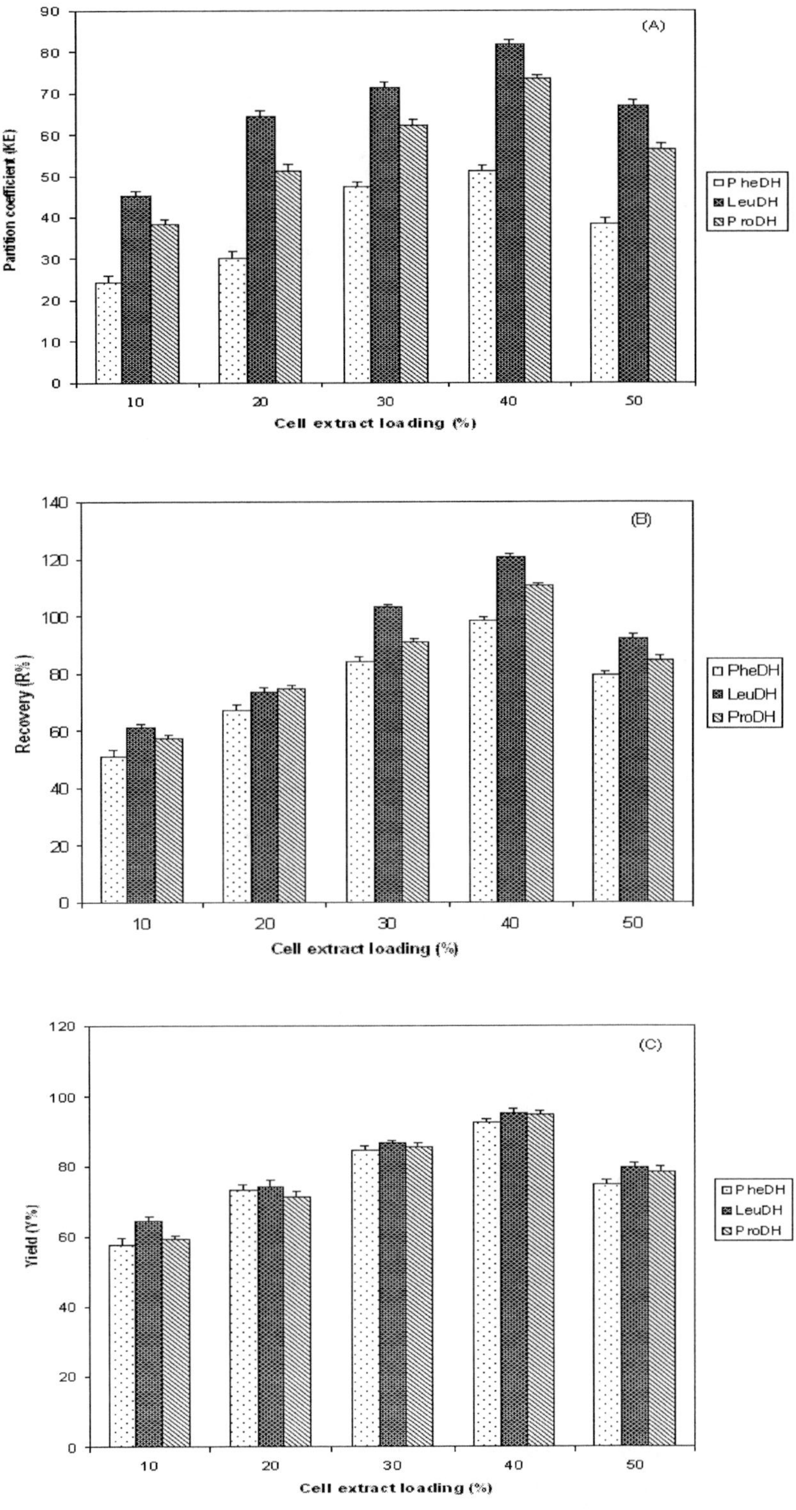

Figure 11. Effect of Cell mass loading on partition coefficient (A), recovery (B) and yield (C) of PheDH, LeuDH and ProDH in PEG-4000 9.0% (w/w) and 18.0% (w/w) K_2HPO_4-KH_2PO_4 (pH 8.0) system. The partition experiments were carried out in triplicate to estimate experimental errors.

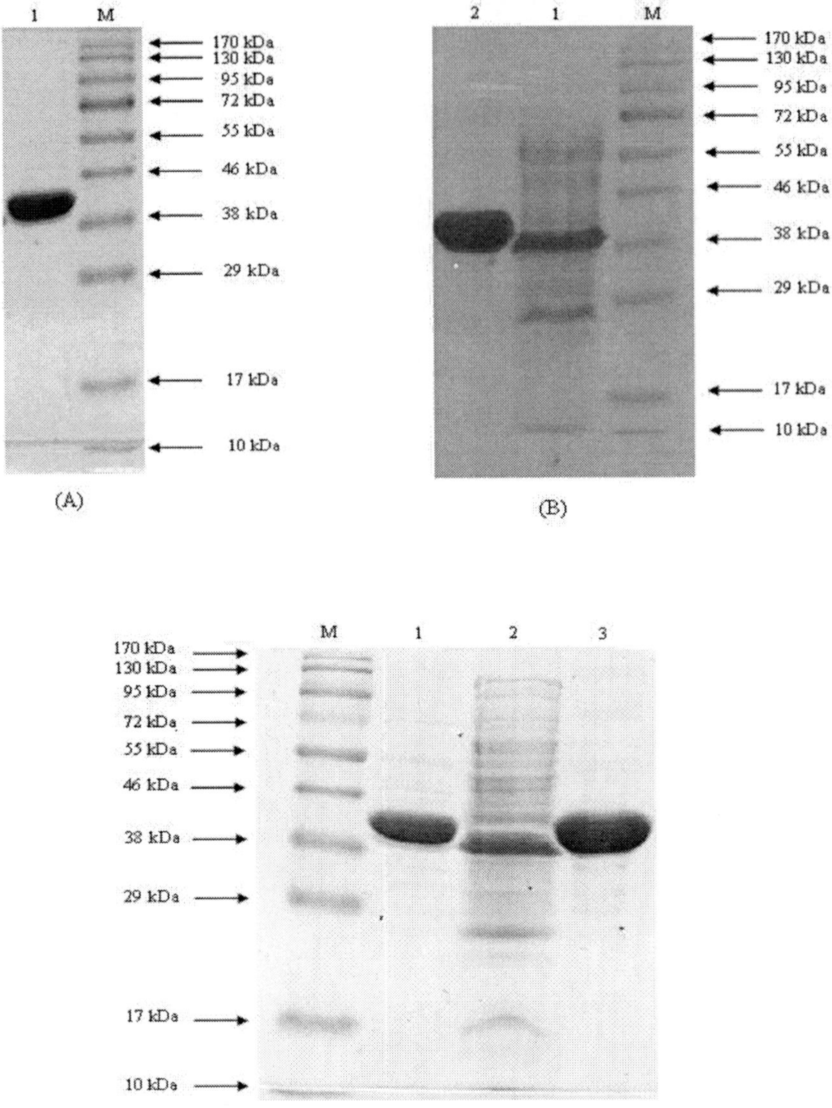

Figure 12. SDS-PAGE gel electrophoresis of PheDH, LeuDH and ProDH purified employing the ATPS containing 9% (w/w) PEG-4000, 18% (w/w) K_2HPO_4-KH_2PO_4 and 8% (w/w) NaCl. A) Partitioning and recovery of PheDH. M: molecular standard markers, Lane 1: Top phase of system. The protein band was stained with Coomassie Brilliant Blue R-250. B) Partitioning and recovery of LeuDH. M: molecular standard markers, Lane 1: top phase of system. The protein bands were stained with Coomassie Brilliant Blue R-250.C) Partitioning and recovery of ProDH. M: molecular standard markers, Lane 1: pellet of the cell lysate, Lane 2: supernatant of the cell lysate. Lane 3: top phase of system. The protein bands were stained with silver staining protocol.

Generally, when the temperature is raised, water molecules are driven from the PEG phase to the salt phase, so the PEG phase becomes more concentrated and the salt density in the bottom phase decreases. This caused the hydrophobicity difference would increase the strength hydrophobic interactions between the interested enzyme and the PEG molecules and lead to effective partitioning. This is also known as the salting-out phenomenon which resulted in higher extraction efficiency values as the temperature increases. As a matter of fact, in PEG/salt ATPS, temperature increases biomolecule partitioning into the PEG-rich phase and the effect of raising temperature can be considered similar to the increasing of salting-out effect, as both result in increasing K_E [3 , 49]. There 49].There was a good agreement between this partition feature of recombinant model enzymes (PheDH, LeuDH and ProDH) and the reported results in other literatures [23, 24]. However, with increasing temperature from 25 to 55 °C, the partitioning of PheDH towards the PEG-rich phase decreased. In this respect, it may be concluded that the salting out effects in the bottom phase become less dominant with elevating the temperature from 25 to 55 °C [24]. In this research, all extraction experiments were done within 20-30 °C.

3.1.9. Effect of Cell Extract Loading and Phase Volume Ratio

In our work, to assess the applicability of ATPS for downstream processing of recombinant AADHS, the effect of cell extract supernatant loading was tested as well. In this section, a set of experiments was done by varying the lysate load from 10% (w/w) to 40% (w/w). The results are shown in Figure 11. As can be found, the partition parameters describing the extraction efficiency (K_E, R, Y) enhanced when the cell extract loading was elevated to 40%. However, the extraction efficiency decreased with increasing extract loading at 50% since the limits of solubility for PheDH were exceeded. These data indicated that there was an optimal ratio for the enzyme loading in ATPS. Therefore, the cell extract could be loaded as high as 40 % (w/w). This feature was advantageous from the viewpoint of industrial scale since larger amounts of enzyme could be processed by the two-phase partitioning systems. Also, more interestingly the volume ratio in the present study remained practically constant (0.25) with the variation in mass loading. Hence, it can be inferred that the phase diagram is not influenced by the variation in cell-extract loading. This finding was in agreement to the theoretical expectations that partition parameters remain constant along the same tie-line [50, 51]. It is necessary to mention that the all experiments in previous sections were carried out with 20% (w/w) lysate load.

3.2. Purity Analysis of Model Enzymes

In ATPS studies, when the proteins or enzymes to be separated from others, partitioning can be carried out successfully. For this reason, the efficiency of partitioning behavior should be finally analyzed by SDS-PAGE method. Therefore, in order to assess the separation performance of the presented method, the partitioning behavior of model recombinant AADHs in the final optimized system (9% (w/w) PEG-4000, 18% (w/w) K_2HPO_4-KH_2PO_4 and 8% (w/w) NaCl) was further tested by SDS-PAGE method. Figure 11 shows the SDS-PAGE analysis of the purified model enzymes by ATPS. As can be observed in Figure 12, purified model AADHs (PheDH, LeuDH and ProDH) in the system composed of 9% (w/w) PEG-4000, 18% (w/w) K_2HPO_4-KH_2PO_4 and 8% (w/w) NaCl at pH 8.0 had considerable purity after partitioning and recovery in ATPS. As expected from the partition behavior previously described, the purified PheDH, LeuDH and ProDH from ATPS method appeared as single bands on stained SDS-PAGE gel. The apparent molecular mass of the PheDH, LeuDH and ProDH were estimated to be 41, 40 and 40. kDa, . These results were in agreement with the observations of Asano et al. [18]. The observed purity of the investigated model enzymes confirmed the soundness of the experimental findings. In fact, it can be concluded that partitioning of AADHs in PEG-4000/K_2HPO_4-KH_2PO_4/NaCl ATPS can be an effective way for purifying them.

Similar results have been reported in the case of using PEG-salt two-phase process for downstream processing of other enzymes such as, laccase [52], invertase [53], and lectin [54].

CONCLUSION

In the present study, we designed and evaluated PEG-4000/salt two-phase partitioning for the development of a process for separation and recovery of AADHs. Partitioning behaviors of three models of recombinant AADHs, PheDH, LeuDH and ProDH were experimentally analyzed. The influence of different factors such as polymer MW, type and concentration of salt, pH, phase volume ratio (V_R), tie-line length (TLL), temperature, addition of inorganic salt and cell extract loading were evaluated in order to establish the optimal conditions. It was shown that the PEG MW, phase compositions, pH and salt type influenced the separation behavior of target enzymes. Addition of sodium chloride to the ATPS proved to be suitable to increase the recovery of AADHs whereas potassium chloride was unsuitable for increasing the yield. The final optimized system in terms of partition coefficient, recovery and yield was obtained using 9.0% (w/w) PEG-4000 and 18.0% (w/w) K_2HPO_4-KH_2PO_4 and 8% (w/w) NaCl at pH 8.0 and TLL 52.3% (w/w). Model AADHs were obtained in the top phase of PEG 4000/K_2HPO_4-KH_2PO_4 ATPS. The cell and other protein contaminants were exclusively partitioned to the bottom phase. The investigated model enzymes showed similar partitioning pattern. Nevertheless, partitioning parameters of these model enzymes were very one-sided and dependent on the tie-line length. In this sense, it can be concluded that partition features of these models AADHs could offer relevant information about partitioning of other enzymes of this family. In fact, the obtained partitioning data from these enzymes provide a clear picture of partition trends of other AADHs in PEG/K_2HPO_4-KH_2PO_4 ATPS and facilitate the selection of an appropriate ATPS for AADH purification process. Therefore, the separation conditions established for the PEG/potassium phosphate ATPS extraction can be an alternative for the potential recovery of any member of AADHs family. In resume, the strategy proposed in this study is more effective than the reported methods for recovering AADHs, and represents a scalable and economically visible alternative for production processes of the family of AADHs.

ACKNOWLEDGMENTS

This research project was financially supported by the Food and Drug Deputy, Ministry of Health, Treatment and Medical Education of Iran. The authors wish to gratefully acknowledge the Biochemistry Dept., Pasteur Institute of Iran.

REFERENCES

[1] Albertsson, P. A. (1986). Partition of Cell Particles and Macromolecules. 3rd ed., New York: John Wiley and Sons.

[2] Hatti-Kaul, R. (1999). Methods in Biotechnology: Aqueous two-phase systems: Methods and Protocols. Humana Press Inc., Totowa, NJ.

[3] Diamond, A. D. and HSU, J. T. (1992). Aqueous two-phase systems for biomolecule separation. *Advances in Biochemical Engineering Biotechnology*, 47: 89-135.

[4] Bensch, M., Selbach, B. and Hubbuch, J. (2007). High throughput screening techniques in downstream processing: preparation, characterization and optimization of aqueous two- phase systems. *Chemical Engineering Science*, 62: 2011-2021.

[5] Raghavarao, K. S. M. S., Rastogi, N. K., Gowthaman, M.K. and Karanth, N.G. (1995). Aqueous tow phase extraction for downstream processing of enzyme/proteins. *Advances in Applied Microbiology*, 41: 97-171.

[6] Ratio-Palomares, M. (2004). Practical application of aqueous two-phase partition to process development for the recovery of biological products. *Journal of Chromatography* B, 807: 3-11.

[7] Albertsson, P. A. Cajarville, A., Brooks, D. E. and Tjerneld, F. (1987). Partition of proteins in aqueous polymer two-phase systems and the effect of molecular weight of the polymer. *Biochimica et Biophysica Acta*, 926: 87-93.

[8] Zaslavsky, B.Y. (1995). Aqueous two-phase partitioning: physical chemistry and bioanalytical applications. Marcel Dekker Inc, New York.

[9] Madeira, P., Xu, X., Teixeira, J. A. and Macedo, E. A. (2005). Prediction of protein partition in polymer/salt aqueous two-phase systems using the modified Wilson model. *Biochemical Engineering Journal*, 24: 147-55.

[10] Lin, D-Q, Wu, Y-T, Mei, L-H, Zhu, Z-Q and Yao, S-J. (2003). Modeling the protein partitioning in aqueous two-phase systems: influence of polymer concentration and molecular weight. *Chemical Engineering Science*, 58: 2963-2972.

[11] Ooi, C. W., Hii, S.L., Kamal, S. M. M., Ariff, A. and Ling, T.C. (2011). Extractive fermentation using aqueous two-phase systems for integrated production and purification of extracellular lipase derived from *Burkholderia pseudomallei*. *Process Biochemistry*, 46: 68-73.

[12] Rosa, P. A. J., Azevedoa, A. M., Sommerfeld, S., Mutter, M., Aires-Barros, M.R. and Backer, W. (2009). Application of aqueous two-phase systems to antibody purification: a multi-stage approach. *Journal of Biotechnology*, 139: 306-313.

[13] Saravanan, S., Raghava Rao, J., Unni Nair, B. and Ramasami, T. (2008). Aqueous two-phase poly(ethylene glycol)–poly(acrylic acid) system for protein partitioning: influence of molecular weight, pH and temperature. *Process Biochemistry*, 43: 905-911.

[14] Yan-Min, L., Yan-zaho, Y., Xi-Dan, Z. and Chuan-Bo, X. (2010). Bovine serum albumin partitioning in polyethylene glycol (PEG)/potassium citrate aqueous two-phase systems. *Foods and Bioproducts processing*, 88: 40-46.

[15] Gu, Z. and Glatz, C. E. (2007). Aqueous two-phase extraction for protein recovery from corn extracts. *Journal of Chromatography* B, 845: 38-50.

[16] Shahbaz Mohammadi, H. and Omidinia, E. (2009). Comparison between polyvinyl pyrrolidone/Na_2SO_4 aqueous two-phase systems and chromatographic methods for purification of recombinant phenylalanine dehydrogenase. 20 (4): 311-317.

[17] Nalinanon, S., Benjakul, S., Visessanguan, W. and Kishimura, H. (2008). Partitioning of protease from stomach of albacore tuna (Thunnus alalunga) by aqueous two-phase systems. *Process Biochemistry*, 44: 471-476.

[18] Rahimpour, F., Mamoa, G., Feyzi, F., Maghsoudi, S. and Hatti-Kaul, R. (2007). Optimizing refolding and recovery of active recombinant *Bacillus halodurans* xylanase in polymer–salt aqueous two-phase system using surface response analysis. *Journal of Chromatography* A, 1141: 32-40.

[19] Shahbaz Mohamadi, H. and Omidinia, E. (2007). Purification of recombinant phenylalanine dehydrogenase by partitioning in aqueous two-phase systems. *Journal of Chromatography* B, 854: 273-278.

[20] Omidinia, E., Shahbaz Mohammadi, H., Dinarvand, R. and Taherkhani, H.-A. (2010). Investigation of chromatography and polymer/salt aqueous two-phase processes for downstream processing development of recombinant phenylalanine dehydrogenase. *Biosystem and Bioprocess engineering*, 33 (3): 317-329.

[21] Asano, Y. Phenylalanine dehydrogenase. In: Encyclopedia of bioprocess technology: fermentation, biotechnology and bioseparation. (Flinckinger, M.C. and Drew, S.W. eds.), Vol. 2. John Wiely and Sons, Inc., New York, 1999, pp 1955.

[22] Tachibana, S., Suzuki, M. and Asano, Y. (2006). Application of an enzyme chip to the microquantification of L-phenylalanine. *Analytical Biochemistry*, 359:72-78.

[23] Busca, P., Paradisi, F., Moynihan, E., Maguire, A. R. and Engel, P. C. (2004). Enantioselective synthesis of non-natural amino acids using phenylalanine dehydrogenases modified by site-directed mutagenesis. *Organic and Biomolecular Chemistry*, **2:** 2684-2691.

[24] Weiss, D. J., Dorris, M., Loh, M. and Peterson, L. (2006). Dehydrogenase based reagentless biosensor for monitoring phenylketonuria. *Biosensors and Bioelectronics*, 22: 2436-2441.

[25] Baban, B. A., Vinod, M. P., Tanner, J. J. and Becker, D. F. (2004). Probing a hydrogen bond pair and the FAD redox properties in the proline dehydrogenase domain of *Escherichia coli* PutA. *Biochimica et Biophysica Acta*, 1701: 49–59.

[26] Katoh, R., Nagata, S. and Misoni, H. (2003). Cloning and sequencing of the leucine dehydrogenase gene from *Bacillus sphaericus* IFO 3525 and importance of the C-terimianl region for the enzyme activity. *Journal of Molecular catalysis* B: Enzymatic, 23: 239-247.

[27] Shahbaz Mohammadi, H., Omidinia, E. and Dinarvand, R. (2007). Evaluation of recombinant phenylalanine dehydrogenase behavior in aqueous two-phase partitioning. *Process Biochemistry,* 42: 1296-1301.

[28] Bradford, M. M. (1976). Rapid and sensitive method for the quantification of microgram quantities of protein utilizing the principles of protein-dye binding. *Analytical Biochemistry*, 72: 248-254.

[29] Sambrook, J., Fritsch, E. F. and Maniatis, T. (1994). Molecular Cloning: A laboratory manual. 2nd ed., Cold Spring Harbor Laboratory press, Cold Spring Harbor. New York, USA. pp 1847-1857.

[30] Chen, X., Xu, G., Li, X., Li, Z. and Ying H. (2008). Purification of an a-amylase inhibitor in a polyethylene glycol/fructose-1, 6-bisphosphate trisodium salt aqueous two-phase system. *Process Biochemistry,* 43: 765-768.

[31] Yang, S., Huang, Z., Jiang, Z. and Li, L. (2008). Partition and purification of a thermostable xylanase produced by *Paecilomyces thermophila* in solid-state fermentation using aqueous two-phase systems. *Process Biochemistry,* 43: 53-61.

[32] Klomklao, S., Benjakul, S., Visessanguan, W., Simpson, B. K. and Kishimura, H. (2005). Partitioning and recovery of proteinase from tuna spleen by aqueous two-phase systems. *Process Biochemistry,* 40: 3061-3067.

[33] Yang, L., Huo, D., Hou, C., He, K., Lv, F., Fa, H. and Luo, X. (2010). Purification of plant-esterase in PEG1000/NaH2PO4 aqueous two-phase system by a two-step extraction. *Process Biochemistry*, 45: 1664-1671.

[34] Chen, X., Xu, G., Li, X., Li, Z. and Ying, H. (2008). Purification of an a-amylase inhibitor in a polyethylene glycol/fructose-1,6-bisphosphate trisodium salt aqueous two-phase system. *Process Biochemistry*, 43: 765-768.

[35] Forciniti, D., Hall, C.K. and Kula, M.R. (1991). Protein partitioning at the isoelectric point: influence of polymer molecular weight and concentration and protein size. *Biotechnology and Bioengineering*, 38: 986-994.

[36] Bim, M. A. and Franco, T. T. (2000). Extraction in aqueous two-phase systems of alkaline xylanase produced by *Bacillus pumillus* and its application in kraff pulp bleaching. *Journal of Chromatography* B, 743: 349-56.

[37] Guo-qing, H., Xiu-yan, Z., Xing-jun, T., Qi-he, C. and Hui, R. (2005). Partitioning and purification of extracellular β-1, 3-1, 4-glucanase in aqueous two-phase systems. *Journal of Zhejiang University Sciences*, 6: 825-31.

[38] Reh, G., Spelzini, D., Tubfo, G., Pico, G. and Farruggia, B. (2010). Partition features and renaturation enhancement of chymosin in aqueous two-phase systems. *Journal of Chromatography* B, 860: 98-105.

[39] Gautam, S. and Simon, L. (2006). Partitioning of ß-glucosidase form Trichoderma reesei in poly (ethylene glycol) and potassium phosphate aqueous two-phase systems: influence of pH and temperature. *Biochemical Engineering Journal*, 30: 104-108.

[40] Patil, G. and Raghavarao, K. S. M. S. (2001). Aqueous two phase extraction for purification of C-phycocyanin. *Bimolecular Engineering*, 17: 167-182.

[41] Da Silva, A. A. S., Coimbra, J. S. R., Rpjas, E. E. G. and Teixeira, J. A. C. (2009). Partitioning of glycomacropeptide in aqueous two-phase systems. *Process Biochemistry*, 44:1213-1216.

[42] Zafarani-Moattar, M. T. and Sadeghi, R. (2005). Effect of temperature on the phase equilibrium of aqueous two-phase systems containing polyvinylpyrrolidone and disodium hydrogen phosphate or trisodium phosphate. *Fluid Phase Equilibrium*, 238: 129–135.

[43] Brunhuber, N. M. and Blanchard, J. S. (1994).The biochemistry and enzymology of amino acid dehydrogenases. *Critical Reviews in Biochemistry and Molecular Biology*, 29: 415-67.

[44] Wongmongkol, N. and Prichanont, S. (2006). Partition of alkaline protease in aqueous two-phase systems of polyethylene glycol 1000 and potassium phosphate. *Korean Journal of Chemical Engineering,* 23: 71-76.

[45] Saravanana, S., Raghava Raoa, J., Murugesanb, T., Unni Nair, B. and Ramasami, T. (2007). Partition of tannery wastewater proteins in aqueous two-phase poly (ethyleneglycol)-magnesium sulfate systems: effects of molecular weights and pH. *Chemical Engineering Journal* 62: 969-978.

[46] Hemavath, A. B. and Raghavarao, K. S. M. S. (2011). Differential partitioning of ß-galactosidase and ß-glucosidase using aqueous two phase extraction. *Process Biochemistry*, 46: 649-655.

[47] Oliveria, M. C. D., Filho, M.A. N. D. A. and Filho, P. D. A. P. (2007). Phase equilibrium and protein portioning in aqueous two-phase systems containing ammonium carbamate and block copolymers PEO-PPO-PEO. *Biochemical Engineering Journal* 37: 311-318.

[48] Mokhtaran, B., Karimzadeh, R., Amini, M. H. and Darvish Manesh, S. (2008). Partitioning of Ciprofloxacin in aqueous two-phase system of poly(ethylene glycol) and sodium sulphate. *Biochemical Engineering Journal*, 38: 241–247.

[49] Malpiedi, L.P., Pico G.A.., Loh, W., Nerli, B.B. (2010). Role of polymer–protein interaction on partitioning pattern of bovine pancreatic trypsinogen and alpha-chymotrypsinogen in polyethyleneglycol/sodium tartrate. *Journal of Chromatogarphy* B, 878: 1831-1836.

[50] Selvakumar, P., Ling, C., Walker, S. and Lyddiatt, A. (2010).Redefinition of working aqueous two-phase systems: A generic description for prediction of the effective phase chemical composition for process control and biorecovery. *Journal of Chromatogarphy* B, 878: 1784-1790.

[51] Tubio, G., Pico, G. A. and Nerli, B. B. (2009). Extraction of trypsin from bovine pancreas by applying polyethyleneglycol/sodium citrate aqueous two-phase systems. *Journal of Chromatography* B, 877: 115-120.

[52] Mayolo-Deloisa, K., Trejo-Herna, M. D. R. and Rito-Palomares, M. (2009). Recovery of laccase from the residual compost of *Agaricus bisporus* in aqueous two-phase systems. *Process Biochemistry*, 44: 435-439.

[53] Yücekan, I. and onal S. (2011). Partitioning of invertase from tomato in poly(ethylene glycol)/sodium sulfate aqueous two-phase systems. *Process Biochemistry*, 46: 226-232.

[54] Soaresa, P. A. G., Nascimentob, C. O., Portob, T. S., Correia, M. T. S., Porto, A. L. F. and Carneiro-da-Cunha, M. G. (2011). Purification of a lectin from Canavalia ensiformis using PEG–citrate aqueous two-phase system. *Journal of Chromatography* B, 879: 457-460.

In: Polymer Phase Behavior
Editors: T. P. Ehlers and J. K. Wilhelm, pp. 265-293
ISBN: 978-1-61324-336-7
© 2011 Nova Science Publishers, Inc.

Chapter 7

PHASE SEPARATION, PHASE DISSOLUTION AND CRYSTALLIZATION IN POLY(ε-CAPROLACTONE)/ POLY(STYRENE-CO-ACRYLONITRILE) BLENDS

Petr Svoboda*
Centre of Polymer Systems, Faculty of Technology,
Tomas Bata University in Zlin,
Zlin, Czech Republic

ABSTRACT

A blend of poly(ε-caprolactone) (PCL) and poly(styrene-co-acrylonitrile) (SAN) containing 27.5 wt% of acrylonitrile having the critical composition (80/20 PCL/SAN) was studied. This PCL/SAN blend having a lower critical solution temperature (LCST) phase boundary at 122°C offered an excellent opportunity to investigate, firstly the kinetics of phase separation above LCST (125-180°C), and secondly the kinetics of phase dissolution below LCST (50-115°C). The blend underwent a temperature-jump above LCST where spinodal decomposition (SD) proceeded, yielding a regularly phase-separated structure (SD structure). Then, it was quenched to the temperatures below LCST when the phase dissolution proceeded. Optical microscopy was used to observe the spinodal decomposition qualitatively while light scattering was used to characterize the phase separation and phase dissolution quantitatively. It was found that during phase dissolution the peak maximum moved towards a smaller angle (wavelength of concentration fluctuations increases) while the peak intensity decreased. This behavior was explained by a model. Also it was found that the fastest phase dissolution kinetics at 80°C, which was characterized by an apparent diffusion coefficient, was about 10 times slower than the kinetics of phase separation at 180°C. Crystallization after various levels of spinodal decomposition was observed by optical microscopy. Order parameter of the lamellae inside the spherulites was evaluated with the help of H_v light scattering.

* Nam. T.G.Masaryka 5555, 760 01 Zlin, Czech Republic. Tel. 420–576 031 335, fax. 420–577 210 172. E-mail: svoboda@ft.utb.cz.

Transmission electron microscopy revealed interesting lamellar structure after spinodal decomposition.

Keywords: Spinodal Decomposition; Phase Dissolution; Crystallization; Polymer Blends; PCL; SAN

INTRODUCTION

Poly(ε-caprolactone) (PCL) is a linear semicrystalline aliphatic polyester that became available commercially following efforts at Union Carbide to identify synthetic polymers that could be degraded by microorganisms. The high solubility of PCL, its low melting point (59-64°C), and its exceptional ability to form blends has stimulated research on its application as a biomaterial. PCL degrades at a slower pace than PLA (polylactic acid) and can therefore be used in drug delivery devices that remain active for over 1 year [1].

Industrially produced copolymer of styrene and acrylonitrile (SAN) usually contains 15 to 35 wt% of acrylonitrile and is prepared by radical polymerization. The higher the acrylonitrile content, the greater the heat and chemical resistance, impact strength, toughness, scratch resistance and barrier properties. SAN is frequently used in households as well as in industry.

Quite many scientists have studied PCL/SAN blends, several examples are listed here [2-17]. PCL and SAN are miscible only when SAN contains 8 ~ 28 wt% of acrylonitrile. Near the boarders of this "miscibility window", the lower critical solution temperature (LCST) behavior was observed. In this study SAN with 27.5 wt% of acrylonitrile (AN) was used.

Being driven by thermal noises, local concentration in a polymer blend fluctuates continually. When temperature is changed, the distribution of wavelength of concentration fluctuations changes until it reaches a new equilibrium. If a temperature change exceeds a certain limit, concentration fluctuations grow enormously and the blend undergoes phase separation through spinodal decomposition (SD) mechanism. Dynamics of fluctuation growth during SD has been extensively investigated in the last decades because of its theoretical as well as practical importance, e.g. by Hashimoto et al. [18-22].

When the temperature of a blend undergoing SD is changed to a value in the single-phase region, concentration fluctuations that had grown at one time in the spinodal region, gradually attenuate. Dynamics of such attenuation of concentration fluctuations is expected to follow the same mechanism as the growth of fluctuations during SD [23]. However, only a few experiments have been made to investigate the dynamics of attenuation of concentration fluctuations [23-29], which makes a marked contrast with the number of experimental studies performed to clarify the dynamics of fluctuation growth during the phase separation. Therefore, our research focused on this not so thoroughly understood field.

In this study PCL/SAN-27.5 blend exhibiting LCST behavior with critical composition (80/20) was investigated. Firstly the blend underwent phase separation at various temperatures above LCST giving us an opportunity to study the spinodal decomposition by optical microscopy and time-resolved light scattering (TRLS). Then the phase-separated blend was quenched to various temperatures below LCST where we could study the process of phase dissolution. Dynamics of attenuation of concentration fluctuations was experimentally investigated by light scattering. Also crystallization after spinodal decomposition was studied by light scattering and by optical microscopy.

EXPERIMENTAL

The PCL was a commercial polymer of Union Carbide Co. (PCL-767; M_w =40,400 g mol^{-1}, M_w/M_n=2.61). SAN containing 27.5 wt% of acrylonitrile (SAN-27.5) was synthesized at 60°C using ethylbenzene as a solvent and azoisobutyronitrile (AIBN) at a concentration of 0.02 mol/l as an initiator; M_w=169,000 g mol^{-1}, M_w/M_n=2.09.

The solution cast method was used to prepare PCL/SAN blends with various compositions. PCL and SAN-27.5 were dissolved at 5 wt% of total polymer in 1,2-dichloroethane. The solution was cast onto a cover glass and the solvent was allowed to evaporate for 24 h at a room temperature. The dried film was annealed at 80°C (below LCST that is at around 122°C) for 1 min to achieve a single-phase state, and then it was subjected to a temperature jump to 125–180°C (>LCST) to allow isothermal SD for the desired time (3–240 min). The decomposed sample was subjected to a temperature drop to temperatures below LCST (50–115°C) where it was isothermally annealed to observe the phase dissolution.

For the optical microscopy observations, the specimen was initially annealed on first hot stage at desired temperature. The melted specimen was then placed onto a LINKAM hot stage of the microscope. Structural development during the isothermal annealing was observed under both the optical and the polarizing microscope (Olympus BH-2) equipped with a video recording system and exposure control unit (Olympus PM-20).

For the TEM analysis, the specimen was microtomed to an ultrathin section about 70 nm thick using a Reichert-Jung ultracryomicrotome with a diamond knife at -80°C and then the section was stained with RuO_4 vapor at room temperature for 2 h. The structure was observed by electron microscope, JEM 100CX (100 kV) or HITACHI 700.

Real time analyses of the structure developments above and below LCST were carried out by light scattering method. A polarized He-Ne gas laser of 632.8 nm wavelength was applied vertically to the film specimen. The scattered light passed through an analyzer and then onto a highly sensitive CCD camera with 567 x 382 pixels in a sensor of dimension 13.3 x 8.8 mm (Princeton Instruments, Inc.). This allowed the time-resolved measurement of a one-dimensional angular distribution of scattered light with 576 points in a time scale of 0.14 s. The input data from the CCD camera was digitized by a ST-13X controller. The digitized data was stored in a personal computer equipped with Dyna-100 software for further analysis.

RESULTS AND DISCUSSION

Figure 1 shows the phase diagram of PCL/SAN-27.5 blend. All specimens of PCL/SAN-27.5 blends were optically clear, and no structure was observed by the microscope at 80°C (above the melting temperature of PCL). When the specimens were heated above LCST, they became opaque, and a two-phase structure was observed.

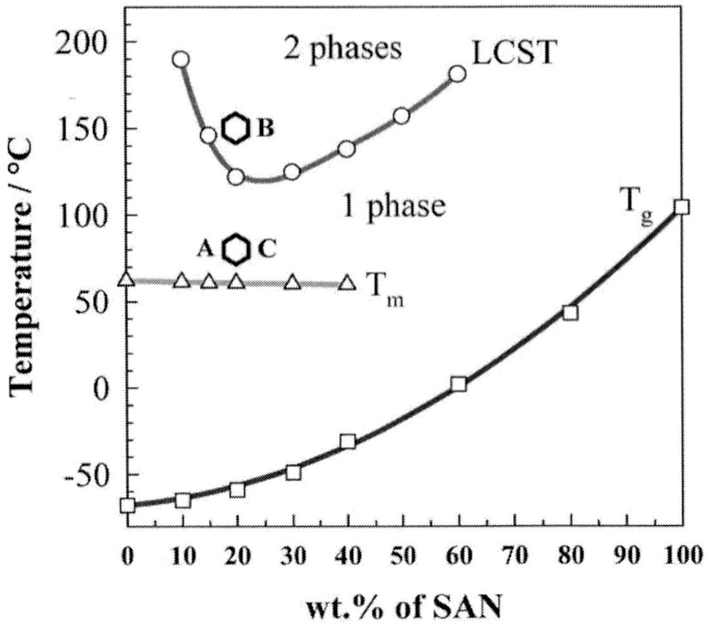

Figure 1. Phase diagram of PCL/SAN-27.5 blend.

As described in the experimental section, the blend was first annealed below LCST in one-phase region (point A). Then the sample was transferred onto another hot stage to various temperatures set above LCST (point B) and annealed isothermally there. Phase separation proceeded to various levels depending on temperature and duration of annealing at the point B. Finally, the phase-separated sample was quenched to various phase dissolution temperatures (50-115°C) below LCST (point C).

In Figure 2, the structural development in 80/20 PCL/SAN-27.5 after the temperature jump from 80 to 130°C is shown. Initially no structure was seen. After 5 min, a two-phase structure with a unique periodicity and phase connectivity was detected. The contrast of the structure became higher with the time of annealing (30 min). Then the structure became coarser and larger with time (60, 90 and 120 min).

Figure 3 shows a typical example of the change in the scattering profile with annealing (demixing) time t at 130°C. Here, the scattered light intensity I_{Vv} is shown as a function of the scattering vector q, given by [18]

$$q = \frac{4\pi}{\lambda}\sin\left(\frac{\theta}{2}\right) \qquad (1)$$

where λ is the wavelength of the light in the specimen and θ is the scattering angle. At $t = 0$ min the scattering intensity is very weak and the profile has hardly any angle dependence, suggesting a homogeneous mixture. The homogeneous mixture starts to phase separate by annealing at 130°C, as shown by the increase in the scattering intensity. The existence of a scattering peak that is moving to lower values of angle suggests that demixing proceeds by spinodal decomposition (SD) mechanism.

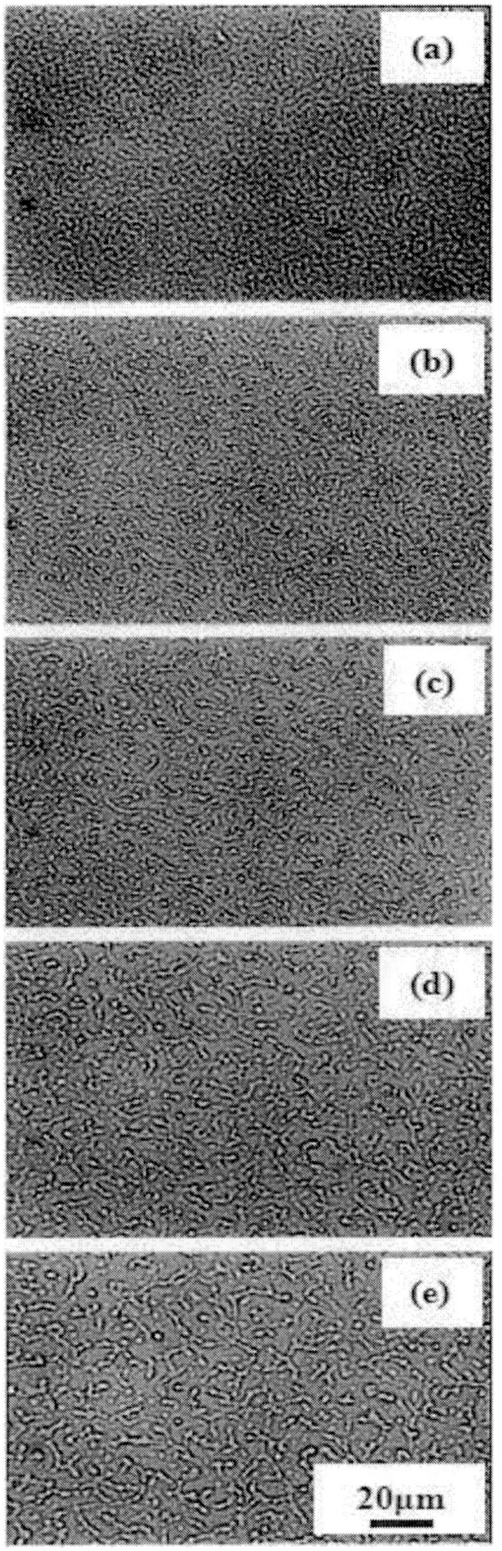

Figure 2. Optical micrographs of PCL/SAN-27.5 (80/20) blend showing phase separation at constant temperature 130°C developing in time: (a) 5 min, (b) 30 min, (c) 60 min, (d) 90 min, (e) 120 min.

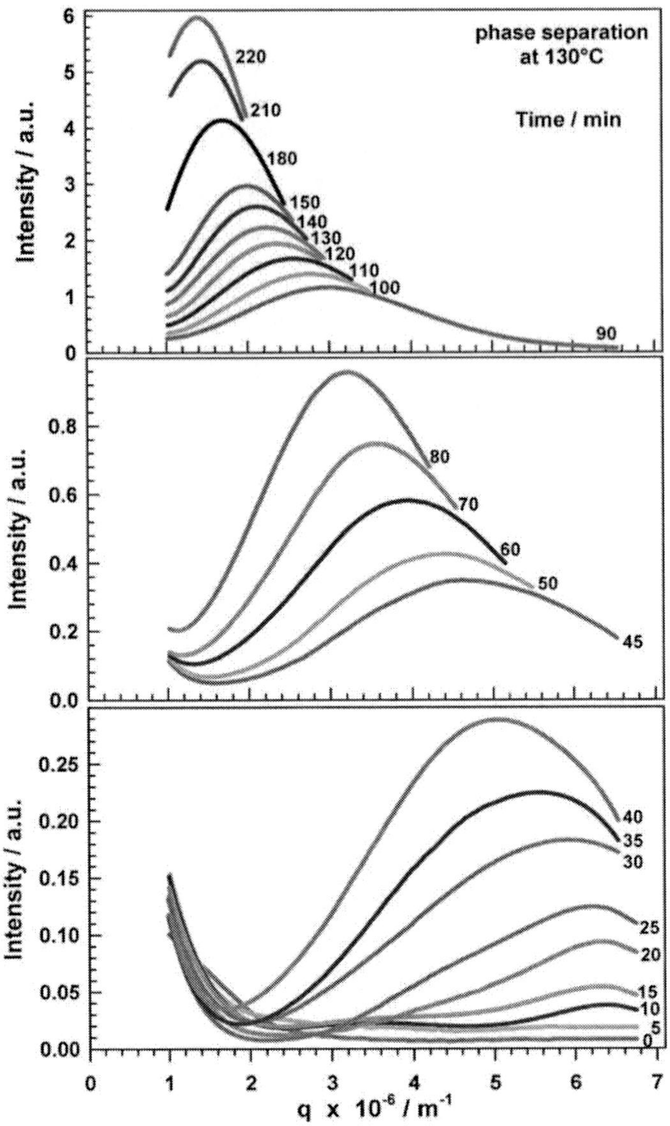

Figure 3. Change of light-scattering V_v profile of PCL/SAN-27.5 (80/20) blend during isothermal phase separation at 130°C.

The SD process in *near-critical* mixtures can be divided into (i) early stage, (ii) intermediate stage, and (iii) late stage [31]. In the early stage the time evolution of the concentration fluctuations can be approximated by the linearized theory, first proposed by Cahn [32]. The characteristic wavenumber $q_m(t,T)$, which is related to the domain spacing $\Lambda_m(t,T)$,

$$\Lambda_m(t) = \frac{2\pi}{q_m(t)} \tag{2}$$

of the dominant mode of the concentration fluctuations at time t and phase-separation temperature T is independent of t. The intensity of the Fourier mode of the fluctuations with wavenumber q, $I(q,t,T)$, grows exponentially with the time. The intermediate stage is characterized by a decrease of $q_m(t,T)$ as well as a further increase of the intensity due to a nonlinear nature in the time evolution of the concentration fluctuations. In the late stage the two phases reach equilibrium composition and the interface between the two phases narrows to equilibrium thickness. The domains, however, $\Lambda_m(t,T)$ are still increasing in order to lower the interfacial free energy of the system [31].

In order to observe the whole process of SD it is convenient to plot the position of the maximum q_m and intensity at the maximum I_m (from Figure 3) as a function of time in a double logarithmic plot, see Figure 4. In this way one is able to divide the SD process into three stages. In the early stage the position of the maximum q_m stays almost constant. Then in the intermediate and late stages a scaling analysis is being used. Binder and Stauffer [33] proposed scaling rules, assuming that clusters aggregate by a diffusion process and coalesce into larger clusters and described the self-similar growth of the structure by the time evolution of $q_m(t,T)$ and $I_m(t,T)$ as

$$q_m(t,T) \sim t^{-\alpha} \tag{3}$$

and

$$I_m(t,T) \sim t^{\beta} \tag{4}$$

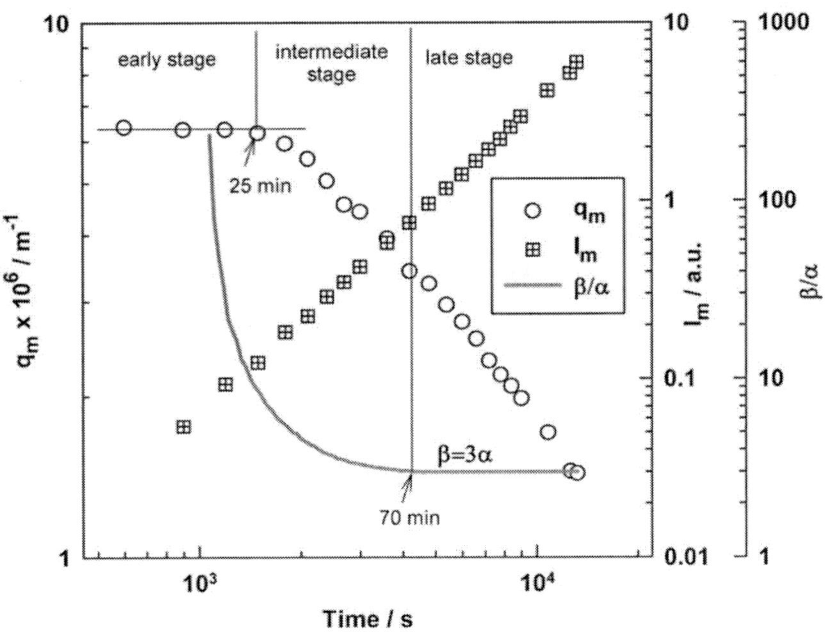

Figure 4. Time evolution of the characteristic wavenumber at maximum q_m, intensity at maximum I_m and β/α ratio for PCL/SAN-27.5 (80/20) blend during isothermal phase separation at 130°C.

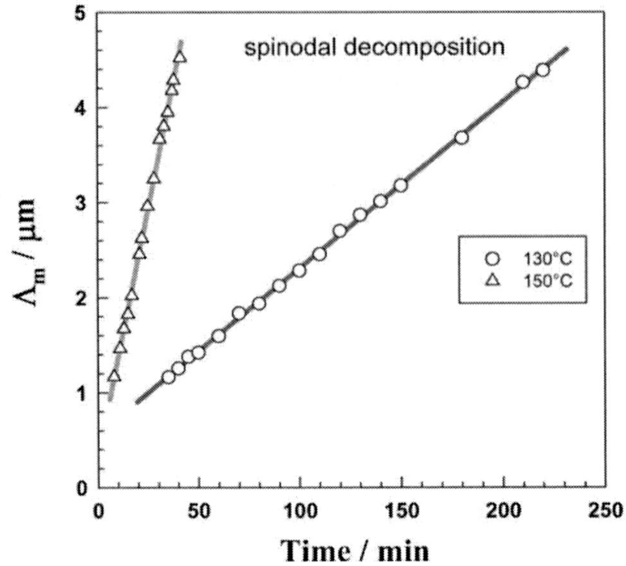

Figure 5. Wavelength of concentration fluctuations Λ_m (from V_v scattering) vs. time for PCL/SAN-27.5 (80/20) blend. Samples were heated to 80°C for 1 min (one-phase region) and then transferred to another hot stage at 130°C or 150°C (two-phase region).

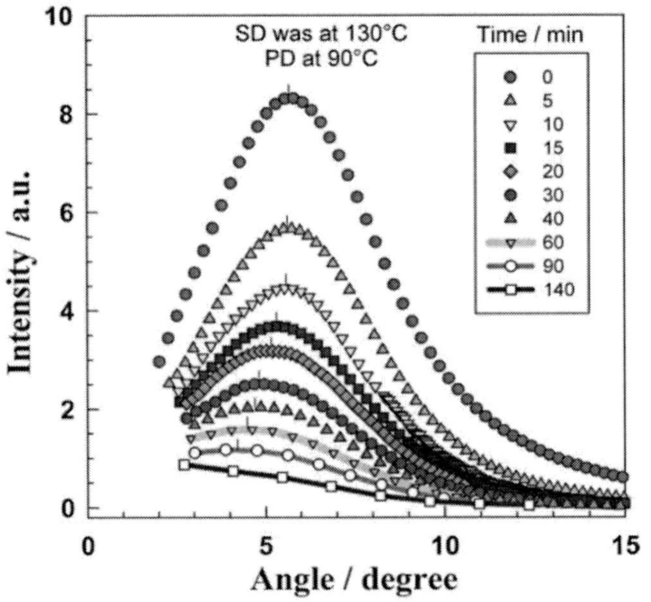

Figure 6. Change of light scattering profile of PCL/SAN-27.5 (80/20) blend during isothermal phase dissolution at 90°C after phase separation at 130°C for 220 min.

Various models and various α and β exponents were also reported, e.g. by Langer, Baron, Miller [34] or Siggia [35]. In the intermediate stage a relation β>3α is observed. The late stage is reached when β=3α [31,36-40]. In the late stage the two phases attain equilibrium compositions, and self-similar growth of the structure can be observed. Summary of the analysis presented in Figure 4:

1) In the *early stage*, there is no increase in wavelength Λ_m, (q_m stays constant) concentration fluctuations $\Delta\phi(t)$ (ϕ is volume concentration) are increasing and the slope of the line (α) is 0. The duration for our case was found 0 – 25 min. This is in agreement with observation by optical microscopy (see Figs. 2a and 2b) when there is only very small increase in the wavelength Λ_m between 5 and 30 min.
2) In the *intermediate stage*, both the amplitude of the concentration fluctuations $\Delta\phi(t)$ and wavelength $\Lambda_m(t)$ of the fluctuations grow with time, and the position of maximum q_m in Figure3 is decreasing to lower angles, the slope of the line α is gradually increasing. Because β is almost a constant, β/α is gradually decreasing until it reaches value 3. The duration was found to be 25 – 70 min.
3) In the *late stage*, where the value $\Delta\phi$ reaches an equilibrium value $\Delta\phi_e$ determined by the coexistence curve and the phase separation temperature (in Figure1), only Λ_m grows with time. The duration of the late stage was 70 – 220 min and the ratio β/α=3.

The variation of Λ_m in time for PCL/SAN-27.5 (80/20) blends at two different temperatures 130 and 150°C is plotted in Figure 5. The values of Λ_m can be used for designing the desired size of domains. And indeed we used this plot to prepare two different samples with apparently the same wavelength of the concentration fluctuations Λ_m being about 4.3 μm, the first one prepared by spinodal decomposition at 130°C for 220 min and the other at 150°C for 38 min. These two supposedly same samples were then subjected to phase dissolution (PD) at 90°C.

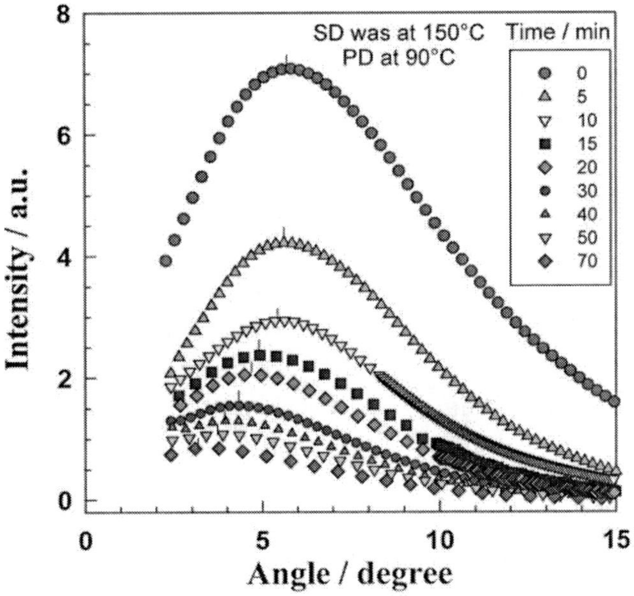

Figure 7. Change of light scattering profile of PCL/SAN-27.5 (80/20) blend during isothermal phase dissolution at 90°C after phase separation at 150°C for 38 min.

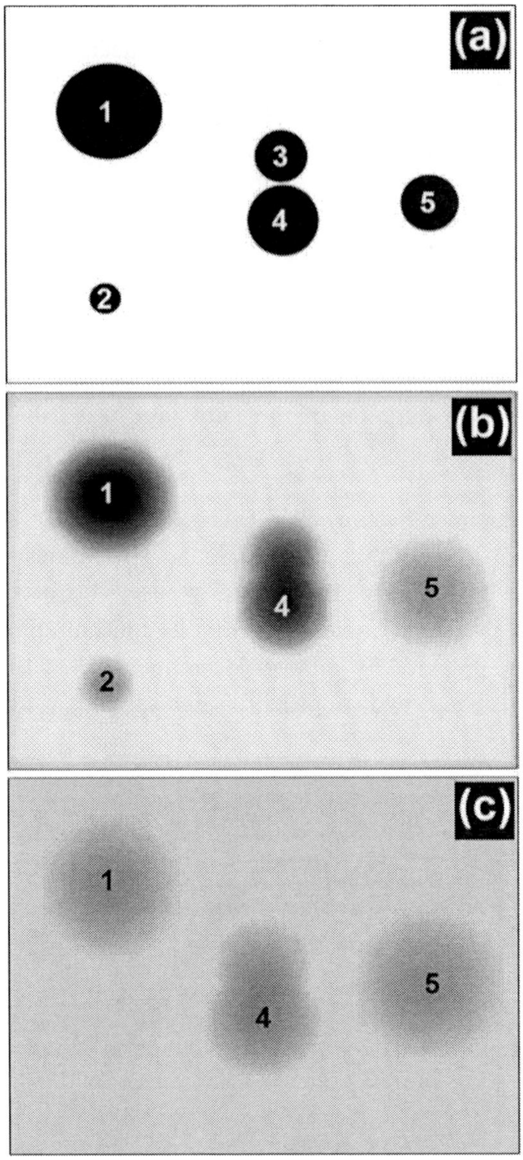

Figure 8. Model of phase dissolution development in time derived from light scattering data shown in Figures 6 and 7.

The results are shown in Figures 6 and 7. The light scattering machine was set to measure precisely the intensity at smaller angles (compared to Figure 3). At first please note that the initial peak at 0 min is a little broader in the case of the sample phase separated at 150°C compared to that separated at 130°C. This indicates quite narrow particle size distribution of the phase-separated structure created at 130°C and a coexistence of larger and smaller particles created at 150°C.

In both cases, the light scattering peak is decreasing and the position of the maximum is shifting towards the lower angle. There are several possible interpretations of the shift of the peak maximum in Figs. 6 and 7, one of which is the faster decay of the high-q Fourier components in the concentration pattern. Only the slower components with long wavelengths (small q) survive at longer times. This results in an effective peak shift towards small q (small angles) during the dissolution process.

This is schematically shown in Figure 8, particles 1 and 2. While the large particle no. 1 survives up to Figure 8c particle no. 2 is in Figure 8c already dissolved. Light scattering at late stage of phase dissolution detects only the remaining particle no. 1. Another possible interpretation of the peak shift towards smaller angle is coalescence of close particles no. 3 and 4. The actual growth of the particles during phase dissolution is possible too, as it was observed directly by TEM by Okada [22,23] and calculated by using a model by Cheng [26]. This is schematically shown by particle no. 5. In our particular PCL/SAN (80/20) system the particles are formed by SAN-rich phase. SAN has rather high T_g (about 103°C) while for PCL T_g=–66°C and T_m≅60°C. Also molecular weight of SAN (M_w=169,000 g mol^{-1}) is higher compared to PCL (M_w =40,400 g mol^{-1}). Above mentioned facts might contribute to much higher mobility of PCL compared to SAN at phase dissolution temperature (90°C). Thus it is possible that PCL molecules diffuse into SAN particle faster than SAN molecules are leaving. This would result in swelling of SAN particles.

Concerning *phase separation*: the existence of scattering peak that is moving to lower values of scattering angle (already shown Figure 3) indicates that the demixing proceeds by the spinodal decomposition (SD) mechanism. To confirm this point, we analyzed the early stage on the basis of the linearized Cahn-Hilliard theory [32, 41].

Cahn in 1965 [32] laid a basis for mathematical treatment of spinodal decomposition. He started with expression for free energy of inhomogeneous solution whose composition everywhere differs only slightly from the average composition, and with small composition gradients.

$$F = \int [f(c) + \kappa(\nabla c)^2] dV \qquad (5)$$

Here *f(c)* is the free-energy density of homogeneous material of composition *c*, $\kappa(\nabla c)^2$ is the additional free-energy density if the material is in a gradient in composition. In the development of the theory it became convenient to consider the Fourier components of the composition rather than the composition. Because of the orthogonality of the Fourier components, ΔF is the sum of contributions from each Fourier component separately.

The kinetics of the initial stages of phase separation can be obtained by solving the diffusion equation. We define a mobility *M* which is (minus) the ratio of diffusional flux to the gradient in chemical potential

$$J_B = -J_A = M\nabla(\mu_A - \mu_B) \qquad (6)$$

Simple thermodynamic considerations show that *M* must be positive if diffusion which results spontaneously from the chemical potential gradient is to result in a decrease in free energy. We obtain μ_A-μ_B from the variational derivative of *F* in Eq. (5),

$$\mu_A - \mu_B = \left(\frac{\delta F}{\delta c_A}\right) = \left(\frac{\partial f}{\partial c_A}\right) - 2\kappa\nabla^2 c_A + \textit{ terms higher in order and degree} \quad (7)$$

We may discard the higher order terms if we are only interested in the initial stages. By substituting Eq. (7) into Eq. (6) and taking the divergence, and keeping only first degree terms, we obtain the following diffusion equation:

$$\frac{\partial c}{\partial t} = M\left(\frac{\partial^2 f}{\partial c^2}\right)\nabla^2 c - 2M\kappa\nabla^4 c \quad (8)$$

Since the coefficient of ∇^2c may be identified with a diffusion coefficient we see that at the spinodal the diffusion coefficient changes sign. The solution to Eq. (8) is

$$c - c_0 = e^{[R(\beta)t]}\cos(\beta\,\mathbf{r}) \quad (9)$$

where β is wavenumber and R(β) is given by

$$R(\beta) = -M\left(\frac{\partial^2 f}{\partial c^2}\right)\beta^2 - 2M\kappa\beta^4 \quad (10)$$

Cahn has developed his model for small molecules in metallurgy. His equations were slightly modified for polymer blends. These days we are using *g* instead of *f* and *q* instead of β. Light scattering became popular in quantitative evaluation of spinodal decomposition in polymer blends. Here, in the initial stage of phase separation (or phase dissolution), the intensity of scattered light (*I*) is expected to increase (or decrease) exponentially with time *t*.

$$I_{Vv}(q,t) \propto e^{2R(q)t} \quad (11)$$

where *R(q)* is the growth rate (or decay rate) of concentration fluctuation at the scattering vector *q*. Today *R(q)* is given by (essentially the same as Eq. 10 in 1965)

$$R(q) = -Mq^2\left(\frac{\partial^2 g}{\partial c^2} + 2\kappa q^2\right) \quad (12)$$

where *M* is the mobility, *g* is the free-energy density of the blend with composition *c*, and *κ* is the gradient energy coefficient.

Cahn started with describing the Gibbs free energy of fluctuating system by a sum of the local free energy of mixing and the square of concentration gradient term. The gradient term remains in Eq.12 as the second term. The first term of Eq.12 is the second derivative of g which is the thermodynamic driving force for fluctuation growth (i.e., the degree of instability of homogeneously mixed state). The second term of Eq.12 suggests that a very sharp gradient with large q (short wavelength of fluctuation) suppresses the fluctuation growth. In contrast, the fluctuation with small q (long wavelength) grows slowly, because the molecules should diffuse long distance for the fluctuation growth. Then, there should exist a peak of rate constant $R(q)$ at a particular q.

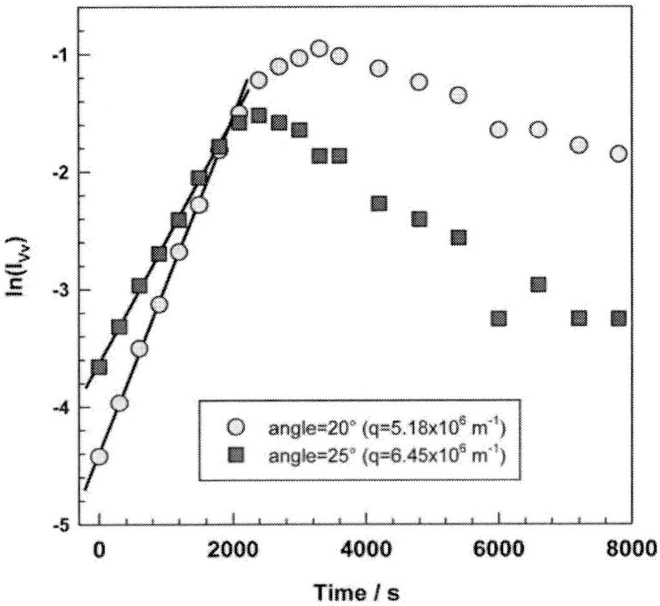

Figure 9. Change of ln(I_{Vv}) in time during spinodal decomposition at 130°C (from light scattering). Evaluation of slopes for various angles.

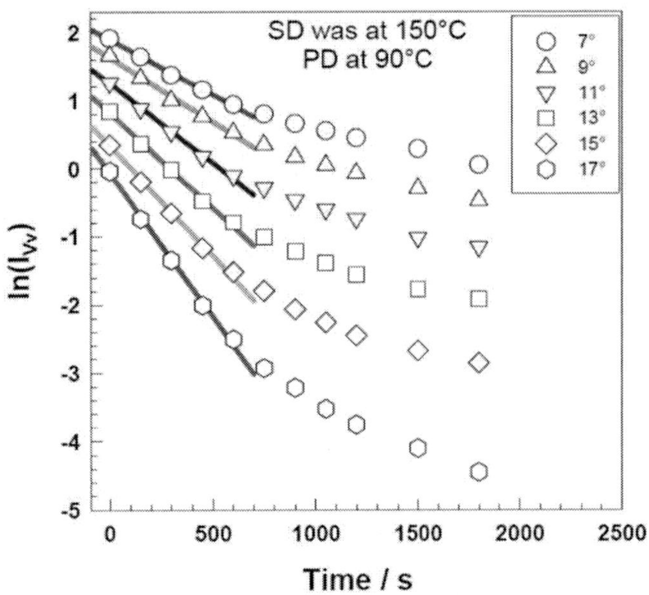

Figure 10. Change of ln(I_{Vv}) in time during phase dissolution at 90°C (from light scattering). Evaluation of slopes for various angles.

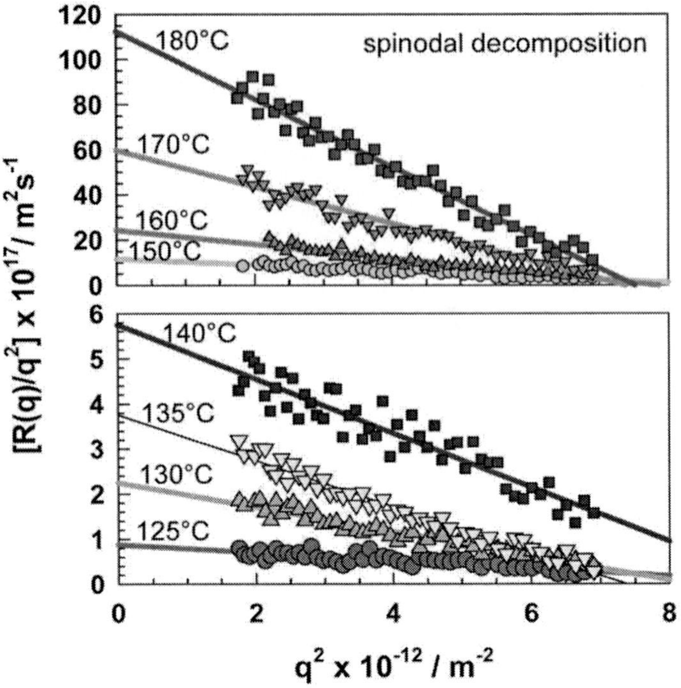

Figure 11. Plot of $R(q)/q^2$ vs. q^2 of PCL/SAN-27.5 (80/20) blend: spinodal decomposition at 125-180°C. *R(q)* values were obtained by analysis of slopes shown by Figure 10.

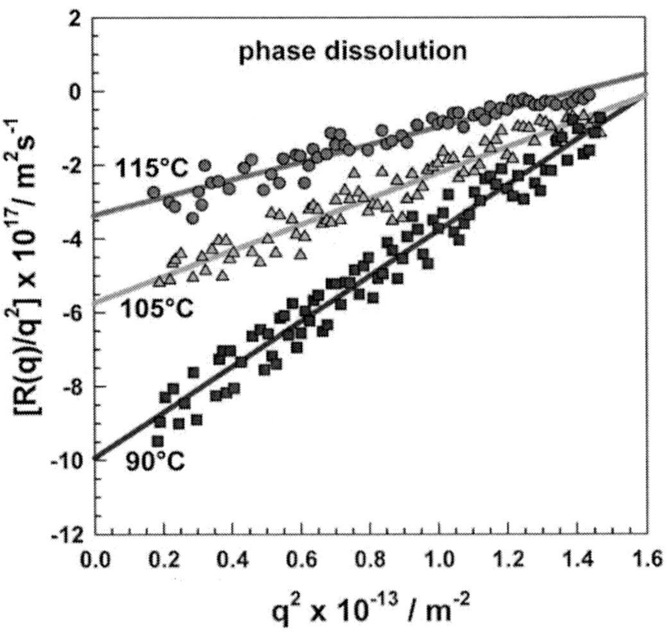

Figure 12. Plot of $R(q)/q^2$ vs. q^2 of PCL/SAN-27.5 (80/20) blend: phase dissolution at 90, 105 and 115°C after previous spinodal decomposition at 150°C for 15 min. *R(q)* values were obtained by analysis of slopes shown in Figures 11 and 12.

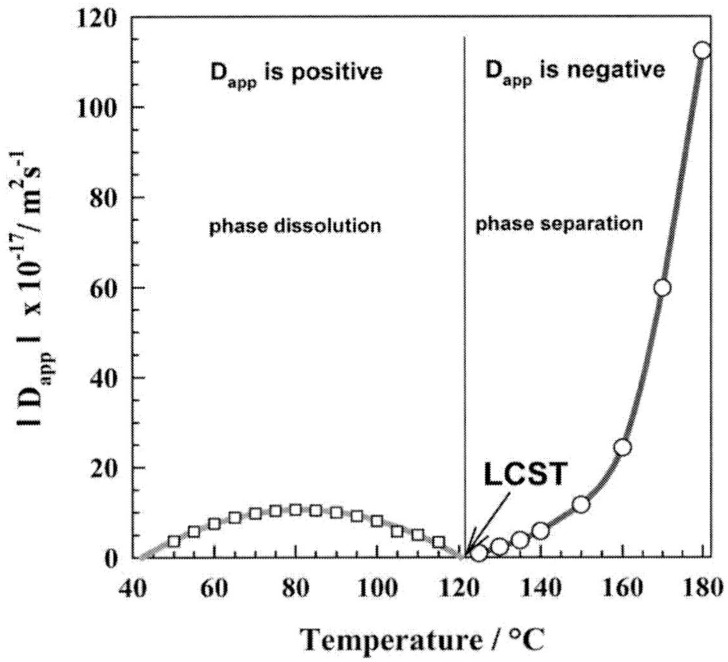

Figure 13. Apparent diffusion coefficient for phase dissolution (squares) and phase separation (circles) as a function of annealing temperature. Lower critical solution temperature (LCST) is shown by arrow.

As it was mentioned earlier and described by Eq. 11 in the early stage of spinodal decomposition the intensity should increase exponentially with time. This was indeed observed and is shown in Figure 9. We used natural logarithm of intensity and therefore the plot yields straight lines in the initial stage. Then one can estimate the growth (or decay) rate $R(q)$ of concentration fluctuations from the initial slope of a $\ln(I_{Vv})$ vs. t plot (in Figure 9 for phase separation and in Figure 10 for phase dissolution) for different angles (or q values).

After such detailed analysis, it is possible to calculate the $R(q)/q^2$ vs. q^2 plot as it is shown in Figure 11 for SD and Figure 12 for PD for various temperatures. The apparent diffusion coefficient D_{app} is defined by

$$D_{app} = -M\left(\frac{\partial^2 g}{\partial c^2}\right) \tag{13}$$

and can be obtained from $R(q)/q^2$ vs. q^2 plot as the intercept of the $R(q)/q^2$ axis in Figures 11 and 12.

The apparent diffusion coefficient D_{app} obtained in this way was positive for phase dissolution and negative for phase separation. It is interesting to compare the absolute values of D_{app} for SD and PD in one plot, see Figure 13.

The D_{app} for phase dissolution was found to have values between 3 and 11 x 10^{-17} m^2s^{-1} in the temperature range 50-115°C. Closer to LCST the values are decreasing and in range 50-65°C they are low as well. The diffusion should be slower at lower temperature due to a lower mobility. In our case, also the presence of a virtual upper critical solution temperature (UCST) around 40°C, found earlier for similar systems [42], should cause the decrease in

D_{app}. Comparable values of D_{app} for phase separation (just with opposite sign – negative values) were found in the temperature range 125-150°C. By a further increase of the temperature the spinodal decomposition proceeded much faster, an example being value of D_{app} at 180°C which is more than 10 times higher than the fastest phase dissolution. From the exponential-like curve one could assume that this difference would be probably even much higher (such as 100x or even 1000x) if we increased the temperature to e.g. 200 or 220°C.

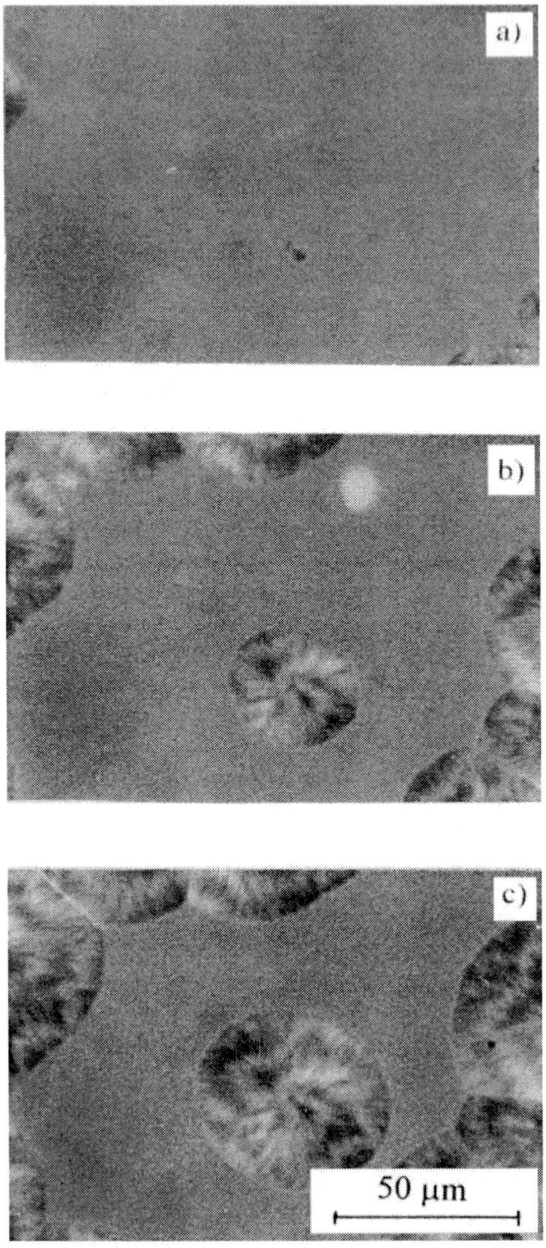

Figure 14. Optical micrographs of PCL/SAN-27.5 (80/20) blend that was heated in the one-phase region (100° for 1 min) and then quenched to the crystallization temperature (40°C). Spherulite growth times (a) 14 min, (b) 26 min, (c) 34 min.

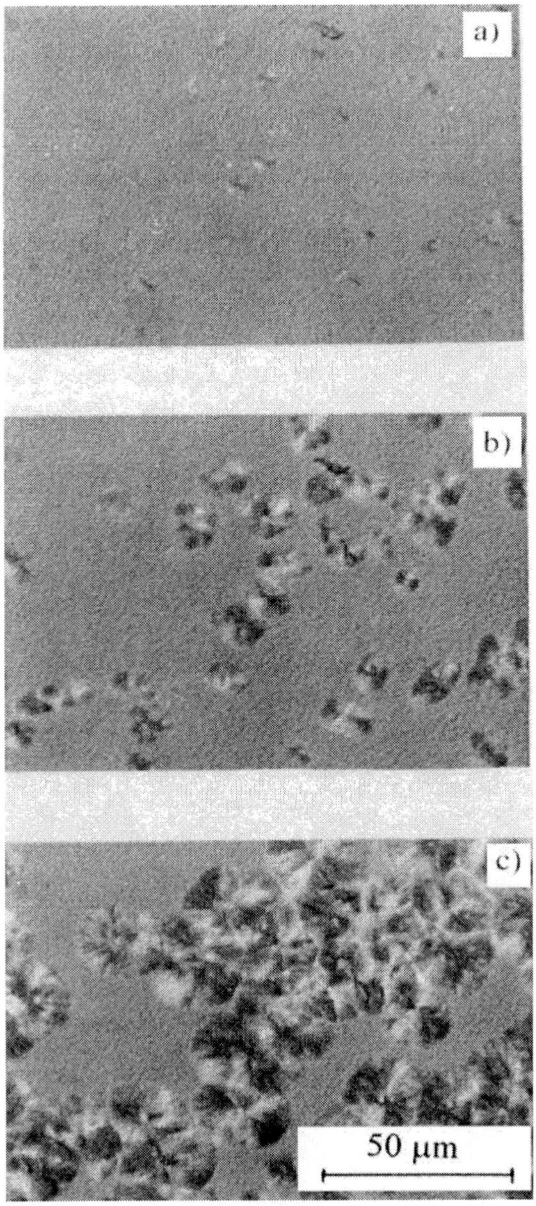

Figure 15. Optical micrographs of PCL/SAN-27.5 (80/20) blend that was heated in the one phase region (100°C for 1 min), then phase-separated 40 min at 150°C and finally crystallized at 40°C. Time of development: (a) 2 min, (b) 4 min, (c) 8 min.

Crystallization Kinetics by Optical Microscopy

The specimens were forced to undergo demixing into different levels, that was quantitatively evaluated using the V_v light scattering technique described above; then they were crystallized by quenching to a temperature below T_m. The starting point of the crystallization kinetics analysis was crystallization of PCL/SAN-27.5 (80/20) from the homogeneous state (one phase) attained by annealing the sample at 100°C for 1 min. After this annealing, the sample was transferred onto another hot stage at 40°C, where the time development of spherulites was observed by optical microscopy. An example of growth of spherulites in time is given in Figure 14. The radius of the spherulites vs. time of crystallization for this case is plotted in Figure 16. As the size and shape of spherulites slightly differ, about three spherulites were usually averaged. As can be seen, a linear dependence was always obtained, so the slope of the line is equivalent to the spherulite growth rate G.

In addition the samples were forced to undergo SD to different levels, represented by different times of SD at 150° and then quenched to crystallization temperature of 40°C. An example of growth of spherulites from these phase-separated structures is given in Figure 15. The first observation leads to the conclusion that after SD the number of small spherulites is higher, implying more nucleation centers. The size of the spherulites is clearly larger than the size of the domains. This means that crystallization proceeds through the PCL-rich matrix somehow ignoring the obstacles represented in this case by SAN-rich domains. Quantitative and more detailed conclusions are difficult to obtain using just optical microscopy. Therefore, light scattering and TEM techniques were used for this purpose, as will be described later. The variations in the radius of the spherulites are plotted in Figure 16. Again the growth is linear with time. Apparently, the growth rate is accelerated by SD. The further SD proceeds, the faster is the crystallization. Even though crystallization does not proceed in SAN-rich domains, the crystallization is not stopped at the SAN-rich borders because of interconnectivity of the PCL-rich phase. It proceeds in the PCL-rich phase. As shown in Figure 17, crystallization kinetics is accelerated in the early and middle stage of SD. The increase of the G may be ascribed to the increase in the concentration of PCL in PCL-rich region with the demixing time of SD of the evolution of concentration fluctuations. On the other hand, G is almost constant with demixing time in the late stage of SD. This is due to no further evolution of the concentration fluctuations in the late stage of SD (only Λ_m increases with demixing time).

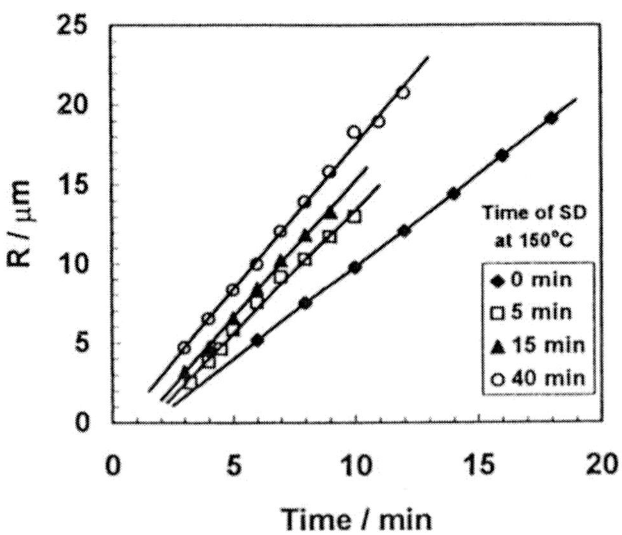

Figure 16. Variation of spherulite radius R with crystallization time for PCL/SAN-27.5 (80/20) at 40°C by optical microscopy. Comparison is made for sample crystallized from the one-phase region (0 min) with samples that have undergone different times of spinodal decomposition at 150°C.

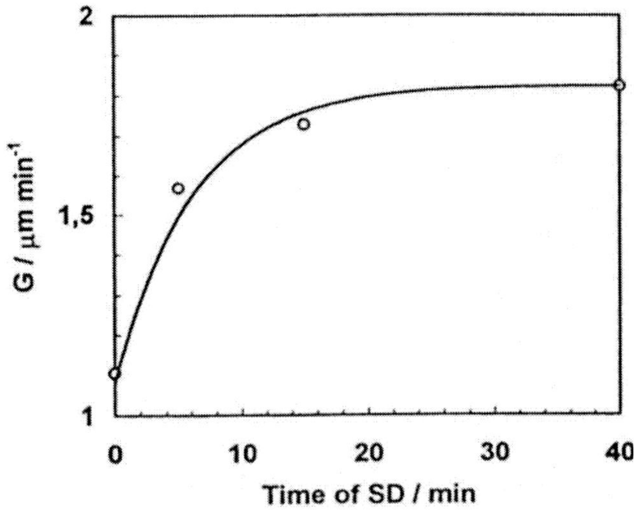

Figure 17. Growth rate of spherulites, G, at 40°C vs. time of spinodal decomposition at 150°C.

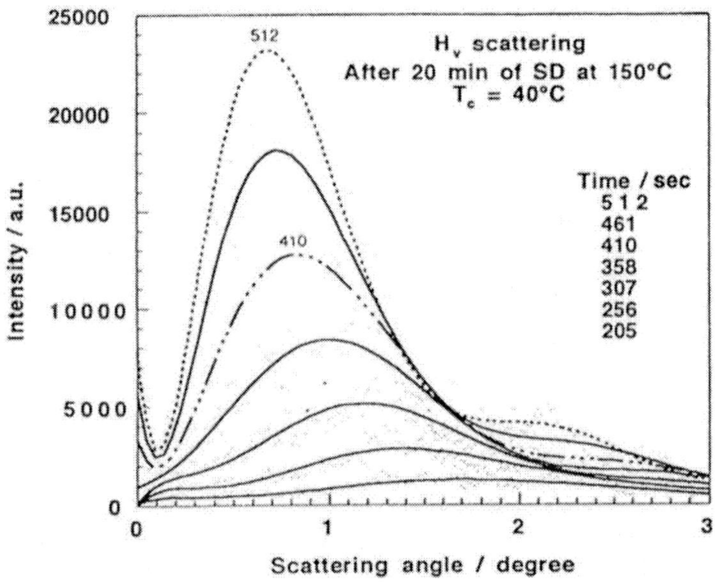

Figure 18. Change of H_v scattering profiles with time after quenching the PCL/SAN-27.5 (80/20) blend to 40°C after 20 min of spinodal decomposition at 150°C, azimuthal angle μ=45°.

An interesting phenomenon is also the ordering of the PCL lamellae inside the spherulites. A deeper insight into this problem is discussed in the next two sections.

Crystallization by H_v Light Scattering

The H_v light scattering is a powerful technique for investigation of crystallization in polymer blends. Compared to optical microscopy, where always has to choose from limited number of spherulites, H_v light scattering gives information about the whole system.

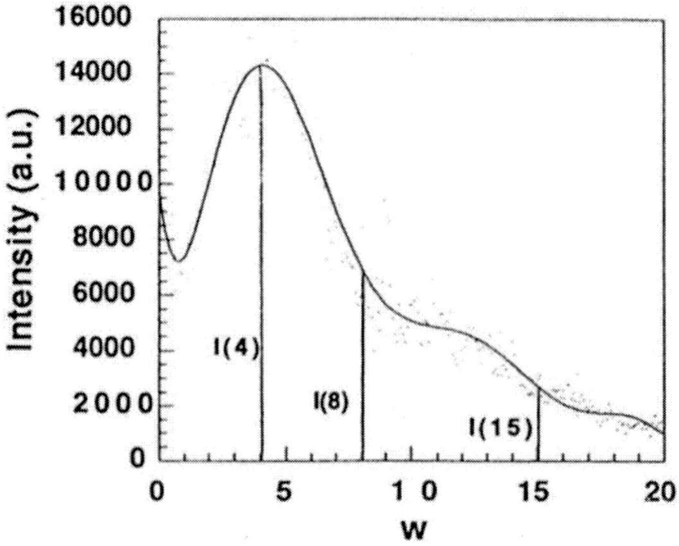

Figure 19. Schematic picture of determination of order parameter I(4)/I(8) and I(4)/I(15).

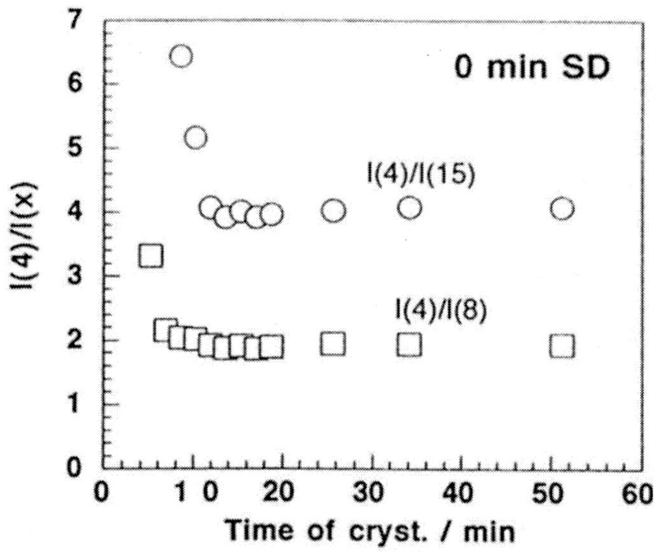

Figure 20. Order parameter vs. time of crystallization at 40°C for (80/20) blend quenched from homogeneous state (100°C for 1 min).

For this study a crystallization temperature of 40°C was chosen. The observed scattering patterns were of the four-leaf-clover type. The clover patterns became smaller with time of crystallization. Figure 18 shows the one dimensional H_v scattering profiles at an azimuthal angle of 45° for a sample crystallized after 20 min of SD. The average radius of the spherulites R_{Hv} can be calculated from the position of the maximum in scattering intensity, θ_m, by using this equation:

$$4.09 = 4\pi(R_{Hv}/\lambda)sin(\theta_m/2) \quad (14)$$

In order to discuss the time variation of the internal spherulitic structure, it is convenient to employ the scattering profiles at an azimuthal angle 45° using a reduced scattering angle w.

$$w = \left(\frac{2\pi}{\lambda}\right) R_{Hv} sin\theta \quad (15)$$

Using w, the scattering profiles are corrected to the reduced profiles with the peak maximum at w=4 as shown in Figure 19. The sharpness of the peaks is different for the samples with different time of SD. Increased radial disorder leads to increased relative I_{Hv} scattered intensity at angles larger than that of the maximum. To compare quantitatively the order inside the spherulite, it is convenient to define a so-called order parameter (Pr).

$$Pr = \frac{I(4)}{I(x)} \quad (16)$$

where I(4) represents intensity at w=4 and I(x) is the intensity at w=x. In the literature two values of x were used: 8 and 15. The situation is schematically shown in Figure 19. How the order parameter is changing during crystallization is shown in Figure 20. At the beginning of crystallization the ordering inside the spherulite is higher and then it decreases and remains at an almost constant level. This behavior was observed for the sample crystallized from the homogeneous phase and also for samples crystallized after different times of SD. This phenomenon can be explained in this way. At the beginning of the crystallization, lamellae of PCL grow from the center of the spherulite in the radial direction, and the order parameter is high. After a short time branching starts in the regions where the concentration of PCL is lower. This branching decreases the order parameter. Then the radial growth and the branching reach an equilibrium, and the order parameter becomes constant.

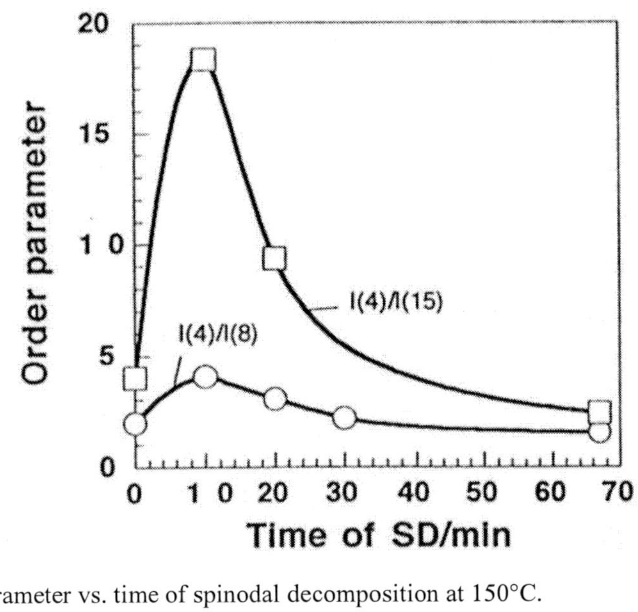

Figure 21. Order parameter vs. time of spinodal decomposition at 150°C.

This constant value was used to compare samples crystallized from different stages of SD, as shown in Figure 21.

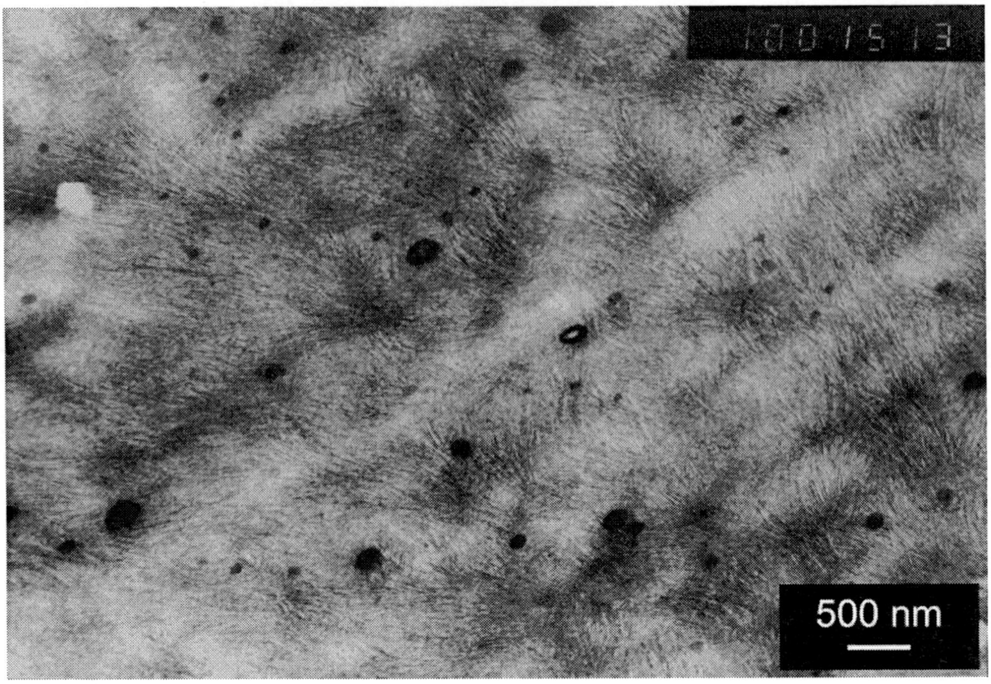

Figure 22. TEM micrograph of PCL/SAN-27.5 (80/20) sample crystallized at 40°C for 14 h after this history: 80°C for 1 min.

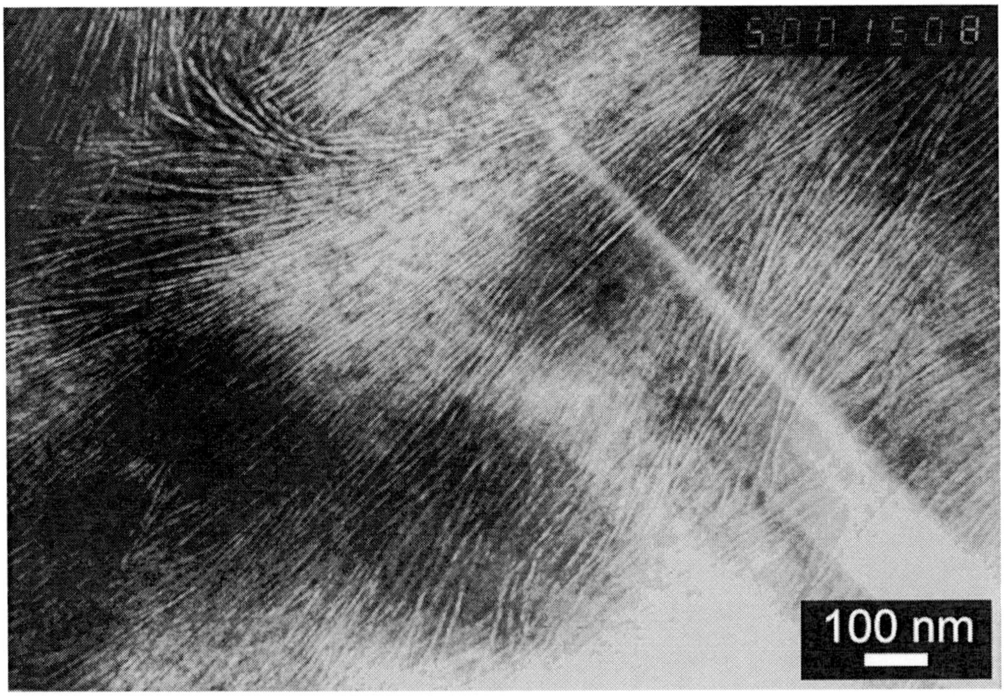

Figure 23. TEM micrograph of PCL/SAN-27.5 (80/20) sample crystallized at 40°C for 14 h after this history: 100°C for 1 min, 150°C for 5 min.

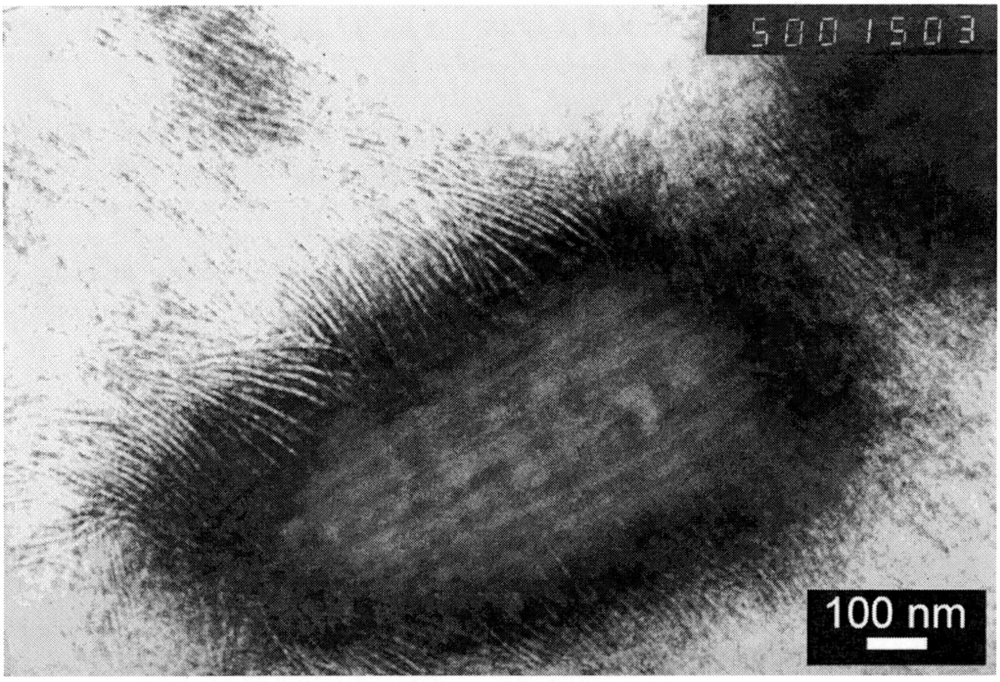

Figure 24. TEM micrograph of PCL/SAN-27.5 (80/20) sample crystallized at 40°C for 14 h after this history: 100°C for 1 min, 150°C for 15 min.

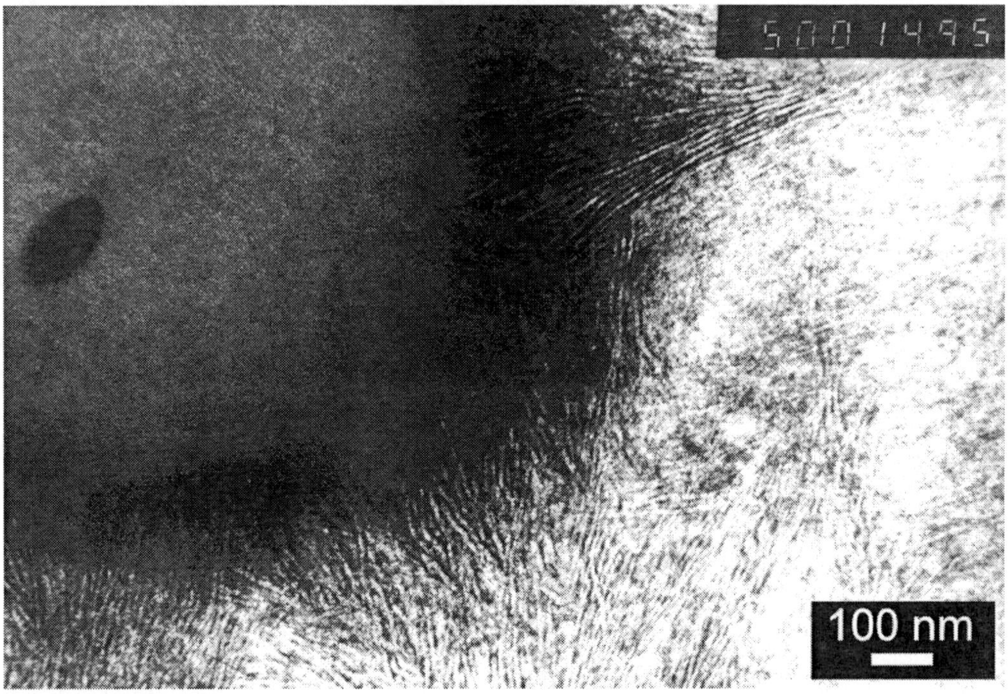

Figure 25. TEM micrograph of PCL/SAN-27.5 (80/20) sample crystallized at 40°C for 14 h after this history: 100°C for 1 min, 150°C for 40 min.

In this graph, two order parameters [I(4)/I(8) and I(4)/I(15)] are plotted as a function of time of SD at 150°C. All samples were crystallized at constant temperature 40°C. Regardless of which order parameter is considered, an interesting phenomenon appeared. At the beginning the order parameter increases (about 10 min) and then slowly decreases with increasing time of SD. By the V_v light scattering analysis at 150°C one can find a change from the middle to late stages of SD at about 10 min. Then from Figure 21, it is apparent that ordering of the spherulites increases for the samples crystallized after SD in the early and middle stages. Coming to the late stage, the ordering parameter decreases continually with increasing time of SD. In order to understand this phenomenon fully, it was necessary to make a TEM analysis.

TEM Analysis

For the transmission electron microscopy (TEM) analysis, the following samples were chosen. All the samples have the same blend composition PCL/SAN-27.5 (80/20) and were crystallized at 40°C for 14 h after different prior history. The first sample (Figure 22) was crystallized from the homogeneous phase that was attained by annealing the sample at 80°C for 1 min. The lamellae are short and oriented in many directions, differing from the radial growth of spherulites. It looks like the SAN is partly segregated from the growing front and partly remains in an amorphous mixture with PCL between the PCL lamellae. As RuO_4 stains benzene rings, that is, SAN, the broad dark region might be the areas where the concentration of SAN is increased due to segregation during crystallization. Many times the lamellae end in these areas and start to grow in new directions. This could explain a low order parameter. Three additional samples with different degrees of phase separation (Figure 23 - 5 min, Figure 24 - 15 min, Figure 25 - 40 min) were also examined. A clear two phase system can be seen in all three cases. Figure 23 shows a micrograph for 5 min of SD (middle stage). The lamellae are highly oriented in one direction. This corresponds with a high order parameter. An interesting phenomenon in this case is that lamellae grow through the SAN-rich areas (dark) without change of direction. The number of lamellae in these areas is, as expected, smaller because of the smaller concentration of PCL in these SAN-rich areas. The next micrograph (Figure 24) shows the system phase separated for 15 min before crystallization (late stage). The concentration inside the SAN-rich domains has already reached the equilibrium value. The difference is also in the size of SAN-rich domains that are in this case larger. As the concentration of PCL inside the SAN-rich domains crossed the critical level that would allow crystallization, the lamellae are not able to penetrate these domains. This is probably caused by the mobility effect. The T_g of these domains is probably higher than the crystallization temperature and the mixture of PCL and SAN is frozen inside these domains. The alignment of lamellae in PCL-rich matrix is rather high, which corresponds with a high order parameter. The growth of spherulite continues through the bridges between SAN-rich domains. The last sample (Figure 25) had undergone 40 min of SD (again late stage). The size of the domains had gradually increased. The difference compared with Figure 24 is that the alignment of lamellae is much worse; lamellae grow in all directions. This corresponds with a lower order parameter. In both cases (Figure 24 and 25), the surface of the SAN-rich domain is darker, suggesting that in this interface the concentration of SAN is higher than inside the domain. This could be explained by crystallization of almost all the PCL from this area while

SAN remains. Inside the domains the concentration of amorphous PCL is higher than at the surface of domains.

In SAN-rich phase, the crystallization growth rate is much slower (or the crystallization can be prevented because of the mobility effect) than in the PCL-rich phase. This is due to increasing T_g of the blend when the content of SAN is increasing (T_g of pure PCL is about -60° and T_g of SAN is about 105°C). However, the PCL-rich phase is highly interconnected, so lamellae are growing in this phase much faster, somehow ignoring the slow SAN-rich domains. At this blend composition (PCL/SAN 80/20), the growth of large spherulites is not prevented by spinodal decomposition. By the formation of the PCL-rich phase during spinodal decomposition, the concentration of SAN in this phase is gradually decreasing down to an equilibrium value; thus the T_g of this phase is decreasing, which finally results in a higher crystallization rate.

CONCLUSION

PCL/SAN-27.5 blend gave us an excellent opportunity to investigate the phase dissolution kinetics and to add some experimental data (that are still quite rare) to this subject. Firstly, we focused on the phase separation above LCST (being at 122°C), and studied it qualitatively by optical microscopy (at 130°C), and then quantitatively by light scattering at various temperatures (125-180°C). The spinodal decomposition mechanism of phase separation was confirmed by the detailed data analysis based on Cahn-Hilliard theory. Secondly, the phase separated structure was quenched to various temperatures below LCST (50-115°C) to study the phase dissolution. The phase dissolution was affected by the temperature of previously performed spinodal decomposition. Based on light scattering data, a model of phase dissolution mechanism was created. There are several possible interpretations of the shift of the peak maximum during phase dissolution, one of which is the faster decay of the high-q Fourier components in the concentration pattern. Only the slower components with long wavelengths (small q) survive at longer times. This results in an effective peak shift towards small q (small angles) during the dissolution process. Again, for the PD, we were able to use linearized theory to obtain an apparent diffusion coefficient D_{app}. Finally, the D_{app} values were compared for SD and PD with the result showing that the fastest PD was about 10 times slower than the fastest SD in measured temperature range.

By the use of optical microscopy, the crystallization kinetics was observed for samples crystallized after decomposition into different stages of SD. A major acceleration of crystallization kinetics was observed for the samples that had undergone SD in the early and middle stages. This can be explained by a gradual change of concentration in the PCL-rich domains. However, in the late stage, the concentration reaches an equilibrium value and the crystallization kinetics reach almost constant level.

The H_v light scattering technique enabled a quantitative evaluation of the ordering of PCL lamellae inside the spherulites which, in our case, were always bigger than the size of the phase-separated domains. The result of this analysis showed that the radial ordering of lamellae is increasing for samples crystallized after SD to the early and middle stages and then gradually decrease for samples separated to the late stage of SD.

TEM analysis explained rather surprising results of the previous analysis. In the sample crystallized from the homogeneous phase, the ordering of lamellae is rather poor. During crystallization SAN is probably partly segregated. When the amount of segregated SAN exceeds a certain level, lamellae start to grow in new directions. For the samples crystallized after separation to the middle stage of SD, the ordering of lamellae is much higher. Lamellae grow through the SAN-rich areas without changing direction. For the samples crystallized after SD to the late stage, the lamellae of PCL cannot penetrate the SAN-rich domains. Crystallization stops at the borders of the SAN rich domains. As the PCL-rich phase is interconnected, crystallization can proceed through bridges between SAN-rich domains. Then the growth of lamellae continues by branching, which, of course, decreases the radial ordering of lamellae inside the spherulite. The more decomposition proceeds, the less ordered are the lamellae.

Review

This work was reviewed by Professor Jörg Kressler from University Halle Wittenberg, Fachbereich Chemie, Institut für Physikalische Chemie, D-06099 Halle (Saale), Germany.

Acknowledgment

This work has been supported by the Ministry of Education of the Czech Republic as a part of the project No. VZ MSM 7088352102 and also with support of Operational Programme Research and Development for Innovations co-funded by the European Regional Development Fund (ERDF) and national budget of Czech Republic within the framework of the Centre of Polymer Systems project (reg.number: CZ.1.05/2.1.00/03.0111).

References

[1] Ratner BD, Hoffman AS, Schoen FJ, Lemons EJ. Biomaterials Science. New York: Elsevier Academic Press, 2004.

[2] Madbouly SA, Abdou NY, Mansour AA. Isothermal crystallization kinetics of poly(ε-caprolactone) with tetramethyl polycarbonate and poly(styrene-co-acrylonitrile) blends using broadband dielectric spectroscopy. *Macromolecular Chemistry and Physics* 2006;207(11):978-986.

[3] Kiersnowski A, Piglowski J. Polymer-layered silicate nanocomposites based on poly(ε-caprolactone). *European Polymer Journal* 2004;40(6):1199-1207.

[4] Men Y, Strobl G. Critical strains in poly(ε-caprolactone) and blends with poly(vinyl methyl ether) and poly(styrene-co-acrylonitrile). *Macromolecules* 2003;36(6):1889-1898.

[5] Aoki Y, Arendt O Melt rheology of poly(ε–caprolactone)/ poly(styrene-co-acrylonitrile) blends. *Journal of Applied Polymer Science* 2001;82(8):2037-2041.

[6] Kim SW, Jo WH, Lee MS, Ko MB, Jho JY. Preparation of clay-dispersed poly(styrene-

co-acrylonitrile) nanocomposites using poly(ε-caprolactone) as a compatibilizer. *Polymer* 2001;42(24):9837-9842.

[7] Hong Z, Shaw MT, Weiss RA. Structure evolution and phase behavior of polymer blends under the influence of shear. *Polymer* 2000;41(15):5895-5902.

[8] Keroack D, Zhao Y, Prud'homme RE. Molecular orientation in crystalline miscible blends. *Polymer* 1999;40(1):243-251.

[9] Wang Z, Wang X, Yu D, Jiang B. The formation of ring-banded spherulites of poly(ε-caprolactone) in its miscible mixtures with poly(styrene-co-acrylonitrile). *Polymer* 1997;38(23):5897-5901.

[10] Oudhuis AACM, Thiewes HJ, van Hutten PF, ten Brinke G. A Comparison between the morphology of semicrystalline polymer blends of poly(ε-caprolactone)/poly(vinyl methyl ether) and poly(ε-caprolactone)/(styrene-acrylonitrile). *Polymer* 1994; 35(18): 3936-3942.

[11] Li W, Prud'homme RE. Orientation and miscibility of poly(ε-caprolactone)/ poly(styrene-co-acrylonitrile) mixtures. *Polymer* 1994;35(15):3260-3267.

[12] Gan PP, Paul DR, Padwa AR. Phase behaviour of blends of homopolymers with α-methylstyrene/acrylonitrile copolymers. *Polymer* 1994;35(7):1487-1502.

[13] Janathanan V, Kressler J, Karasz FE, Macknight WJ. Intramolecular interactions in Poly(styrene-co-acrylonitrile)/Poly(ε-caprolactone) blends. *Journal of Polymer Science, Part B: Polymer Physics* 1993;31(8):1013-1017.

[14] Kressler J, Kammer HW, Silvestre C, Di Pace E, Cimmino S, Martuscelli E. Crystallization kinetics in blends of poly-ε-caprolactone and poly(styrene-ran-acrylonitrile). *Polymer networks and blends* 1991;1(4):225-233.

[15] Shah VS, Keitz JD, Paul DR, Barlow JW. Miscible ternary blends containing polycarbonate, SAN, and aliphatic polyesters. *Journal of Applied Polymer Science* 1986; 32(3):3863-3879.

[16] Chiu SC, Smith TG. Compatibility of poly(epsilon-caprolactone) (PCL) and poly(styrene-*co*-acrylonitrile) (SAN) blends – II. The influence of the AN content in SAN copolymer upon blend copatibility. *Journal of Applied Polymer Science* 1984; 29(5):1797-1814.

[17] Schulze K, Kressler J, Kammer HW. Phase behaviour of poly(ε-caprolactone)/ poly(styrene-ran-acrylonitrile) blends exhibiting both liquid-liquid unmixing and crystallization. *Polymer* 1993;34(17):3704-3709.

[18] Hashimoto T, Tsukahara Y, Kawai H. Dynamic Small-Angle X-ray Scattering Studies on Diffusion of Macromolecules in Bulk. 2. Principle and Preliminary Experimental Results. *Macromolecules* 1981;14(3):708-715.

[19] Hashimoto T, Kumaki J, Kawai H. Time-Resolved Light Scattering Studies on Kinetics of Phase Separation and Phase Dissolution of Polymer Blends. 1. Kinetics of Phase Separation of a Binary Mixture of Polystyrene and Poly(vinyl methyl ether). *Macromolecules* 1983;16(4):641-648.

[20] Inaba N, Sato K, Suzuki S, Hashimoto T. Morphology Control of Binary Polymer Mixtures by Spinodal Decomposition and Crystallization. 1. Principle of Method and Preliminary Results on PP/EPR. *Macromolecules* 1986;19(6):1690-1695.

[21] Nakai A, Shiwaku T, Hasegawa H, Hashimoto T. Spinodal Decomposition of Polymer Mixtures with a Thermotropic Liquid Crystalline Polymer as One Component.

Macromolecules 1986;19(12):3008-3010.
[22] Inaba N, Yamada T, Suzuki S, Hashimoto T. Morphology Control of Binary Polymer Mixtures by Spinodal Decomposition and Crystallization. 2. Further Studies on Polypropylene and Ethylene-Propylene Random Copolymert. *Macromolecules* 1988; 21(2): 407-414.
[23] Okada M, Tao J, Nose T. Attenuation of concentration fluctuations after a quench to a temperature in the single-phase region. *Polymer* 2002;43(2):329-335.
[24] Okada M, Tao J, Nose T. Dissolution of phase-separated domains after a jump to a temperature in the single-phase region. *Polymer* 2002;43(26):7429-7432.
[25] Akcasu AZ, Bahar I, Erman B, Feng Y, Han CC. Theoretical and experimental study of dissolution of inhomogeneities formed during spinodal decomposition in polymer mixtures. *J. Chem. Phys.* 1992;97(8):5782-5793.
[26] Cheng MH, Nauman EB. Non-linear diffusion with concentration-driven flows in miscible systems. *Polymer* 2003;44(21):6707-6712.
[27] Okamoto M, Shiomi K, Inoue T. LCST-type phase behaviour and structure development during melt processing in a polycarbonate/poly(styrene-co-acrylonitrile) blend. *Polymer* 1995; 36(1):87-91.
[28] Sato T, Han CC. Dynamics of concentration fluctuation in a polymer blend on both sides of the phase boundary. *J. Chem. Phys.* 1988;88(3):2057-2065.
[29] Takagi Y, Ougizawa T, Inoue T. Phase dissolution in polymer blends. Kinetics of dissolution and related problems in rubber technology. *Polymer* 1987;28(1):103-108.
[30] Svoboda P, Svobodova D, Chiba T, Inoue T. Competition of phase dissolution and crystallization in poly(ε-caprolactone)/poly(styrene-co-acrylonitrile) blend. *European Polymer Journal* 2008;44(2):329-341.
[31] Ribbe AE, Hashimoto T. Real Space Analysis of the Structural Evolution of a Polymer Blend via Spinodal Decomposition. *Macromolecules* 1997:30(14):3999-4009.
[32] Cahn JW. Phase Separation by Spinodal Decomposition in Isotropic Systems. *J. Chem. Phys.* 1965;42(1):93-99.
[33] Binder K, Stauffer D. Theory for the Slowing Down of the Relaxation and Spinodal Decomposition of Binary Mixtures. *Physical Review Letters* 1974:33(17):1006-1009.
[34] Langer J, Baron M, Miller HD. New computational method in theory of spinodal decomposition. *Physical Review* A 1975:11(4):1417-1429.
[35] Siggia ED. Late stages of spinodal decomposition in binary mixtures. *Physical Review* A 1979:20(2):595-605.
[36] Fujita H, Hashimoto T, Takenaka M. Time change in the maximum of scattering intensity during spinodal decomposition. *Macromolecules* 1989:22 (12):4663-4664.
[37] Hashimoto T, Izumitani T. Effect of a block copolymer on the kinetics of spinodal decomposition of polymer blends. 1. Nonuniversality in scaled characteristic quantities versus reduced time. *Macromolecules* 1993:26(14):3631-3638.
[38] Izumitani T, Hashimoto T. Effect of a Block Copolymer on the Kinetics of Spinodal Decomposition of Polymer Blends. 2. Scaled Structure Factor. *Macromolecules* 1994: 27(7):1744-1750.
[39] Takeno H, Iwata M, Takenaka M, Hashimoto T. Combined Light Scattering and Laser Scanning Confocal Microscopy Studies of a Polymer Mixture Involving a Percolation-to-Cluster Transition. *Macromolecules* 2000:33(26):9657-9665.
[40] Jinnai H, Hasegawa H, Hashimoto T, Han CC. Time-Resolved Small-Angle Neutron

Scattering in Intermediate and Late-Stage Spinodal Decomposition of DPB/HPI Blends. *Macromolecules* 1991:24(1):282-289.

[41] Cahn JW, Hilliard JE. Free Energy of a Nonuniform System. I. Interfacial Free Energy. *Journal of Chemical Physics* 1958:28(2):258-267.

[42] Svoboda P, Kressler J, Chiba T, Inoue T, Kammer HW. Light-scattering and TEM analyses of virtual upper critical solution temperature behaviour in PCL/SAN blends. *Macromolecules* 1994;27(5):1154-1159.

In: Polymer Phase Behavior
Editors: T. P. Ehlers and J. K. Wilhelm, pp. 295-301
ISBN: 978-1-61324-336-7
© 2011 Nova Science Publishers, Inc.

Chapter 8

THERMO- AND PH-SENSITIVITY OF POLY(*N*-VINYLPYRROLIDONE) IN WATER MEDIA

N. I. Pakuro,* A. A. Arest-Yakubovich,† B. I. Nakhmanovich and F. Kh. Chibirova

Karpov Institute of Physical Chemistry,
Moscow, Russia

ABSTRACT

In recent years, a number of polymers that undergo phase separation in water solutions on temperature rising are studied. These polymers are characterized by lower critical solution temperatures (LCST). Poly(*N*-vinylpyrrolidone) (PVP) is not thermo- or pH - sensitive under usual conditions. However, since this polymer is widely used, especially in medicine, several studies are dedicated to the problem of making this polymer stimuli-responsive, too. In the review, the phase behavior of PVP in water solutions under various conditions is covered. The phase behavior of PVP-containing copolymers and hydrogels are described. The effect of the addition of salts, including transition metal ones, on the PVP phase separation temperature is considered, the attention being paid to the different influence of anions and cations on this value. It is known that PVP readily forms complexes with many organic and inorganic compounds. Examples of such complex formation effects on cloud points of the polymer solutions are given. The phase behavior of PVP is compared with that of poly(*N*-vinylcaprolactam), a PVP close analog, which is a well-known thermosensitive polymer.

---

* Per. Obukha 3-1/12, str. 6, 105064 Moscow, Russia. e-mail: pakuro@cc.nifhi.ac.ru.

† e-mail: arest@cc.nifhi.ac.ru.

INTRODUCTION

Stimuli-responsive water-soluble polymers are the subject of wide research because of their possible applications in medicine (drug release), biotechnology, and many other fields. Temperature responsive polymers undergo phase separation on increasing the solution temperature, i.e., they have a lower critical solution temperature (LCST). Poly(N-isopropylacrylamide) (PNIPAM), other polyacrylamides, poly(N-vinylcaprolactam) (PVCL) are typical examples of such polymers. A number of reviews on the subject are published [1-5]. Some thermosensitive polymers, for example, poly(N-vinylcaprolactam) - co methacrylic acid, are also pH responsive [6].

Poly(N-vinylpyrrolidone) (PVP) is the closest analog of PVCL, but while the latter phase separates at temperatures of 32-39 °C, PVP exhibits thermosensitivity only under pressure higher than 1 kbar [7]. It is interesting that in the pressure range 2-4 kbar, both LCST, and upper critical solution temperature (UCST) are observed. PVP is widely used in medicine and pharmacology due to its biocompatibility and very low toxicity. Nowadays, it finds a new application as a stabilizer of metal and metaloxide sols in the sol-gel nanotechnology [8-10]. In this connection, tries have been taken at making PVP stimuli-responsive, too.

THERMOSENSITIVITY OF PVP SOLUTIONS, CONTAINING SALTS

The most usual way of controlling the temperature of phase separation $T_{ph.s}$ of thermosensitive polymers is addition salts into their solutions. Influence of inorganic salts on properties of PVCL water solutions has been studied by Kirsh [3, 4, 11]. Addition of various salts was shown to result both in an increase and decrease in polymer $T_{ph.s}$, depending on the nature and concentration (c) of the salt. Sulfates, carbonates, and phosphates decrease $T_{ph.s}$ of PVCL most effectively. This salting out effect can be explained by weakening hydrogen bonds in the polymer - hydrate complex that increases hydrophobic interactions in the polymer chain and enhances coil-to-globule transition. Some authors believe that this phenomenon is caused by a change in the water structure upon salt addition [3, 4].

The same effects have been observed for water solutions of PVP. $T_{ph.s}$ of this polymer (M = 4×10^4) was found to linearly decrease with increasing alkali metal salts concentrations in 1% polymer solutions [12]. The influence of anions corresponds to the following row: F^- (1-2) < $H_2PO_4^-$ (0.8-1.4) < CO_3^{2-} (0.4-0.7) < SO_4^{2-} (0.17-0.27). Concentration ranges (mol/l) where $T_{ph.s}$ falls down from 80 to 25 °C are given in brackets. Addition of NaCl or NH_4NO_3 does not result in phase separation even at their concentrations higher than 3 mol/l.

These data are in agreement with results of viscometric measurements [13] where addition of sulfates, diphosphates, and tetraborates is shown to decrease the intrinsic viscosity of PVP solutions more strongly than NaCl and $NaClO_4$ do. This indicates the greater compactization of macromolecules in the first case that facilitates phase separation.

θ-temperatures for PVP solutions, containing these salts and mixtures of phenolic and some other organic compounds, have been obtained from their cloud point temperatures [14]. Extrapolation of the $T_{ph.s}$ - c plots to zero concentration gives $T_{ph.s}$ value close to 170 °C which is in agreement with that obtained earlier [3].

For PVCL, it is shown that a change in $T_{ph.s}$ of its solutions is determined much more by the nature of anions rather than that of cations. Apparently, this is the case for PVP, too. The effectiveness of salts with the same anion (sulfates) is found to depend on the cation charge not sequentially and increase slowly in the following row: Zn^{2+}(0.63-0.9) < Mg^{2+} (0.55-0.87) < Al^{3+}(0.23-0.37) < Na^{+}(0.17-0-27) [15]. In brackets, the concentration ranges (mol/l), where $T_{ph.s}$ values of PVP solutions decrease linearly from 80 down to 25 °C, are presented. This effect is known for other stimuli-responsive polymers, but is not theoretically considered.

$T_{ph.s}$ values of PVP solutions are shown to depend on polymer concentration that differs this polymer from PVCL. For the latter, the difference in $T_{ph.s}$ for 1 and 20% solutions is not greater than 1-2 °C while for PVP of the same concentrations in 1.5 M KF solution, $T_{ph.s}$ = 52 and 28 °C, respectively [12].

Effect of transition metal chlorides on the properties of PVP and PVCL water solutions is also studied [16]. Addition of $ZnCl_2$ to PVP solutions is shown to result in polymer phase separation upon heating. In the presence of $CdCl_2$ this phenomenon is absent, but on addition of HCl into PVP solutions, containing $CdCl_2$, phase separation also takes place. A very sharp decrease in $T_{ph.s}$ is observed if NaCl (KCl) and a small amount of HCl are introduced into PVP - $CdCl_2$ solution. Addition of $CuCl_2$ to PVP solution is effective in the presence of alkali metal chlorides, the effect being more pronounced if HCl is also added. The same phenomena are observed for PVCL. In the authors opinion, the effects described are connected with complex anions $MtCl_4^{2-}$ which are known to form in the transition metal water solutions, containing HCl and/or alkali metal halogenides [17]. The authors [16] believe that in complex systems formed, which include bivalent anions $MtCl_4^{2-}$, complexes of transition metal cations with PVP, and free ions, screening electrostatic interaction between such chains, bivalent anions are responsible for binding chain units with participation of H_2O molecules and cations, that facilitates polymer globulization and phase separation on heating. If HNO_3 is used instead of HCl, the effects described above are absent that confirms this suggestion.

PHASE SEPARATION OF PVP SOLUTIONS IN THE PRESENCE OF ORGANIC COMPOUNDS CAPABLE TO FORM COMPLEXES WITH THE POLYMER

Complex formation is one of the ways of polymer hydrophobization that facilitates phase separation in their solutions on heating. PVP forms complexes with many organic compounds, including phenols and other aromatic compounds, a number of alcohols and organic acids [18]. The effects of these compounds on PVP solution properties have been studied in comparison with those of PVCL. Low-molecular-mass organic acids, such as formic, acetic, and oxalic ones, are shown to decrease $T_{ph.s}$ of PVCL [3, 11], but are ineffective in the case of PVP. However, addition of isovaleric and isobutyric acids, which have longer hydrocarbon chains, results in phase separation in PVP solutions [19], indicating an important role of hydrophobic interactions in these systems. Hydrogen bonding between OH groups of acids and C=O groups of polymer lactam rings also takes place. Apparently, these two types of interaction violate the polymer-hydrate complex that facilitates phase separation.

Even the stronger effect on the behavior of PVP water solutions is observed if trichloroacetic acid (TCA), which is also known to form complexes with PVP [18], is used as an additive [19]. In the TCA concentration range 0.17-0.30 mol/l, $T_{ph.s}$ drops from 80 to 10 °C. This system turned out to be also pH-responsive. HCl introduction into solutions, containing TCA, causes even more drastic decrease in $T_{ph.s}$ The same is shown for PVCL while HCl addition into PVCL solutions in the absence of TCA slightly increases $T_{ph.s}$ These phenomena are supposed to be connected with the fact that non-ionized form of TCA takes part in the complex formation via hydrogen bonds between COOH groups of TCA and C=O groups of PVP (PVCL).

To prove this suggestion, the dependence of $T_{ph.s}$ on the concentrations of non-ionized TCA $c(1-\alpha)$ in PVP-TCA solutions was elucidated. The dissociation degree α was calculated for cases of either the presence or absence of HCl in the solution, using the known value of the TCA dissociation constant K_a. All $T_{ph.s.}$ vs. [HCl] plots obtained at various TCA concentrations and $T_{ph.s}$ vs. c plot for the case of HCl absence turned out to merge when replotted on the coordinates $T_{ph.s}$ vs.c $(1-\alpha)$. The same is shown to be true for PVCL. These results corroborate the supposition that complexes between the polymers studied and the non-ionized form of TCA cause phase separation in these systems.

The behavior of PVP in water - trifluoroethanol (TFE) mixtures has been studied [20]. It was found that at TFE concentrations of 1.3-1.6 mol/l (solution of alcohol in water), transitions of LCST type are observed, i.e., the polymer precipitates on heating. At TFE concentrations in the range 5-6 mo/l (solution of water in alcohol), UCST transitions take place, i.e., the polymer dissolves on heating. In the intermediate field, the system is shown to consist of two phases at any temperatures. Thus, in this system, the phenomenon of so called cononsolvency is observed, i.e., the polymer entirely dissolves both in water and TFE taken separately, but does not dissolve in their mixtures. TFE and water are completely mixable in any proportions. The cononsolvensy phenomenon was reported earlier for PNIPAM [21], where it was explained by stronger interaction between water and cononsolvent than that between the polymer and any solvent, resulting in decomposition of polymer-hydrate complex and polymer sedimentation. For PVCL it was not reported.

PHASE SEPARATION IN WATER SOLUTIONS OF PVP COPOLYMERS AND DERIVATIVES

Copolymerization of VP with less hydrophylic monomers is a means to obtain thermoresponsive polymers. Random copolymer of VP and VCL ($M = 30 \times 10^4$) has been synthesized by the method of free radical polymerization [22]. Addition of 20% VCL units in the PVP chain is shown to decrease $T_{ph.s}$ of its solution from 170 °C (the calculated value [3]) down to 70 °C. The same effect is observed for copolymers of VP and 2-NN-dimethylamino ethylmethacrylate [23].

Effect of temperature on the structural characteristics of VP-VCL copolymer molecules has been investigated in very dilute aqueous solutions [24]. Relaxation times τ_{IMM}, characterizing the intramolecular mobility of macromolecules, were determined with the use of polarized luminescence at stationary light exitation. An increase in τ_{IMM} value indicates an increase in intramolecular hindrance, i.e., macromolecule compactization. It was found that the greater the proportion of VCL units in the copolymer, the greater is an increase in τ_{IMM}. The highest increase in τ_{IMM} was observed for the copolymer, containing 85 mol. % VCL units at 43-48 °C. This phenomenon was explained by disturbing syndiotactic sequences of VCL units in the presence of some VP units in the macromolecule. It is interesting that the copolymers were shown to exhibit the higher complexing ability with phenol than PVP and PVCL homopolymers with the maximum for the above copolymer composition.

Many of thermosensitive water-soluble polymers have LCST near 30 °C. Introduction of VP units in the chains of such polymers increases their LCST. For example, this is true for VP-VCL copolymers described above. The behavior of water solutions of VP-NIPAM copolymers, containing 5 and 10% VP units, was also studied [25]. $T_{ph.s}$ of such copolymers was found to be somewhat higher than that of PNIPAM. Formation of monomolecular globules of these copolymers stable in very weak solutions was shown to take place under similar conditions [26]. VP-NIPAM copolymers turned out to be pH-responsive, too. An increase in LCST of the copolymer as compared with that of PNIPAM was found to be higher at pH = 4 than at pH = 7.4 because of the basic character of VP [27].

The $T_{ph.s}$ dependence of random VP and N,N-dimethylacrylamide copolymer solutions on the amide units content is reported [28]. It is shown that $T_{ph.s}$ of the copolymers in 1% water solutions, containing Na_2SO_4 and Na_2CO_3 (0.55M), increase linearly from 7 to 45-49 °C with increasing the PVP content from 0 to100%.,

Recently, PVP derivatives, which have n-C_2H_6 and n-C_4H_9 groups in lactam rings, have been synthesized [29]. These polymers turned out to also exhibit reversible temperature-dependent water solubility. The polymers were applied as stabilizers in formation of "smart" thermoresponsible Au nanoparticle catalysts.

STIMULI-RESPONSIVE HYDROGELS ON THE BASE OF PVP COPOLYMERS

PVP-g-PNIPAM hydrogel was prepared by NIPAM grafting from crosslinked PVP derivative hydrogel beads. The polymerization was initiated by PVP-Br, which was prepared through bromination of pendant allylic groups of the PVP derivative with N-bromosuccinimide. PVP-g-PNIPAM hydrogels are reported to show more rapid temperature-responsive properties as compared with those of hydrogels based on conventional VP-NIPAM copolymers owing to the rapid dehydration of the freely mobile graft chains [30].

Crosslinked pH-sensitive copolymers of VP and acrylic acid (AA) were prepared by free radical copolymerization of VP and AA in the presence of ethylenglycole dimethacrylate [31].

Preparation of the pH-sensitive semi-interpenetrating network (semi-IPN) hydrogel based on hydrogen bond between chemically crosslinked PVP and linear PAA has been reported [32]. Crosslinked PVP beads were obtained by suspension polymerization of VP with AIBN in the presence of NN'- methylene-bis-acrylamide. Physical complexation occurred between PAA and porous PVP beads in the water media. The semi-IPN hydrogel was shown to have excellent pH-sensitivity in the range of pH from 2.2 to 4.0. COOH groups of PAA partly dissociate that results in the chain repulsion and formation of hydrogels of higher swelling degree. It is noted that addition of NaCl does not affect much the pH-sensitivity of the hydrogel.

REFERENCES

[1] Dimitrov, I.; Trzebika, B.; Müller, A. H. E.; Dworac, A.; Tsvetanov, C. B. *Prog. Polym. Sci.* 2007, vol. 32, 1205-1237.
[2] .Aseyev, V. O.; Tenhu ,H.; Winnik, F. M. *Adv. Polym. Sci.* 2006, vol. 196, 1-85.
[3] Kirsh, Yu. E. Water-soluble Poly-(*N*-vinylamides): Synthesis and Physico-Chemical Properties; John Wiley and Sons: Chichester, 1998.
[4] Kirsh, Yu. E. *Prog. Polym. Sci.* 1993, vol. 18, 519-542.
[5] Kumar, A.; Srivastava, A.; Galaev, I. Yu.; Mattiasson B. *Prog. Polym. Sci.* 2007, vol. 32, 1205-1237.
[6] Makhaeva, E. E.; Tenhu, H.; Khokhlov, A. R. *Macromolecules* 2002, vol. 35, 1870-1876.
[7] Sun, T.; King, H. E.; *Phys. Rev.* E 1996, vol. 54, 2696-2703.
[8] Sivudu, K. S.; Shailaja D. *Mater. Lett.* 2007,vol. 61, 2167-2169.
[9] Chen, C.; Wang, L.; Jiang, G.; Yu, H. *Res. Adv. Mater. Sci.* 2006, 11, 1-18.
[10] Si, R.; Zhang, Y.-W.; You L.-P.; Yan, C. H. *J. Chem. Phys. Ser. B* 2006, vol. 110, 5994-6000.
[11] Kirsh, Yu. E.; Yanul, N. A.; Kalninsh, K. K.; Maslov, V. G. *J. Molec. Liq.* 1999, vol. 82, 117-130.
[12] Nakhmanovich, B. I.; Pakuro, N. I.; Akhmet'eva, E. I.; Litvinenko, G. I.; Arest-Yakubovich, A. A. *Vysokomolek. Soedin. Ser. B* 2007, vol. 49, 941-944; *Polym. Sci. B* 2007, vol.49, 136-138.
[13] Güner, A. J. *Appl. Polym. Sci.* 1996, vol. 62, 785-788.
[14] Güner, A.; Ataman, M. *Colloid. Polym. Sci.* 1994, vol. 272, 175. Kinci B.; Güner, A. *Eur. Polym. J.*2001, vol. 37, 361-365. DOI:1016/S00143057(00)00106-3. Güner, A.; Kara M. *Polymer* 1998, vol.39, 1569-1572. DOI:10.1016/S0032-3861(97)00501-6. Kavlak S.; Güner, A. *J. Appl. Polym. Sci.* 2000, vol. 78, 507-510. DOI:10.1002/1097-4628(20001017)78:3<507::AID-APP40>3.0C0;2-8.
[15] Nakhmanovich, B. I.; Arest-Yakubovich A. A. (to be published).
[16] Pakuro, N. I.; Nakhmanovich, B. I.; Pergushov, D. V.; Chibirova, F. Kh. *Vysokomolek.Soedin. Ser. A*, 2011, vol. 53, 9-14; *Polym. Sci. A* 2011, vol. 53, 6-11. DOI:10.1134/S0965545X11010044.
[17] Colton, R.; Canterford, J. H. Halides of the First Row Transition Metals; Wiley and Sons: London, 1969.
[18] Molyneux, P. Water-Soluble Synthetic Polymers: Properties and Behavior; CRC Press: Boca Raton. Fl.: 1984, Vol. 2.
[19] Pakuro, N. I.; Yakimansky, A. V.; Chibirova, F. Kh.; Arest-Yakubovich, A. A. *Polymer* 2009, vol. 50, 148-153.
[20] Nakhmanovich, B. I.; Arest-Yakubovich, A. A. (to be published).
[21] Costa, R. O. R.; Freitas, R. F. S. *Polymer* 2002, vol. 43, 5879-5885.

[22] Popkov, Yu. M.; Nakhmanovich; B. I.; Chibirova F. Kh.; Bune, E. V.; Arest-Yakubovich, A. A. *Polym Sci. B* 2007, vol. 49, 155-158.
[23] Nakhmanovich B. I.; Arest-Yakubovich, A. A. (unpublished results).
[24] Anufrieva, E. V.; Gromova, R. A.; Kirsh, Yu. E.; Yanul, N. A.; Krakoviak, M. G.; Lushchik, V. B.; Pautov, V. D.; Sheveleva, T. V. *Eur. Polym. J.* 2001, vol. 37, 323-328.
[25] Siu, M. H.; He, C; Wu, C. *Macromolecules* 2003, vol. 36, 6588-6592.
[26] Siu, M. H.; Zhang, G. Z.; Wu, C. *Macromolecules* 2002, vol. 35, 2723-2727
[27] Dincer, S.; Rzaev, Z. M. O.; Piskin, E. J. *Polym. Res*. 2006, vol. 13, 121-131.
[28] Alencar de Queiroz, A. A.; Gallardo, A.; San Román, *J. Biomaterials*. 2000, vol. 21, 1631-1643.
[29] Yan, N.; Zhang, J.; Yuan, Y.; Chen, G.-T.; Dayson, P. J.; Li, Z.-Ch.; Kou, Y. *Chem. Commun.* 2010, vol. 46, 1631-1633.
[30] Jin, S.; Liu, M.; Chen, S.; Gao, C. *Eur. Polym. J.* 2008, vol. 44, 2162-2170.
[31] Devine, D. M.; Higginbottom, C. L. *Eur. Polym. J.* 2005, vol. 41, 1272-1279.
[32] Jin, S; Liu, M.; Zhang, F.; Chen, S.; Niu, A. *Polymer* 2006, vol. 47, 1526-1532.

INDEX

C

D

E

F

G

I

J

K

L

M

N

O

P

Q

R

T

U

V

W

X

Y